Ecology of Desert Systems

Ecology of Desert Systems

Walter Whitford

RESEARCH ECOLOGIST (EMERITUS)
USDA-ARS JORNADA EXPERIMENTAL RANGE
PO BOX 30003, MSC 3JER
NEW MEXICO STATE UNIVERSITY
LAS CRUCES, NM 88003-8003, USA

Illustrations by
Elizabeth Ludwig Wade

SCOTTSDALE, ARIZONA, USA

ACADEMIC PRESS
An Elsevier Science Imprint

San Diego   San Francisco   New York
Boston   London   Sydney   Tokyo

Cover Illustration by Glynn Gorick.

The cover illustration conveys ecological properties and processes, from the fine scale of soil microflora and mesofauna living in the plant rhizosphere and soil pores, to the landscape scale. Subterranean termites and harvester ants foraging in the vegetation show the importance of social insects in patch dynamics and ecological processes in deserts. The patch scale depicts the three photosynthetic pathways of desert plants: C_3 shrubs (creosotebush, *Larrea tridentata*), C_4 grasses, and CAM succulents (prickly-pear cactus, *Opuntia* spp.). The landscape behind the vegetation portrays the patchy nature of desert terrain.

This book is printed on acid-free paper.

Academic Press
An Elsevier Science Imprint
Harcourt Place, 32 Jamestown Road, London NW1 7BY, UK
http://www.academicpress.com

Academic Press
An Elsevier Science Imprint
525 B Street, Suite 1900, San Diego, California 92101-4495, USA
http://www.academicpress.com

ISBN 0-12-747261-4

Library of Congress Catalog Number: 2001097955

A catalogue record for this book is available from the British Library

Typeset by M Rules, London
Printed and bound in Great Britain by MPG Books Ltd, Bodmin, Cornwall

02   03   04   05   06   07   MP   9   8   7   6   5   4   3   2   1

Dedication

To Linda C. G. Whitford, who, despite aversion to foreign travel has accompanied me to several of the world's deserts and has graciously hosted many of my colleagues and visitors and to the memory of our daughter, Colleen A. 'Coco' Whitford (November 5, 1974 – January 26, 1981). She shared our lives and our love for the desert for all too short a time.

Contents

Foreword   xi

Preface   xv

Chapter 1 Conceptual Framework and Paradigms   1
  1.1 Pattern and Process   2
  1.2 Definition of Ecosystem   6
  1.3 Landscape Units   9
  1.4 Questions   10
  1.5 Defining Deserts   11
  1.6 The Pulse-Reserve Paradigm and the Autecological Hypothesis   12
  1.7 Temporally Uncoupled Resources   14
  1.8 The Soil Resource   14
  1.9 Ecosystem Processes   16
  1.10 Problems of Scaling   18
       References   18

Chapter 2 Landforms, Geomorphology, and Vegetation   21
  2.1 Desert Mountains and Hillslopes   22
  2.2 Piedmonts, Alluvial Fans, and Bajadas   24
  2.3 Ephemeral Streams   30
  2.4 Basins and Flatlands   33
  2.5 Desert Pavements and Stony Surfaces   36
  2.6 Rivers and Floodplains   37
  2.7 Sand Dunes and Sand Features   37
  2.8 Landform Stability   40
       References   40

Chapter 3 Characterization of Desert Climates   43
  3.1 Seasonality   48
  3.2 Spatial Effects   52
  3.3 Predictability   53
  3.4 Intensity, Duration, Frequency, and Return Time   55
  3.5 Microclimate   58
       References   61

Chapter 4 Wind and Water Processes   65
4.1 Wind Erosion   65
4.2 Redistribution of Rainfall   71

    4.2.1 Interception and Stem Flow   73
    4.2.2 Throughfall   76

4.3 Splash Erosion-kinetic Energy of Raindrops   77
4.4 Infiltration and Run-off   78
4.5 Exchanges Among Landscape Units   84
4.6 Exchanges Within Landscape Units   86
4.7 Episodic Events   88
4.8 Ephemeral Ponds and Lakes   90
    References   93

Chapter 5 Patch–Mosaic Dynamics   97
5.1 Seeds: Germination–Establishment Sites   97
5.2 Single Species Patches   102
5.3 Intraspecific Morphological Variation   103
5.4 Mosaics and Multi-species Patches   104
5.5 Competition versus Facilitation   106
5.6 Animal-produced Patches   110
5.7 Temporal Dynamics and Feedbacks   117
    References   119

Chapter 6 Adaptations   123
6.1 Avoidance of Extremes   123

    6.1.1 Annual Plants   124
    6.1.2 Drought Deciduous Perennials   125
    6.1.3 Amphibians   126
    6.1.4 Reptiles   129
    6.1.5 Mammals   129
    6.1.6 Birds   131
    6.1.7 Arthropods   131

6.2 Physiological and Morphological Adaptations   133

    6.2.1 Perennial Plants   133
    6.2.2 Reptiles   141
    6.2.3 Invertebrates   142
    6.2.4 Large Mammals   143
    6.2.5 Small Mammals   148
    6.2.6 Carnivores   150
    6.2.7 Birds   150

    References   151

Chapter 7 Primary Production   157
  7.1 Measurement of Net Primary Production   157
  7.2 Comparisons of Production Estimates with Mesic Ecosystems 160
  7.3 Rain Use Efficiency   161
  7.4 Below-ground Productivity   162
  7.5 Models of AGNPP   163
  7.6 Landscape Relationships   166
  7.7 Productivity in Extreme Deserts   170
  7.8 Productivity Linked to Rainfall   171
  7.9 Productivity of Species and Functional Groups   172
     References   178

Chapter 8 Consumers, Consumption, and Secondary Production   181
  8.1 Secondary Production   181
  8.2 Foliage Chewers and Browsers: Vertebrate and Invertebrate 185
  8.3 Canopy Insect Communities   192
  8.4 Surface Active Arthropods   198
  8.5 Large Herbivores: Grazers   202
  8.6 Granivory   206
  8.7 Impacts of Rodents and Ants on Ecosystem Structure and Processes 213
  8.8 Vertebrate Predators   215
  8.9 Below-ground Food Webs   220
 8.10 Ephemeral Ponds and Lakes   222
     References   226

Chapter 9 Decomposition and Nutrient Cycling   235
  9.1 Nutrient Limitations   235
  9.2 Decomposition   237
     9.2.1 Sources and Characteristics of Decomposable Material   238
     9.2.2 Rates and Mechanisms of Mass Loss (Surface Litter)   239
     9.2.3 Spatial Variation   244
     9.2.4 Decomposition of Buried Litter and Roots   245
     9.2.5 The Role of Microfauna in Decomposition and Mineralization 247
     9.2.6 The Role of Termites   249
     9.2.7 Conceptual Model of Decomposition Processes in Deserts 251
  9.3 Landscape Patterns of Nutrient Distribution   252
  9.4 Resorption   257
  9.5 Nitrogen Cycle   258
     9.5.1 Nitrogen Fixation   258
     9.5.2 Nitrogen Losses   262
  9.6 Other Potentially Limiting Nutrients   266
  9.7 Role of Mycorrhizae   266
     References   269

Chapter 10 Desertification   275
10.1 History of Desertification   277
10.2 Ecosystem and Landscape Consequences of Desertification   282
10.3 Resistance and Resilience of Desertified Landscapes   289
10.4 Desertification and Climate Change   290
10.5 Social and Economic Consequences of Desertification   291
10.6 Effects of Desertification on Biodiversity   293
10.7 Rehabilitation of Desertified Landscapes   295
      References   301

Chapter 11 Monitoring and Assessment   305
11.1 Definitions   306
11.2 Existing Assessment and Monitoring Systems   306
11.3 Data for Assessment and Monitoring   309
11.4 Ecosystem Functions   310
11.5 Indicators   310
      References   316

Chapter 12 Desert Ecosystems in the Future   319
      References   325

Index   327

Foreword

That the desert here is mildly austere is certainly true, and yet neither the plants nor the animals live under what is, for them, painfully difficult conditions. The vegetation flourishes in its own way. For the desert birds and the desert animals this is not an unfavorable environment. They have had to make their adaptations to the heat and dryness of their land, just as the animals of other climates made other adaptations to theirs. For the jack rabbit and the ground squirrel, as well as for the dove and cactus wren, this is obviously a paradise and there is no paradox in the smile which the face of the desert wears. Only to those who come from somewhere else is there anything abnormal about the conditions which prevail.

The Desert Year, Joseph Wood Krutch

Deserts are often perceived as extreme, barren landscapes that are seemingly inhospitable and capable of sustaining only the most rudimentary plant and animal life. Hence, deserts have long captivated the imagination of both layperson and scientist alike. Such was the case for author, naturalist, and philosopher Joseph Wood Krutch who, in the early 1930s, made his first excursion to the American Southwest, a trip that would forever change his life. Krutch, a New Englander and self-proclaimed 'city man,' acknowledged that this journey was taken 'without much enthusiasm,' so he was undoubtedly contemplating considerable misgivings as his train pulled into the dusty depot at Lamy, New Mexico. However, during the next several weeks, as he crisscrossed New Mexico, Arizona, and parts of California and southern Utah, he became enamored of the enormous diversity of vistas, landforms, plants, animals and human cultures of these drylands. Krutch amassed a volume of enduring impressions of these arid landscapes, and yearned for more: '. . . I was still only a traveler or even only the traveler's vulgar brother, the tourist. No matter how often I looked at something I did no more than look. It was only a view or a sight. It threatened to become familiar without being really known and I realized that what I wanted was not to look at but to live with this thing whose fascination I did not understand.' Years later he returned, to spend a single year in the Sonoran Desert, fulfilling a desire to gain a degree of intimacy with this alien land. His delightful book, *The Desert Year*, chronicles this quest.

Several decades later, another New Englander and 'city man,' made his first trek to the American Southwest. Walter Whitford, trained as an animal physiologist in Rhode Island, was about to start his first job as assistant professor of biology at New Mexico State University. When he arrived in Las Cruces, Whitford was most likely entertaining his own set of misgivings. While embracing the opportunity to study a system so unlike the eastern forest and coastal ecosystems to which he was accustomed, this stark and unfamiliar land undoubtedly made him feel uncertain, much like 'the traveler's vulgar brother, the tourist.' However, like Krutch, the desert would forever change his life. As a naturalist, Whitford immediately took to the field (often at high speeds in old trucks), seeking to observe and understand the geology, flora, and fauna; as a scientist, he conducted experiments and measured everything he thought made deserts distinctive. Over the next forty years, Whitford would amass an impressive volume of work. This book, *Ecology of Desert Systems*, is witness to his duel role of scientist and naturalist, an anthology of a life's work striving to gain an intimacy with desert ecosystems.

Initially, Whitford viewed the varied organisms of the northern Chihuahuan Desert through the lens of a classically trained physiologist. However, in the 1960s, ecology was undergoing a revolution of sorts with the emergence of ecosystem ecology as a discipline. The International Biological Program (IBP) was launched in a bold attempt to understand how entire ecosystems function and, by modeling the behavior of these systems, predict how they would respond to environmental stresses. Whitford organized an IBP study to understand water and nutrient dynamics in a small desert watershed north of Las Cruces and, drawing from his intuitive understanding of natural systems, adopted a holistic approach – combining empirical observations with theoretical constructs – to forge a systems view of the complex workings of deserts, one of the very first attempts to apply this perspective to aridlands. He followed his curiosity and expanded his own research queries to an extraordinarily broad set of topics, including nutrient cycling, plant productivity, microbial ecology, and rodent population dynamics, all in the context of a systems understanding. This remarkable breadth is at the core of the *Ecology of Desert Systems*.

Considering that nearly 20% of the world's population and one-third of its land surface is located in arid and semiarid regions, linking ecological knowledge to the economic and social consequences of climate change and other anthropogenic disturbances – and to concerns about sustainable land management – is crucial. Land degradation in arid systems, which involves many complex interactions and feedbacks between human social structures, soils, vegetation, and biogeochemical and hydrological processes, is a complex phenomenon that eludes simple characterizations. Nevertheless, it is clear that ecologists will only be able to make substantial contributions to the debate if armed with strong conceptual models of how arid lands work and how people change them. As suggested by the title, the *Ecology of Desert Systems* is one such conceptual model, written by an expert with a naturalist's vision that also embraces a systems perspective, and provides a dynamic 'blueprint' for the understanding of the ecology of arid lands. Whitford provides a set of simple but powerful guidelines for managing aridlands: that short-term dynamics are not necessarily indicative of longer-term phenomena; that the behavior of desert ecosystems is wonderfully complex and all too often difficult or impossible to predict; and that humans have major impacts on these 'simple' ecosystems, often in unknown and unpredictable ways.

During his year living in the desert, Krutch recognized both the beauty of his surroundings and their subtle complexities:

> There is all the difference in the world between looking at something and living with it. In nature, one never really sees a thing for the first time until one has seen it for the fiftieth. It never means much until it has become part of some general configuration, until it has become not a 'view' or a 'sight' but an integrated world of which one is a part; until one is what the biologist would call part of a biota.

The *Ecology of Desert Systems* is an affirmation of the difference between 'looking at something and living with it.'

James F. Reynolds
Duke University
Durham, NC

Preface

No wind. The heat bears down.
It has not rained for one year.
I love this dry land
Am caught even by blowing sand, reaches
Of hot winds. I am not the desert
But its name is not so far from mine.

Extract from 'Desert Cenote' by Keith Wilson in *Bosque Redondo: The Encircled Grove*, Pennywhistle Press. (Cenote is Spanish-Aztec for 'water hole, oasis')

If I had the talent of my friend Keith Wilson, I would have expressed my love and fascination with deserts in poetry. Since I do not possess that talent, I have resorted to narrative of a scientific nature to communicate a picture of the harsh reality of life in the wonderful dry places of the planet. I was encouraged to write this book by several colleagues and students. Students enrolled in my desert ecology classes provided critique's of early drafts. Without their encouragement, I would never have attempted the daunting task that this book represents. This book was written for them and for future generations of students. It is hoped that this book will encourage those students to address many of the fascinating questions that this ecologist could not address in his lifetime.

'Ecology of Desert Systems' is the culmination of nearly 40 years of fascination by deserts and the organisms that reside therein. That fascination has maintained my enthusiasm for research on all aspects of desert ecology to the present day. The completion of the book was accomplished after I was no longer bound by the requirements of gainful employment. The support for continuing research and collaboration with colleagues and students postretirement has been afforded me by the US Department of Agriculture, Agricultural Research Services's Jornada Experimental Range in cooperation with New Mexico State University. I could not have completed this book without that support.

Most of my research for the past 38 years has been conducted on the Jornada Experimental Range. The Chihuahuan Desert of southern New Mexico has been my frame of reference in my work in other deserts of the world. The Jornada Experimental Range has provided the baseline for comparisons and contrasts of the Chihuahuan Desert with other arid lands. Although I have attempted to write this book in a global context, my familiarity with the Chihuahuan Desert has undoubtedly dominated my thinking about deserts. Despite this, I hope that the reader will find this a suitable springboard for their own research and thinking about deserts.

The ideas and concepts developed in the 'Ecology of Desert Systems' result from collaboration and discussions with many desert scholars. Many of the authors of papers cited in the book are colleagues, postdoctoral associates and students from whom I learned a great deal at conferences, in the laboratory, and in the field. My mentors and students led me to research topics in desert ecology in which I had no formal training. My colleagues in the Biology Department at New Mexico State University, William Dick-Peddie, Ralph J. Raitt, and James R. Zimmerman provided much needed postdoctoral mentoring by generously allowing me to tag along on field trips and patiently identified the plants, animals, and landscapes which were all new to me. Visitors and colleagues who had a large impact on me and served as role models in my career include: John Cloudsley-Thompson, George Ettershank, Robert Chew, William Nutting, Clifford Crawford, Claude Grenot, John A. Ludwig, Gary L. Cunningham, Mark Westoby, M. Timm Hoffman, Susan Milton, W. Richard Dean, Graham I. H. Kerley, and John A. Wallwork.

I owe special thanks to David Tongway of CSIRO Wildlife and Ecology, who spent hours discussing the relationships between patch dynamics and landscape structure and landscape processes in the laboratory and while driving the hundreds of kilometers between the CSIRO laboratory in Deniliquin to Lake Mere. He influenced my thinking about desert ecosystems and landscapes more than any other individual. David made my visits to the Lake Mere research site in northwestern New South Wales intellectually stimulating and enjoyable.

Chapter 1 | Conceptual Framework and Paradigms

Nearly one-third of the land area of this planet is classified as arid or semiarid. These lands are areas where rainfall limits productivity and/or is so unpredictable that cropping is not feasible. Of the approximately 37,000,000 km² classified as arid and semiarid lands, 25,560,000 km² are used as rangelands: lands used by pastoralists for domestic livestock production (Verstraete and Schwartz, 1991). Despite these limitations, humans have inhabited these lands for millennia, using the limited and varied productivity to support pastoralism. In the 20th century, nomadic pastoralism has been replaced by pastoral industries in many areas of the world. Commercial livestock production has had very different impacts on arid lands than nomadic pastoralism. Only the hyper-arid regions (those areas receiving less than 80 mm of rainfall per year (UNESCO, 1977) are excluded from the desert rangelands. Nearly coincident with the development of commercial ranching or pastoral industry, was the realization that these lands were fragile and that when degraded, recovery was slow or did not occur (Dregne, 1986). Along with the recognition that degradation of rangelands was occurring, came the realization that we lacked sufficient knowledge about how these rangelands worked to develop sustainable management strategies. During the last half of the 20th century, a considerable literature has developed on most aspects of the ecology of deserts and of desert organisms. Despite this explosion of information, problems of sustainable use and management of arid lands remain nearly as intractable as they were a half century ago.

In economically developed regions of the world, desert lands are viewed by humans not only as places for the production of harvestable animal products but also as regions that support diverse landscapes and floras and faunas. These lands are considered resources that provide opportunities for recreation and escapes from the hectic pace of cities. These changes in attitude toward 'wastelands' have generated conflicts among different interest groups. Therefore the problems of managing and sustaining desert lands have not disappeared nor are they any more tractable now than they were a half-century ago. Buzz words like sustainable ecosystems have entered the lexicons of many people but what they mean and how these words will lead to achieving that desired state still eludes us. A conceptual framework that incorporates the information from population to ecosystem levels of organization at appropriate temporal and spatial scales may provide a blueprint for understanding arid lands.

1.1 PATTERN AND PROCESS

Studies of ecological systems must focus on both pattern and process. Pattern may be structural or temporal. Processes are mechanisms, many of which are similar in ecological systems that are structurally different. Ecological systems are in many ways analogous to an automobile. Understanding how an automobile moves down a highway involves understanding the structural components in relationship to each other and the processes of combustion of fuels and expansion of gasses that move pistons within a cylinder. Other processes involve the transfer of energy to the drive shaft and axles via gears in order to provide sufficient torque to overcome the resistance of the mass of the vehicle. The structures of a diesel automobile are different but the basic processes (expansion of combusted gasses and gearing) are similar. Included in the efficient functioning of an automobile is the temporal pattern of firing sequence to provide a smooth transfer of energy to the drive chain. In ecological systems, temporal patterns of variables such as rainfall and temperature drive the sequencing of processes such as nutrient cycling and primary production. Although the automobile analogy provides some insights into what is needed to understand ecological systems, the analogy is limited. The patterns and processes of an automobile are sufficiently well understood for predictive models (for design and construction) to work. However, for ecological systems, structural complexity, variable time scales, and the variable spatial scales of important structural features contribute to the difficulty of developing predictive understanding of ecological systems. Understanding the ecology of desert systems requires understanding both pattern and process and their relationships.

Ecology is generally considered at eight levels of organization: cell, organism, population, community, ecosystem, landscape, biome, biosphere. However, our thinking about ecology and ecological relationships must not be constrained by a conventional hierarchy. It is necessary to keep temporal and spatial scale ordering separate from ecological organization constraints. Scale ordering of time and space addresses the physical side of ecological systems. Both scale-dependent physical fluxes and human intellectual constructs such as community, ecosystem and landscape are necessary to address the full range of phenomena that contribute to the ecology of desert systems. Ecological processes do not function in a way that is limited to physical and chemical mechanisms (Allen and Hoekstra, 1990). For example, the behavior of organisms in a landscape are distinctive and dependent on the genetic characteristics of the species population. A landscape feature such as a river-bed may be a corridor for some species, a barrier for other species or a neutral feature for others. These behaviors cannot be understood as physical and chemical mechanisms but are important for understanding ecological systems (Allen and Hoekstra, 1990).

Communities have historically been identified as discernible aggregations of organisms occupying concrete patches on the ground. However, since communities are human intellectual constructs, they are best considered assemblages of organisms that are constantly accommodating and changing in character as the result of natural selection of species to the presence or absence of associates. Communities are the integration of the complex behavior of the biota that produces a cohesive whole. That whole exhibits properties of self-regulation and self-assertiveness that can modify the physical environment. Populations consist of only one species and communities involve many species (Allen and Hoekstra, 1990). Populations are tangible and can be identified with particular places on the landscape at an instant in time.

Ecosystems involve integration of water and nutrient fluxes over wide ranges of time and spatial scales. Because ecosystems include the influences of abiotic elements and feedbacks between the abiotic and biotic components, defining ecosystems by an area of landscape is of limited value and may be misleading. As a conceptual construct it is important to define carefully the scale at which ecosystem properties and processes are being considered.

Landscapes are composed of distinct, bounded units that are differentiated by biotic and abiotic structure and composition (Pickett and Cadenasso, 1995). Landscape can also be considered as an abstraction representing spatial heterogeneity at any scale. In arid lands, landscape units are generally recognized by the life form and species composition of the dominant species (classical concept of biotic community). For example, creosotebush, *Larrea tridentata*, shrubland may make up one of the distinct units on a watershed. The *L. tridentata* shrub canopies intercept wind and rainfall thereby changing the spatial distribution of the water and modifying the energy and characteristics of the wind at the soil surface. This dominant species creates patches with modified microclimates. Within a landscape unit we may recognize several mosaics, that is combinations of plants that are distributed throughout the unit. An example of different mosaics in *L. tridentata* shrublands are: shrubs with large clumps of the grass, *Muhlenbergia porteri*, around the base of the shrubs, shrubs with a relatively continuous understory of a grass such as *Scleropogon brevifolia*; shrubs within a matrix of several species of bunch grasses and shrubs with no grasses in the understory. Each of these mosaics may be present in a single landscape unit recognized as a creosotebush shrubland. Within mosaics there is a finer scale of landscape unit that can be recognized as patches (Table 1.1). Patches may be unvegetated areas, diggings of animals, contrasting surface materials such as lag gravel or aggregations of stones. Many agents of patch formation are episodic especially in arid regions. Patches may form at different times and places on the landscape thereby producing a shifting mosaic (Pickett and Cadenasso, 1995). At the finest scale, a patch is a single plant and the soil around that plant.

The living component of desert ecosystems is composed of many species that are 'living on the cusp of disaster'. Many, but not all, of the species living in deserts are operating in conditions that are very close to their tolerance thresholds or very close to the maxima or minima for limiting factors. When extreme conditions occur (long periods with little or no rainfall, periods of time with temperatures well above or below the average temperature maxima or minima)

Table 1.1
Definitions and Spatial Relationships of Landscape Units from the Patch Scale to the Regional Scale

Unit type	Definition	Size
Patch	A patch is a function of the size of a single kind of plant growing on a homogeneous soil	$< 1–10 \ m^2$
Mosaic	A unit containing contiguous patches	$< 1 \ ha \ to \ 1 \ km^2$
Landscape	A unit that contains several mosaics	$> 1 \ km^2$
Region	A number of landscapes within a general bioclimatic zone, may be arbitrarily defined by major drainages etc.	Hundreds of km^2

Based on Reynolds *et al.* (1997).

there is often local extinction of species for which the conditions exceed their tolerance limits. These episodic climatic events contribute a temporal component to structural heterogeneity.

The heterogeneity in structure of desert ecosystems and landscapes is also a product of historical forces acting on the geological structures of a region, i.e. pedogenic (soil forming) processes that form soil patches. The soil patches are further modified by the interactions between soil, climate, plants, and animals. The availability of suitable patches may be extremely important for the long-term survival of species during 'crunch' periods of extreme drought or other episodic event that pushes species to their tolerance limits.

The structural characteristics of patches determine how the abiotic environment affects that patch and how the abiotic environment is modified by the structural elements of the patch. For example a patch made up of relatively tall shrubs with large dense canopies interacts with the abiotic environment very differently from a patch composed of three to four grass clumps. The redistribution of rainfall via stemflow and subsequent depth of storage in the soil is very different in these patches, as is capacity to retain litter etc. The structural characteristics affect every process and hence must be considered in any analysis of the functional aspects of patch or ecosystem. Although the patch is frequently the spatial scale at which studies of functional relationships are focused, understanding ecosystem processes requires studies at larger scales. It is not possible, for example, to predict the quantity of water entering an ephemeral stream by aggregating the run-off from the patches that make up the watershed. There are processes at the watershed scale that are unique to that assemblage of patches that make up the cover on the watershed. These processes cannot be derived by simply summing inputs and outputs from the component patches (Table 1.1).

Before we can understand ecosystem processes at the patch to landscape scale, it is necessary to examine the characteristics of the structural components and how these components serve as determinants of ecosystem and landscape properties.

The characteristics of the dominant plant species within a patch, mosaic, or landscape unit determines the way in which that entity interfaces with weather and how the environment is modified by that entity. Life form and morphological characteristics must be considered in terms of general life history characteristics in order to understand the contribution of various plant species to ecosystem and landscape properties. The way in which plants interface with wind, rain, and sunlight thereby modifying the quantity and/or intensity reaching the soil surface is very important in establishing the spatial patterns of soil moisture and air temperature. That interface also affects the physical location of materials such as plant litter and seeds.

In deserts the life forms that are most important in interfacing with weather are perennial species of bunch grasses, subshrubs, shrubs, and small trees. Perennial herbaceous plants are too sparse in most deserts to serve as determinants of ecosystem or landscape properties. Annual or ephemeral plants are generally too intermittent temporally to play an important role as a determinant of landscape properties.

Perennial bunch grasses interface with weather and landscape level processes such as run-off and deposition. Bunch grasses have dense root mats, which are concentrated under the grass canopy. The density of subcanopy roots and their relationships to the vegetative tillers that make up the bunch varies among species. The rate of root growth and root death and physiological characteristics such as quantities and kinds of materials exuded from roots also varies among species. An important contributor to the variation among species is the morphology of the canopy. Canopy morphology variables that affect the quantity of intercepted rainfall that is translocated to the root system include: leaf length, leaf width, and mean slope angle of the leaf (DePloey, 1982; Van Elewijck, 1989). Some bunch grasses have narrow, short leaves, which may be very erect (high slope angles) whereas other bunch grasses may have more prostrate, narrow, short leaves. The proportion of stemflow water that reaches the soil in the immediate neighborhood of the grass roots is immediately available for growth and production. Effective stemflow to dense root mats may stimulate complex carbohydrate exudation from the roots which can increase rates of carbon turnover and nutrient mineralization in the rhizosphere of the plant.

Life history characteristics of bunch grasses have an effect on their importance as determinants of landscape properties. For example, long-lived perennial grasses that reproduce primarily by stolons may have very low seed production. Such grasses may exhibit lags in increasing canopy cover and biomass in response to above average rainfall. If drought is sufficiently prolonged or combined with a stress such as grazing, the plant may die and have produced insufficient seed to be replaced by seedlings during the post-drought period. In such a scenario, the loss of bunch grass canopy of a long-lived grass leaves the landscape unit without perennial cover during the first storms of the post-drought period. The loss of this plant cover exposes the soil surface to the full impact of the raindrops, which initiates soil loss. In mosaics of short-lived and long-lived bunch grasses, some species produce copious seeds (e.g. *Sporobolus flexuosus* in the Chihuahuan Desert). Following drought such a species can dominate a patch in which it was a minor component during the pre-drought period.

1.2 DEFINITION OF ECOSYSTEM

In functional terms, an ecosystem is all of the living organisms in an area interacting with each other and with their physical environment. The interactions include basic ecosystem processes such as energy flow and nutrient cycling, and population processes such as competition and predation. Ecosystem processes and properties also vary as a function of a variety of vegetation structural characteristics. Ecosystem processes are also affected by animal activities and their effects on ecosystem properties such as spatial and temporal heterogeneity in the distribution of essential resources. Interactions such as effects of animals on vegetation structure, soil properties and water distribution are considered as essential modifiers of ecosystem processes. We consider ecosystems as systems because we perceive these units as having properties that are more than a sum of the component parts. System attributes that are more than the sum of the parts develop from the interactions among organisms and interactions between organisms and the physical environment. These interactions produce rate-modifying feedbacks that determine the ultimate responses of component populations through time. Just as mechanical systems require an outside source of energy to move the parts and have that system function, ecosystems require a source of energy from outside the system to drive the processes. If we consider an ecosystem as a functional unit, it has no discrete physical boundaries. We typically select study units from within a reasonably homogeneous piece of landscape (a patch) for our investigations of ecosystem properties, structure, and function. Therefore, most of the studies done to date provide little information about exchanges among distinguishable landscape units and the effects of these exchanges on ecosystem properties and processes.

The ecosystem concept however needs no defined physical boundary. Conceptually an ecosystem can be as simple as a tussock of grass, the soil in which it is rooted and the atmosphere above that plant (Fig. 1.1). The ecosystem processes: energy flow, material cycling, and the inputs, and outputs to and from that system are conceptually similar when considering a grass clump or a larger area of landscape. Considering the simple grass clump as an ecosystem, it is apparent that it is dynamic, that is, it is changing through time. As the grass grows, matures and senesces in response to climate and soil conditions, animals feed on the foliage, flowers, seeds, and roots. At the same time soil microflora and microfauna change the chemical and physical properties of the soil. There are direct and indirect feedbacks within the system that modify the state of that system through time. The magnitude and direction of such feedbacks are the result of biotic interactions such as herbivory, predation, competition, etc. (Fig.

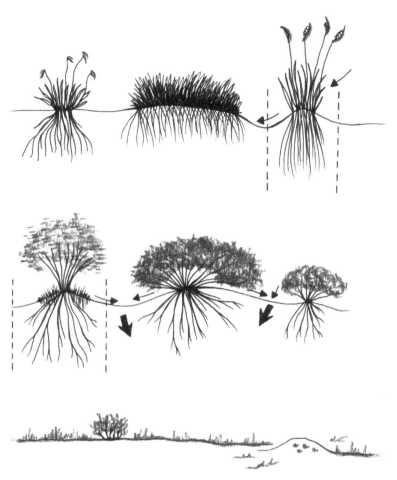

Figure 1.1   Examples of a simple ecosystems or patches (delineated by broken lines) with surrounding patches. Arrows indicate direction and magnitude of inputs and outputs. Inter-patch exchanges occur as the result of differences in microtopography.

1.1). The ecosystem does not function without inputs and outputs. Inputs include atmospheric gasses, precipitation, air-borne dust, and materials brought in by mobile consumers; outputs included gasses, wind and water transported materials and materials carried by consumers.

The focus of most ecosystem research is on ecosystem structure and function. Structure refers to the more static or compositional features of the system and function refers to the dynamic features. Odum (1962) suggests that there are three elements of structure: (1) composition of the biological community; (2) quantity and distribution of the abiotic materials; and (3) range of physical conditions. Odum (1962) lists three elements of function: (1) rate of energy flow; (2)

rate of nutrient cycling; and (3) ecological regulation of organisms by environment and of environment by organisms. Ecosystem ecologists emphasize the 'system' idea focusing on interactions among components and feedbacks that regulate the functions. The ecosystem concept thus is defined as: 'a system is a complex of interacting subsystems, which persists through time due to the interaction of its components. The system possesses a definable organization, temporal continuity and functional properties which can be viewed as distinctive to the system rather than its components' (Reichle et al., 1975). The implications of such a definition are that there are numerous links and feedbacks among the biotic and abiotic elements of the system that are essential for the temporal continuity of that system as a recognizable system. There is a need to study those connections and feedbacks that glue the system together as an integrated whole as well as to examine the connections and feedbacks of the component subsystems. The structure and functions of desert ecosystems (physically patches and mosaics) should provide a general test of the ecosystem hypothesis (sensu Noy Meir, 1979/1980) and provide the functional framework for understanding relationships between landscape units.

Ecosystems are perceived as providing goods and services for the human populations. Exploitation of the goods provided by ecosystems can compromise or change the nature of the services provided by ecosystems. Human society in attempting to understand what is involved in sustaining productive wildlands, especially in arid regions, is faced with inadequate knowledge of the quantitative relationships between ecosystem function and the goods and services needed by society (Table 1.2). The conflicts arise between the segments of society that perceive aridlands as sources of goods and those that perceive arid lands as systems supporting a healthy lifestyle. Understanding ecosystem processes is essential for wise management of ecosystems for either goods or services (Table 1.2).

Table 1.2
The Contributions of Arid Ecosystems to Human Society Divided into General Categories of Goods, Service, and Processes

---

Goods
Foods, Fibers, Fuels, Medicines, Building materials, Industrial products

Services
Modification of climate (albedo, evapotranspiration etc.), Development of soils, Degrading wastes, Storing wastes, Natural control of pathogens and parasites, Places for aesthetic, cultural, and spiritual renewal (recreation and nonconsumptive uses)

Processes
Production of organic matter, Decomposition of organic matter, Nutrient cycling, Redistribution and storage of water (infiltration, run-off, run-on, evapotranspiration), Erosion of soil

---

It is the goal of ecosystem science to predict the state of an ecosystem at some time in the future based on the knowledge of a small but selected set of parameters or variables at time zero and given a specific set of climatic conditions, abiotic inputs, and perturbations imposed by human activities. We need to understand how critical ecosystem processes vary through time and the mechanisms causing that variation. Such understanding will hopefully allow us to address ecosystem properties such as stability, resistance to change and resilience with change. Stability can be viewed as oscillations within boundaries through time such that the essential biological features of the system remain unchanged over long periods of time (decades to centuries). Resistance is a measure of the ability of an ecosystem to maintain its structural and functional properties when subjected to a disturbance. Resilience is a measure of the capacity of a system to return to its predisturbance state after it has been changed by a disturbance. In desert ecosystems, an understanding of stability, resistance and resilience is critical. The desert ecosystems of the world are undergoing rapid changes that affect ecosystem processes. These changes are collectively referred to as desertification. Desertification results in reduced productivity and reduced habitability for humans. Reversing or at least halting desertification will require a better understanding of the basic properties of desert ecosystems.

1.3 LANDSCAPE UNITS

Ecosystems may be studied and modeled without considering fixed geographical boundaries. However, for most ecologists, boundaries between vegetation types or physical boundaries such as land–water interfaces are inherent in the conceptualization of ecosystems. The boundaries of such units may be coincident with boundaries between soil types or may simply reflect changes in cover and composition of the dominant vegetation on the same soil-mapping unit. However, not all boundaries that are structurally important can be identified by such means (Ludwig and Cornelius, 1987). Watersheds and other landscape units may be subdivided into patches or mosaics on the basis of boundaries delineated by suites of structural properties (Table 1.1). Because biotic composition and quantities of abiotic materials are primary determinants of rates of ecosystem processes, these boundaries must be considered if we are to understand processes within ecosystems. Defining the limits of patches is necessary to examine how the transport of materials and movement of biota across such boundaries contributes to the functioning of a landscape or geographic region.

Because the dominant (largest biomass, most frequently occurring species) vegetation is relatively homogeneous within such a unit, it is reasonable to expect that ecosystem processes within the unit may reflect characteristics of the dominant plants, biota and soils associated with those plants. An ecosystem by this definition is equivalent to a 'habitat type' but with the caveat that features of the abiotic environment were integral parameters in characterizing the ecosystem.

However, many of the important features of desert systems involve more than interchanges and feedbacks between the biota and the abiotic environment. The ecology of desert systems must also address population and community processes such as competition and symbiosis. It is important to keep what is allowed by the structures in our observational framework (population processes, community processes) separate from the unbounded world of continuously scaled fluxes (Allen and Hoekstra, 1990).

1.4 QUESTIONS

Questions that focus on mechanisms, i.e. how ecosystems function, are central to ecosystem ecology and ecosystem science. A wide variety of questions evolve from and contribute to this kind of general question. We want to understand ecosystem functioning sufficiently well to be able to predict the state of that ecosystem at some time in the future. That goal is one shared by ecologists whose interest is limited to curiosity about the world with no application in mind and by land managers who may have short-term and long-term goals of commodity production, maintenance of biodiversity or maintenance of ecosystems services as their primary goal. One kind of question asks: what determines the long-term 'stability' of ecosystems that make up a landscape? Stability in the ecosystem sense is generally defined as little change in the rate of ecosystem processes or in the biotic and abiotic components with respect to their equilibrium value (Walker and Noy Meir, 1982). That definition presupposes that undisturbed ecosystems can be characterized by 'equilibrium' values for rates and components and that these may vary within some readily measured limits through time. Although that may not really apply to natural systems, it is a useful working conceptualization for ecosystem study. Indeed, ecosystems may vary around a set of values that changes through time but the short-term result is that of an equilibrium situation. Subsets of the stability question involve what relative proportions of individual species of organisms in the assemblages of plants, birds, mammals, arthropods, etc. will occupy a defined area of land in the future given certain climatic patterns. Related to this, how will the biomass, density etc. of these species or assemblages change through time? We would like to know if assemblages of organisms that make up identifiable and repeating patches on a landscape vary within limits over time and represent 'stable' assemblages or if even in the absence of human intervention the ('n') number of assemblages change in composition through time. A critical consideration is if an ecosystem will retain its stability or sustainability in the face of extinctions and establishment of exotics. Another way of stating that question is 'are all species equal in importance in ecosystem processes and as determinants of ecosystem properties?'

Perhaps more useful than the concept of stability is the concept of resilience. Resilience is more precisely defined as a measure of the magnitude of population perturbations or degree of disturbance a system will tolerate before changing to some qualitatively different dynamic behavior (May, 1976; Walker and Noy Meir, 1982). What are the ecosystem properties that confer resilience to that ecosystem? The notion that complexity of food webs confers greater stability and resilience to ecosystems is not supported by the data for a variety of ecosystem types. May (1976) concluded that mathematically increased complexity makes for dynamic fragility rather than robustness. However, he modified that by stating its dependence upon the general level of random fluctuation in the environment. If trophic complexity confers neither stability nor resilience to an ecosystem, what are the appropriate parameters to examine as contributors to these ecosystem properties?

Arid and semiarid ecosystems must be examined in terms of spatial and temporal heterogeneity of key processes and the organisms involved in these key processes. Arid and semiarid landscapes are complex mosaics of interlocking habitats or units that physically, and/or, by activities of the inhabiting organisms modify the patterns of availability of key resources: water and nutrients within and among units. The stability and resilience of the component ecosystems of desert landscapes must be examined in that larger context if we are to minimally achieve the objectives of predicting future states of the systems given certain climatic fluctuations or imposed management and land use practices. Although there are limited data, it is possible to summarize and synthesize these data to develop a useful conceptual model of arid and semiarid ecosystems.

1.5 DEFINING DESERTS

Before initiating a discussion of the similarities and differences among deserts and more mesic environments, it is necessary to set the limits of the temperature and moisture regimes to be considered and the general framework within which these environments are to be discussed. Considering the importance of extrapolation of results from studies, experiments and models from one 'desert' area to another, it is obvious that we must have objective criteria for making such comparisons. Because of the over-riding importance of climate as the force shaping the physical environment and biological characteristics of deserts, the delimitation and definition of areas referred to as deserts must be based on climatic criteria. I have chosen to follow the UNESCO (1977) definitions and delimitation of arid and semiarid regions based on indices of bioclimatic aridity. Bioclimatic aridity increases as water gained by rainfall falls far below that potentially lost by evaporation and transpiration. The UNESCO Man and The Biosphere (MAB) group that produced the UNESCO (1977) map of arid and semiarid zones of the world chose the ratio P/ET (in which P is mean annual precipitation and ET is the mean annual evapotranspiration) as the index of aridity applied to delimit zones of varying degrees of aridity. They chose P/ET because: (1) it gives the same value for all climates in which potential water loss is proportionally the same in relation to rainfall; (2) it is biologically accurate in climates with highly concentrated seasons and (3) it was used by the FAO (Food and Agriculture Organization) in its study of desertification risk. UNESCO (1977) proposed four classes of aridity:

(1) *Hyper-arid zone* (P/ET < 0.03), areas with very low irregular aseasonal rainfall; perennial vegetation limited to shrubs in riverbeds;

(2) *Arid zone* (0.03 < P/ET < 0.20), perennial vegetation is woody succulent, thorny or leafless shrubs, annual rainfall between 80 mm to as much as 350 mm; interannual rainfall variability is 50–100%;

(3) *Semiarid zone* (0.20 < P/ET < 0.50), steppes, some savannahs and tropical scrub; mean annual rainfall is between 300 mm and 700–800 mm in summer rainfall regimes and between 200 mm and 500 mm in winter rainfall at Mediterranean and tropical latitudes. Interannual rainfall variability is 25–50%.

(4) *Subhumid zone* (0.50 < P/ET < 0.75), primarily tropical savannahs, and chapparal and steppes on chernozim soils. The UNESCO (1977) group included this zone in their map because of the susceptibility to soil and vegetation degradation during droughts.

The deserts in which the processes described in this book are applicable are those defined by Meigs (1953) as arid or semiarid. The areas that are classified as hyper-arid, have little or no perennial vegetation and are seldom used by humans except as a source of mineral resources hence are considered here only for comparative purposes if data are available. The deserts, which are the focus of this book, are characterized by climates where rainfall rarely exceeds potential evaporation and ambient temperatures remain above freezing for more than 6 months a year. In general such areas receive between 50 mm and 400 mm precipitation per year and domestic livestock production is the primary economic use of the land.

1.6 THE PULSE-RESERVE PARADIGM AND THE AUTECOLOGICAL HYPOTHESIS

Noy Meir (1979/80) questioned the applicability of the ecosystem model to deserts suggesting that an alternative and simpler hypothesis might be sufficient for understanding deserts. He offered the autecological hypothesis as the alternative hypothesis. That hypothesis he stated as follows: 'In deserts, the dynamics of each population are determined mainly by its (the population's) independent reaction to the environment, in particular to the highly intermittent availability of water.' He developed a case for the autecological hypothesis using examples for plants both annual and perennial that he identified as 'arido-passive'. He suggested that the 'pulse-reserve' model or paradigm adequately describes the life history and population dynamics of such plants. Noy Meir extends his argument to include herbivores stating their dependence on vegetation for food and water.

These species pass into resistant or dormant stages in dry seasons hence are arid-passive with a 'pulse-reserve' life strategy. The autoecological view is that populations of plants and animals are 'too sparse' and 'too transient' to have much of a modifying effect on microclimate or soil. Thus he concludes 'a sufficient explanation of all phenomena of populations and arid community dynamics in such deserts may consist of the independent autoecological responses of individual species to the weather, without considering interspecific interactions such as competition and predation.'

Noy Meir (1973) listed three obvious attributes of arid ecosystems: (1) precipitation is so low that water is the dominant controlling factor for biological processes; (2) precipitation is highly variable throughout the year and occurs in infrequent discrete events; (3) variation in rainfall is unpredictable. Therefore he defined desert ecosystems as 'water controlled ecosystems with infrequent discrete and largely unpredictable water inputs'. Based on those considerations and definitions, Westoby and Bridges (Noy Meir, 1973) developed the pulse-reserve paradigm as a basic and general model for desert ecosystem function. In that model a rain event triggers a pulse of activity like the growth of vegetation, a variable portion of which is 'lost' to mortality and/or consumption but some part is put into a reserve such as seeds or reserve energy stores in roots or stems (Fig. 1.2). The magnitude of the pulse varies as a function of the 'trigger' event and season of the year as well as the magnitude and duration of the event.

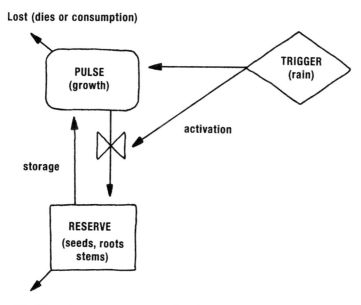

Figure 1.2 The pulse–reserve conceptual model proposed by Noy Meir (1973) for the functioning of desert ecosystems. (From Noy Meir (1973) with permission from *Annual Review of Ecology and Systematics* Volume **4**, by Annual Reviews.) www.AnnualReviews.org

The question that must be asked about this paradigm is: is it of sufficiently general applicability to describe the responses of most organisms living in deserts? If it is not generally applicable, how must it be modified to adequately describe and predict organism responses?

1.7 TEMPORALLY UNCOUPLED RESOURCES

The conventional wisdom concerning deserts is that water is the essential variable and that population responses and processes, such as rates of energy flow, can be understood as direct responses to rainfall. Analyses of plant production data for 4-6 successive years in several arid ecosystems has shown that the direct relationship with rainfall does not always hold (Ludwig and Flavill, 1979; Floret and Pontanier, 1982; Le Houerou, 1984). These investigators and others (Ettershank *et al.*, 1978; West and Skujins, 1978) have suggested that available nitrogen may affect ecosystem processes during periods of adequate moisture. However, since the system components frequently behave as would be expected by the autecological hypothesis, it is logical to conclude that there are a number of important and complex linkages between precipitation and nitrogen availability. This requires a re-evaluation of the autecological hypothesis and modification of the 'pulse-reserve' paradigm. If nitrogen availability regulates composition and quantity of production, are there feedbacks through consumers and decomposers that affect nitrogen availability? In other words do deserts behave primarily as predicted by the autecological hypothesis or as predicted by the ecosystem hypothesis? If deserts behave as ecosystems, then a modified nitrogen-regulated, 'pulse-reserve' paradigm is a more appropriate model for desert ecosystems. The available evidence for these hypotheses will be evaluated in subsequent chapters.

1.8 THE SOIL RESOURCE

As noted above, the pulse-reserve paradigm may not be a model with general applicability to arid ecosystems. The structure and processes of terrestrial ecosystems vary largely as a function of soil properties. The most important factor affecting the structure of vegetation in an ecosystem is soil. Soil is the basic resource of terrestrial ecosystems and arid ecosystems are no exception. Soil is a product of rock weathering. Soil formation is the result of complex interactions between lithological characteristics (geological history of parent rock), climate, and biological activity (Fig. 1.3). These interactions determine the texture and chemical composition of the soil. However, the spatial distribution of soil types in a landscape is the result of the geomorphologic history of the area. Climate variables (wind and water) redistribute soil in addition to contributing to weathering. According to the pulse-reserve paradigm, ecosystem processes in arid and semi-arid ecosystems are triggered by water availability. Rainfall is a poor measure of

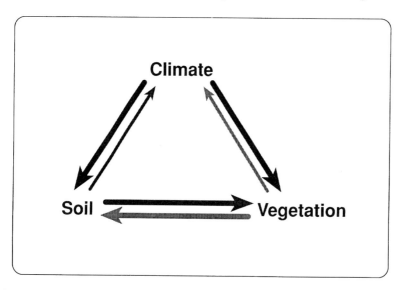

Figure 1.3   The interactions between soil, climate, and biota emphasizing feedbacks that affect the characteristics of each of these key ecosystem elements.

water availability. Plant growth is dependent on the ability of the plant to access soil water. The measure of water availability is not soil water content (percent of mass that is water), it is soil water potential. Soil water potential (matric potential) is a measure of the amount of force necessary to move water out of the soil matrix into a root. Soil water potential is a negative number that is a measure of the vacuum (negative pressure) necessary to extract water from a soil. Water molecules adhere to clay particles by electrostatic force. The relative proportions of sand, silt, and clay therefore determine how tightly water is held by the soil. Water potential is expressed in megapascals (MPa). Soil at field capacity (pore space saturated) has a soil water potential of approximately –0.03 MPa. Wilting point for agronomic species is generally considered to be –0.15 MPa. Desert soils may dry to water potentials of –0.6 MPa (no gravimetric water present). Textural properties determine the differences in the relative availability of water from soils that have the same water content. The supply of nutrients to plants is dependent upon water movement.

The characteristics of desert soils that affect nutrient availability and nutrient cycling include pH, texture, organic matter content, and landscape position. Most desert soils are classified as aridisols, which are divided into those with an argillic (clay) horizon referred to as Argids and those with no argillic horizon, referred to as Orthids. Other soils that occur in deserts are mollisols, which are soils of high base status with a dark A horizon, Entisols which are weakly developed soils and Vertisols, the cracking clay soils. Desert soils may have argillic horizons with accumulated clays, gypsic horizons enriched with calcium sulfate more than

15 cm thick or calcic horizons with calcium carbonate in the form of secondary concretions more than 15 cm thick. Most desert soils are slightly basic to very basic and also tend to be well buffered by a calcium carbonate–bicarbonate system. The pH of desert soils affects phosphorus and micronutrient availability since these are less soluble at pH > 7.0.

Soil texture combined with landscape position determines infiltration and movement of the wetting front. Thus, soil water availability (soil water potential) and aeration vary greatly with soil texture and landscape position because of water run-off–run-on relationships. Soil textural properties are affected by soil organic matter. Low soil organic matter content in general characterizes desert soils. However, there may be some locations within a desert watershed where soil organic matter contents are similar to soils of mesic environments. Soil texture, landscape position and organic matter content have marked effects on the population densities, species richness and biomass of the soil biota. Understanding the relationships between the soil biota and the abiotic features of the soil environment is essential if we are to understand nutrient cycling processes.

Soil textural characteristics are a function of the percentage composition of sand, silt, and clay. Range of particle sizes are: sand, 2–0.05 mm; silt, 0.05–0.002 mm; clay, < 0.002 mm. Soils in the general class sandy loam to sand have high infiltration rates, rapid water percolation, and are well aerated. Loams, silty loams, sandy clay loams and clay loams, have moderate infiltration rates, moderate rates of water percolation and may have water-saturated, anaerobic microsites following heavy rains. Clay soils are sticky when wet, have low rates of infiltration and percolation and are frequently anaerobic when wet. These general characteristics of soils are important determinants of how well the autecological, pulse-reserve paradigm applies to a given landscape unit.

1.9 ECOSYSTEM PROCESSES

The ecosystem processes addressed by most general ecology texts are energy flow and materials cycling. Much of the work on desert ecosystems focused on energy or trophic relationships within the system. The trophic relationship studies formed the basis for the pulse-reserve paradigm (Noy Meir, 1973) which links pulses of productivity to pulses of rainfall. The pulses of productivity provide the resources for the populations of consumers and lead to reproduction in consumer populations. This paradigm is a more formal statement of the widely held view that deserts are water-limited systems and that system responses should be predictable from a knowledge of water inputs. Although material cycling (nutrient cycling) processes have been recognized as keys to understanding forest and grassland ecosystems these processes have received little attention by desert ecologists because of the prevailing view that water is the only important limiting factor in arid ecosystems. Recent work in the Sahel, Tunisia, New Mexico and Australia has indicated the critical importance of nitrogen as a factor interacting with water affecting arid land productivity and trophic relationships. It is the interaction between water and critical materials like nitrogen that must be the focus of efforts to understand the dynamics of desert ecosystems and provide the basis for predicting long-term patterns of productivity and stability of such systems. These are also the critical basic studies needed for designing management schemes for arid lands that will maintain productivity and stable ecosystem structure.

As discussed in the previous section, it is not sufficient to understand processes at only one point in space and time. Other processes affect the patchiness within a definable landscape unit or habitat and exchanges between landscape units. These processes affect patchiness and exchanges vary across space and over time. Processes such as herbivory which result in patches of differential removal of photosynthetic area, patches of fecal matter accumulation and patterns of wastage of plant parts by herbivores must be addressed using the basic currencies that drive the system, e.g. water and nutrients. Consumers must be considered not only as consumers of plant material but also as agents affecting the structure of the system and the fate of the currencies that operate the system. In this approach consumer organisms are viewed in part as regulators of rates as suggested by Chew (1974) and also as effectors of landscape structure and soil characteristics which impact directly on basic ecosystem processes of energy flow materials cycling and resource distribution.

Biotic interactions may modify responses of species to the abiotic environment. For example, competition may be of extreme importance as a determinant of the water–nutrient availability patterns of plants, patterns of seed dispersal and patterns of herbivory that affect carbon allocation patterns and morphologies of plants. It has been suggested that competition is most important as a force organizing community structure only during 'crunch' years, that is, years when critical resources are in short supply (Weins, 1977). The modified pulse-reserve model hypothesizes that competition and predation may uncouple the critical resources threshold from direct linkage with rainfall. Thus, 'community' processes can affect lags in critical ecosystem processes and contribute to the variability of the climate–system response processes.

Before proceeding with a detailed consideration of spatial and temporal variability in important ecosystem processes, it is necessary to examine the structural characteristics of desert ecosystems. Landform (geomorphology) has a direct effect on soils and catenary (watershed) relationships. Soils and catenary relationships directly affect vegetation plus water and materials redistribution. Thus understanding the interaction of the currencies (water and nutrients) requires a basic consideration of desert geomorphology. Rainfall in desert areas is highly variable and responses of desert ecosystems to rain events will vary depending on the characteristics of the particular event. Characteristics of rain events that affect ecosystem processes and landscape processes include seasonality, intensity, quantity, and frequency of recurrence. Since rainfall is the most important driving variable, these characteristics of rainfall in deserts must be considered in detail.

1.10 PROBLEMS OF SCALING

If the ecosystem is the functional unit for study of processes such as energy flow and materials cycling, the problem of scaling then becomes critical. The studies of ecosystem processes and data available in the literature generally report the results for a plot or plots ranging in size from a fraction of a square meter to several square meters. Numerous other studies focus on one species within an ecosystem and report productivity, herbivore utilization, nutrient concentrations and distribution, etc. for that species within a relatively small homogeneous area.

Given that the data are largely based on small, relatively homogeneous bits of landscape, what is the procedure for and reliability of scaling up to an entire landscape or broad geographic areas. These large-scale areas contain numerous subunits, ecosystems, that differ in varying degrees with respect to component species, soil characteristics, topographic relationships etc. The scaling problem is less severe if we can discern some general rules or patterns that will allow us to incorporate the variability within a subunit, into a larger unit. In practice, scaling problems are generally dealt with by considering small areas to be representative of larger areas and simply multiplying by the appropriate factor. Thus the data for 0.1 m^2 plots are summed, averaged and that number multiplied by 100,000 to express the data as numbers ha^{-1} or kg ha^{-1}. There would be no problem with scaling in that manner if the variation were uniform over what appears to be a homogeneous area. Unfortunately there may be considerable variation among patches within an apparently homogeneous area that renders that assumption false.

The ability to scale up from a knowledge of processes at the square meter scale to larger landscape units and eventually to the globe is essential. The relationships between key system processes and geomorphology are beginning to be understood. Efforts to scale up from square meters in uniform vegetation to landscapes within a region necessitate a different perspective of ecological problems. The questions have to be addressed at a different scale. We need to know how patch processes modify the exchanges of resources within a single landscape unit and what resources and populations pass through the boundary *filter* between landscape units. In order to develop the understanding necessary to manage arid lands in a sustainable manner in the future, we will need to understand not only patch dynamics but also the dynamics of inter-patch and inter-landscape unit exchanges.

REFERENCES

Allen, T. F. H. and Hoekstra, T. W. (1990). The confusion between scale-defined levels and conventional levels of organization in ecology. *J. Veget. Sci.* **1**, 5–12.

Chew, R. M. (1974). Consumers as regulators of ecosystems: an alternative to energetics. *Ohio J. Sci.* **72**, 359–370.

DePloey, J. (1982). A stemflow equation for grasses and similar vegetation. *Catena* **9**, 139–152.

Dregne, H. (1986). Desertification of arid lands. In F. El Baz and M. Hassan (eds), *Physics of Desertification*, pp. 4–34. Martinus Nijhoff, Dordrecht.

Ettershank, G., Ettershank, J., Bryant, M., and Whitford, W. G. (1978). Effects of nitrogen fertilization on primary production in a Chihuahuan Desert ecosystem. *J. Arid Environ.* **1**, 135–139.

Floret, C. and Pontanier. R. S. (1982). Management and modelling of primary production and water use in a south Tunisian steppe. *J. Arid Environ.* **5**, 77–90.

Le Houerou, H. N. (1984). Rain use efficiency: a unifying concept in arid-land ecology. *J. Arid Environ.* **7**, 213–247.

Ludwig, J. A. and Flavill, P. (1979). Productivity patterns of *Larrea* in the northern Chihuahuan Desert. In E. C. Lopez, T. J. Mabry, and S. F. Tavison (eds), *Larrea*, pp. 130–150. Centro de Investigacion en Quimica Aplicada. Saltillo, Mexico.

Ludwig, J. A. and Cornelius, J. M. (1987). Locating discontinuities along ecological gradients. *Ecology* **68**, 448–450.

May, R. M. (1976). *Theoretical Ecology: Principles and Applications.* Blackwell Scientific, Oxford.

Meigs, P. (1953). World distribution of arid and semi-arid homoclimes. In *Arid Zone Hydrology*, pp. 203–210. UNESCO. Paris.

Noy Meir, I. (1973) Desert ecosystems: environment and producers. *Annu. Revi. Ecol. Syst.* **5**, 195–214.

Noy Meir, I. (1979/80). Structure and function of desert ecosystems. *Isr. J. Bot.* **28**, 1–19.

Odum, E. P. (1962). Relationships between structure and function in ecosystems. *Jpn. J. Ecol.* **12**, 108–118.

Pickett, S. T. E. and Cadenasso, M. L. (1995). Landscape ecology: spatial heterogeneity in ecological systems. *Science* **296**, 331–334.

Reichle, D. E., O'Neill, R. V., and Harris, W. F. (1975). Principles of energy and material exchange in ecosystems In W. H. Van Dobben and R. H. Lowe-McConnell (eds), *Report of Plenary Sessions of the First International Congress of Ecology, The Hague 1974*, pp. 27–43. W. Junk, The Hague.

Reynolds, J. F., Virginia, R. A., and Schlesinger, W. H. (1997). Defining functional types for models of desertification. In T. M. Smith, H. H. Shugart, and F. I. Woodward (eds), *Plant Functional Types: Their Relevance to Ecosystem Properties and Global Change*, pp. 195–216. Cambridge University Press. Cambridge, UK.

UNESCO (1977). *World Map of Arid Regions.* United Nations Educational, Scientific, and Cultural Organization, Paris.

Van Elewijck, L. (1989). Influence of leaf and branch slope on stemflow amount. *Catena* **16**, 525–533.

Verstraete, M. M. and Schwartz, S. A. (1991). Desertification and global change. *Vegetatio* **91**, 3-13.

Walker, B. H. and Noy Meir, I. (1982). Aspects of the stability and resilience of savanna ecosystems. In B. J. Huntley and B. H. Walker (eds), *Ecology of Tropical Savannas*, pp. 556–590. Springer-Verlag. Berlin.

Weins, J. A. (1977). On competition and variable environments. *Am. Sci.* **65**, 590–597.

West, N. E. and Skujins J. (eds) (1978). *Nitrogen in Desert Ecosystems.* US/IBP Synthesis Series 9. Dowden, Hutchinson and Ross. Stroudsburg, Pennsylvania.

Chapter 2 | Landforms, Geomorphology, and Vegetation

In order to develop conceptual models for the relationships among units of desert landscapes, the processes of transfer among those units and processes within units, it is necessary to consider landforms and their spatial and structural relationships. The base geology/lithology of a region is an important determinant of the geomorphology and landforms that can develop in the region. The geomorphology or landforms of deserts are features which, by interacting with climate and biota, determine soil characteristics. This chapter focuses on those structural features (landforms) that affect ecosystem processes. Landforms have a direct effect on water and materials, i.e. infiltration, run-off, erosion, water storage, salt accumulation etc., and on catenary sequences of soils and nutrients.

The structure of desert landscapes reflects the effects of low rainfall. Despite low rainfall, even in the driest deserts, there are productive patches. The distribution of productive patches in desert landscapes is a function of locations of limiting resource. Noy Meir (1973, 1981) suggested that if soil water was heterogeneously distributed in space, productivity would be higher than if water was evenly spread across the land surface. In order to understand water and soil nutrient distributions, it is necessary to examine the variety of topographic features of arid and semiarid environments and relate these features to the kinds of vegetation assemblages that develop on these features. In addition some topographic features are relatively stable (change little in ecological time of centuries) whereas others are more dynamic.

Mabbutt (1977) groups the world deserts into two general types based on gross relief and major geologic structure: (1) shield-platform deserts and (2) mountain and basin deserts. Broad plains on granitic rocks and table lands and basin lowlands on subhorizontal platform strata characterize shield-platform deserts located in Africa, Arabia, Australia, and India. In these areas mountains or hillslopes are restricted in area to ancient mountains or, in some instances, areas to mountains of recent volcanic activity. Gently sloping erosional plains are well developed and aeolian sands, which occur as sand plains or dunes (Mabbutt, 1977), characterize the lower portions of such landscapes.

In shield and platform deserts the underlying lithology and the geomorphic history of the landscape largely determines the structure and composition of the vegetation communities. For example the vegetation distribution of the Arabian shield desert of north-central Saudi Arabia correlates with six general landscape units that differ in geomorphologic history and lithology (Schulz and Whitney, 1986). The rocky, high relief areas support a tree and shrub association with an understory of bunch grasses and herbs. The surrounding sandy pediments support communities of chenopod shrubs. The ephemeral drainages (wadis) support small trees such as *Acacia* spp. and *Lycium* spp. on the benches of the main drainage channels. The extensive shield plains with fine textured soils have a vegetation that is dominated by three species of shrubs: *Rantherium*, *Fagonia*, and *Astragalus*. Two different shrubs dominate the sand plains: *Anabasis* and *Hammada*. The ephemeral drainages in the sand plains support a mixture of *Acacia* trees and *Panicum* grasses. The ephemeral lake margins support denser stands of the shrubs of the sand plains plus some *Artemisia* shrubs. The sand seas (dunes) support a larger number of species and higher density of plants than the bedrock plains (Schulz and Whitney, 1986).

Mountain and basin or basin and range deserts are characterized by mountain ranges separated by broad alluvial filled valleys. Basin and range deserts are geologically younger than the shield-platform deserts. Basin and range deserts are found in the Americas and in Asia. The alluvial processes tend to create catenary sequences with the course alluvia at the base of the mountains grading into finer sandy soils, sandy loams, clay loams and finally clays of the ephemeral lake basins. Soil salinity may be high in sinks or ephemeral lakes and surrounding soils at the base of such landscapes.

2.1 DESERT MOUNTAINS AND HILLSLOPES

Desert mountains and hills tend to be steep sided and rise abruptly from the gentle slope of the desert plains. There are few slopes of intermediate angle. The nature of hillslopes vary according to the parent materials and geological origin of the hills. Hillslopes on mountains of igneous origin tend to be characterized by debris (boulders, fractured rock) with little if any boulder accumulations at the base. Boulder fields vary in terms of boulder size distribution and physical distribution across the faces of a hill slope. In some areas boulder debris is oriented into debris chains. Boulder fields produce a variety of microsites in which plants establish. Many of the microsites receive run-on from the impervious surfaces of the rocks around the site. Even on hillslopes with an aspect facing the midday sun, microsites within boulder fields may support species of plants that require a relatively constant supply of soil water and moderate ambient temperatures. The shading and crevices in the rock and boulders provide escape sites for many species of invertebrates and vertebrates. Boulder-strewn hillslopes have received little attention from ecologists. However, the unique features of boulder fields on hillslopes which, in some instances, approach mini-oases, deserve attention (Figs 2.1–2.3).

Mountains of sedimentary origin exhibit bedding with little if any boulder field formation. The debris slopes may be simple or complex depending on the

Figure 2.1   Desert mountains, the MacDonald Ranges, in central Australia.

Figure 2.2   A desert mountain with boulder-strewn slopes, the Chihuahuan Desert, New Mexico.

Figure 2.3   An example of arid hillslopes and flat-topped hills flanking a deeply incised ephemeral stream channel in the Negev Desert, Israel.

bedded materials. In some situations rock resistant to weathering forms a cap on a hill resulting in the flat top steep-sided hills (mesas or buttes) characteristic of many arid areas. Hillslopes of sedimentary origin frequently have the form of terraced slopes. The terracing forms with the differential rates of weathering of the different sediment strata and accumulation of soil size particles on weathered benches. The terraces support an array of vascular plants. On the terraces, rainfall is enhanced by run-off water from the surrounding rock. The rock blocks between terraces may be largely devoid of vegetation except for cryptogams (lichens, algae, and moss) or may support vascular plants in crevices and surface fractures.

2.2 PIEDMONTS, ALLUVIAL FANS, AND BAJADAS

Piedmonts are graded slopes of alluvial materials extending outward from the base of mountains, mesas or hills. Piedmont gradients appear to be the result of interactions of coarseness of surface debris, nature of sediment load, magnitude and frequency of run-off across the surface, rate of stream channel cutting and density of perennial vegetation. These same factors plus the chemical nature of the parent rock largely determine the soils developed along piedmont gradients. The characteristic morphologies of drainages that develop on pediments tend to be dendritic moving from rills to small channels into larger channels.

Alluvial fans form at the outlet of mountain valleys (Fig. 2.4). The concentric contours of alluvial fans are formed from alluvia deposited as stream velocity rapidly decreases when released from the confines of the valley walls. Since only part of an alluvial fan receives sediment at any time, the surface is made up of sectors of differing age (Mabbutt, 1977).

Bajadas are compound or coalesced fans that form piedmont plains especially in closed basins. Slope profiles are determined by the slopes of the fans that

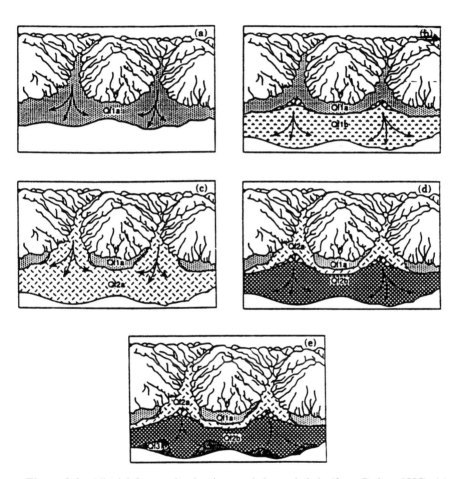

Figure 2.4 Alluvial fans coalescing into a piedmont bajada (from Parker, 1995). (a) Aggradation at canyon mouths, then coalescence of fans into bajadas during the early to mid-Pleistocene. (b) Increasing stream power relative to sediment load causing fanhead entrenchment and a shift in deposition to lower parts of the fan, soil forming on stable interfluves. (c) Deposition of materials on deposited materials (Qf2 on Qf1). (d) Reduction in sediment load relative to stream power (early Holocene) (e) Erosion by sheet wash and deposition on distal fan. (From Parker, 1995 with permission of *Journal of Arid Environments*, Academic Press Ltd.)

coalesced there by forming an undulating surface that gradually grades into slopes of finer alluvia in basin areas that may form ephemeral lakes.

Alluvial fans are one of the most common landforms in intermontane desert regions. Alluvial fans are conical-shaped deposits of fluvial and debris flow sediments that accumulate where stream channels emerge from bedrock-constrained mountain valleys. As the stream channels that are confined by upland rock structure emerge onto the intermontane basins, changes in channel morphology and flow velocity reduce the energy of the stream and its ability to transport sediment. This results in deposition of materials ranging in size from boulders to coarse sand. Where streams drain adjacent valleys fans coalesce over time to form a broad piedmont or *bajada*. Depending on the geological history of the mountains and basin, the deposition of materials may result in a simple gradient where the sorting of particles produces an exponential decrease in downslope particle size (Graf, 1988). Most alluvial fans and bajadas are not formed by simple depositional gradients. Because of a complex geological history, depositional and erosional episodes result in complex surface deposits on what appear to be simple alluvial fans. Each depositional episode may build a different segment of the fan from sediments of different origins and particle sizes (Mabbutt, 1977). Episodic floods rework previously deposited material and, depending on the depth and energy of the flood, carry coarse materials farther downslope where they bury finer materials deposited by earlier small events. This results in complex patterns of alluvial deposits across the fans and bajadas. Episodes of relative stability create surfaces of different ages and different degrees of pedogenic development. Older soils frequently have well-developed clay-enriched argillic horizons or petrocalcic horizons, which are absent in younger soils. The interaction of pedogenic and geomorphic processes produces a complex of spatial mosaics of soil texture rather than a simple linear gradient (Gile and Grossman, 1979).

Mosaics or simple linear gradients may support complex vegetation associations (Parker, 1995). For example, McAuliffe (1991) found large vegetation differences among alluvial terraces with relatively simple geomorphic histories. Other studies have concluded that marked changes in vegetation composition coincide with the boundaries between geomorphic surfaces of different ages (Parker, 1995). Many investigators have suggested that plants respond to variation in water availability governed by soil texture. The coarser soils on the upper fans store run-off water thereby providing more mesic sites than the finer textured soils on the lower slopes (Fig. 2.5). This feature may be a key to the higher diversity of vascular plants observed on upper fans. However, Parker (1995) indicated that other soil physical properties may be more important than surface texture in controlling water availability. These properties include depth to an impenetrable layer, texture of the subsoil, variability in faunal activity (biopedturbation) and the areal extent of surface stone pavement which affects both infiltration and erosion by raindrop splash.

The spatial distribution of various plant life forms is a function of the interaction of several environmental variables: soil texture, elevation, soil nutrients, and

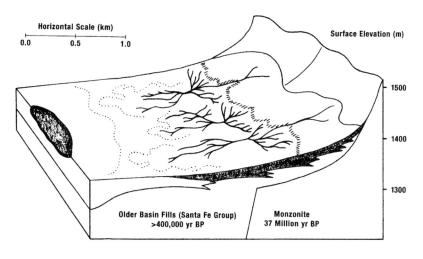

LANDFORM	Major Physiographic Part	Basin Floor		Piedmont Slope				Mountain
	Major Landform	Playa	Alluvial Plain	Fan Piedmont			Alluvial Fan	
	Component Landform			Non-buried Fan Remnant	Alluvial Fan Apron	Erosional Fan Remnant	Fan Collar	

Figure 2.5   Relationship between elevation, slope gradient, soil, and vegetation on a Chihuahuan Desert bajada (From Wondzell *et al.*, 1987). Geomorphic surfaces range in age from Jornada 1 at 250,000–400,000 years BP to Organ II at 1100–2100 years BP.

relative dryness (based on slope angle and aspect). Soil texture and nutrients may be distributed along an elevation gradient or may exhibit a mosaic pattern that is produced by the interaction of run-on, run-off, and perennial vegetation. Vegetation associations reflect the spatial pattern of these variables. For example, in the northern Sonoran Desert, the large columnar cacti exhibit different abundance patterns based on abiotic variables. The saguaro (*Carnegiea gigantea*) reaches maximum densities on coarse granite-derived alluvial soils of flat upper bajadas and basin floors. The organ-pipe cactus (*Stenocereus thruberi*) is most dense on coarse sandy soils of steep, south-facing granitic slopes. The senita (*Lophocereus schottii*) is restricted to very coarse, granite-derived alluvial soils along the banks of washes (ephemeral streams) (Parker, 1988).

Alluvial fans and bajadas of relatively uniform age are also made complex by the movements of stream channels across the fan surfaces over time. These meandering stream channels cause a shifting mosaic of stable and unstable substrates and discontinuities in soil properties. Complex patterns of vegetation assemblages on fans and bajadas are the product of the complexities of geomorphic history and pedogenesis. Understanding ecosystem processes therefore requires

Figure 2.6   A desert watershed toposequence (piedmont grassland (foreground), creosote-bush (*Larrea tridentata*) shrubland, grass–shrub mosaic and ephemeral lake basin) in the Chihuahuan Desert, New Mexico.

an understanding of the geomorphic history and soil properties in the landscape units that make up the fan or bajada.

The lithology of the mountains interacting with the prevailing climatic conditions of the geological past and the ecological present are the most important factors controlling the development of vegetation on the alluvial fans and alluvial plains. In their evaluation of the processes leading to the development of plant communities on a desert watershed that developed from a granitic monzonite mountain, Wondzell *et al.* (1996) list ecological processes and geomorphic processes of the landforms that control the structure of the vegetation community. The percentage slope and soil characteristics determine the erosion/deposition relationships. The fan collar with slope of 7–10% is a depositional area despite the relatively steep slope because of the coarse nature of the soil, which allows high rates of infiltration. The alluvial fan components with slopes of 2–4% may be erosional or depositional features depending on the soil characteristics and the type of vegetation that is developed on that soil. On the lower slopes sediments are transported by sheet flow as suspended sediment and/or bedload. The bedload materials are coarse and are lost from sheetflow when the velocity is reduced. The vegetation of the fan collar is desert perennial grassland that exhibits a relatively sharp transition (ecotone) to a desert shrub community (Fig. 2.6). The lower slopes of the fan were occupied by perennial desert grassland until grazing by domestic livestock during droughts opened these areas to colonization by shrubs. The current vegetation of the lower slopes is a grassland mosaic with small insular shrubland patches.

The importance of lithology as a primary determinant of plant community patterns in relationship to topography has been documented by studies of a gypsum hill – valley region of Spain (Guerrero-Campo *et al.*, 1999). The species that dominate the vegetation in this landscape are gypsophylic species. However, topographic position and aspect determine the life form distribution of the gypsophylic vegetation. Tall perennial grasses dominate the valley bottoms. Scrub vegetation dominates the hillslopes and hill tops. The spatial distribution of the plant communities of this landscape is a function of water availability, soil depth and sediment transport down the hillslopes. These vary with slope and soil characteristics.

The characteristics of vegetation on Sonoran Desert bajadas has been shown to be a function of slope, soil texture and soil salinity (Phillips and MacMahon, 1978). Small trees (paloverde, *Cercidium microphyllum*, ironwood, *Olyneya tesota*, and mesquite, *Prosopis juliflora*) dominate the coarse soils at the top of bajadas. The large columnar cacti, saguaro, *Carnegiea gigantea* are abundant on the coarse soils on the upper slopes and absent on the finer textured soils on the lower slopes. Shrubs are distributed from the top of the bajada to the lowest elevations. Upper slopes and mid-level slopes are dominated by creosotebush, *Larrea tridentata*, and burr sages, *Franseria* spp. On the bajada studied by Phillips and MacMahon (1978) the lower slope soils were somewhat saline and dominated by a chenopod shrub (*Atriplex polycarpa*). The fine-textured soils of the basin were quite saline (electrical conductivity (EC) between 7 and 20, whereas upper slope soils had EC in the range of 0.6–3.4). Leaf succulent shrubs (*Suaeda* and *Sarcobatus*) plus the saltbushes (*Atriplex* spp.) dominated the saline soils.

Although the soil sequences on many alluvial piedmonts (bajadas or alluvial fans) may follow simple gradients in texture and chemistry, many desert areas have a history of climate cycles with wet periods followed by dry periods of varying duration. Historical changes in climate result in changes in the volume and energy of run-off and the erosive power of floods. These historical variations can have dramatic effects on the distribution of soils and vegetation. Erosion can cut away previously deposited alluvium, leaving calcium carbonate deposition layers exposed or buried under a shallow remnant of soil. Some soil horizons (clay-rich, argillic horizons) require tens to hundreds of thousands of years to form (McAuliffe, 1994). Clay horizons and indurated calcium carbonate horizons limit percolation of water and downward transport of nutrients. When argillic horizons form on alluvial piedmonts and are subsequently dissected by later episodes of erosion, a mosaic of patches with and without argillic horizons develops. The remnant patches of piedmont with argillic horizons support a very different mix of plant species than the adjacent eroded areas of alluvium which lack a layer that inhibits water percolation (McAuliffe, 1994) (Fig. 2.7). McAuliffe's work clearly demonstrates the importance of understanding the complex evolution of soils,

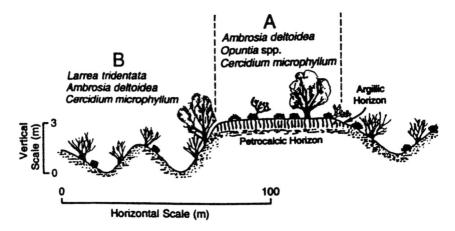

Figure 2.7   A typical landscape cross-section across a portion of an alluvial fan with a complex geological history in the Sonoran Desert, Arizona. Zone A consists of narrow, level areas with preserved argillic horizons; Zone B consists of erosional slopes and ridgelines from which argillic horizons have been completely truncated. Dominant perennials of each landscape position are listed above the landscape unit. (From McAuliffe, 1994. Reprinted with permission of The Ecological Society of America.)

especially in areas with high topographic variability. If we do not understand the soils, we will fail to understand the soil–plant water relationships and fail to understand how various landscape units respond to variations in climate.

2.3 EPHEMERAL STREAMS

Ephemeral streams vary considerably in terms of length and breadth of stream channel, complexity of channel morphology, and the riparian plant communities that develop on channel margins. Ephemeral streams that originate in mountains and that are responsible for the development of alluvial fans and bajadas are different from those streams that drain watersheds with much lower topographic relief. Ephemeral streams originate from anastomosing rills that drain upland areas of a watershed. In areas of shallow soils on steep terrain the small branches at the origin of an ephemeral stream may be steep-sided and cut to the bedrock base. Most of the small origin stream beds are covered with granular sediments.

Desert ephemeral streams (*wadis* or *arroyos*) share several features depending on the position of the drainage channel with respect to the area of watershed draining at that point. In areas where bedrock is close to the surface, ephemeral channels may have eroded channels deep into the bedrock. Here the topography of the ephemeral channel is confined on one side by a steep cliff face and the depositional–erosional benches are on one side only (Batanouny, 1973). In

ephemeral streams where the channels cut through alluvium, the drainage channel is bounded on both sides by one or more deposition–erosion benches. The
active channel is generally sterile but may support widely scattered shrubs that are
both resistant and resilient to torrential flood waters and the debris transported by
flood waters. Small feeder channels are generally shallow. In semiarid deserts the
benches or margins are populated by species of shrubs and grasses that are found
on the upland areas that supply run-off to the channel. The larger channels
develop benches because the drainage channel represents flooding by average
storms and the benches are flooded only by low-frequency episodic events. In
long-established ephemeral streams, benches may have been formed during the
wetter conditions of the Pleistocene.

The deep sediments of ephemeral stream channels are mostly sands and gravels. The sediments are characterized by exceedingly high infiltration rates. Water
entering ephemeral stream channels rapidly infiltrates into the stream bed. This
water is referred to as transmission loss (water that leaves the flowing mass of
water by gravity). Transmission losses are sufficiently high that only the very
large, intense rainstorms produce floods of sufficient depth to reach the terminus
of the stream channel (Fig. 2.2).

The vegetation along the origin stream beds differs from the upland drainage
area vegetation only in terms of cover, not species composition. Middle segment
channels are broad and distinguished by border–margins of obligate and/or
facultative riparian plant species (Fig. 2.8). Middle segment channel sediments

Figure 2.8 An ephemeral stream channel in the Chihuahuan Desert, New Mexico. The
channel margin vegetation includes mesquite (*Prosopis glandulosa*) and the obligate riparian
species *Brickelia lanciniata* in the left foreground.

tend to be coarse sands. Cross-section morphology of middle segment channels may include more than one channel and several benches. The benches of ephemeral streams afford habitats that can be occupied by both ephemeral and perennial vegetation (Batanouny, 1973). In the ephemeral streams in the hyper-arid region of Egypt, the perennial vegetation is sparse even on the stream terraces that are rarely inundated by flood waters (Batanouny, 1973).

In the northern Chihuahuan Desert, a region receiving approximately 250 mm rainfall per year, mid-segment ephemeral stream channels are lined with dense vegetation (close to 100% canopy cover). That vegetation is composed primarily of deep-rooted species such as mesquite (*Prosopis glandulosa*) and desert willow (*Chilopsis linearis*). One of the obligate shrub species, *Brickellia lanciniata*, appears to be both resistant to flood damage and resilient if damaged, and will occasionally establish in the channel. Shrubs in the channel trap sediments and organic debris during floods. This results in island formation that splits the stream channel. Depending on the location of such islands with reference to the sur-rounding drainage area, the stream bed may become braided into several channels that coalesce into a single channel further downstream.

The deep coarse sediments of the mid-segment channels produce the condi-tions for transmission losses. Transmission loss occurs when stream-flow water percolates into the sediments. Transmission losses account for marked reduction in flow volumes over the reaches of desert streams. For example, Mabbutt (1977) reported that a stream in the Sahara that received run-off equal to 80% of the rain-fall on the upstream catchment, was reduced to 15–20% of the rainfall downstream. On a larger ephemeral stream draining a basin of nearly 35,000 km^2 in India, Sharma *et al.* (1984) reported a 70% decrease in flow volume at a gaug-ing station 300 km downstream. Lengths of mid-segment channels are a function of the topography and area of the watershed drained by the channel. The obligate and facultative riparian species bordering mid-segment channels probably reflect the lengths of long-term average flows. The obligate and facultative riparian species have access to the large volume of transmission loss water stored in the sediments. Transmission losses reduce the volume and energy of channel flood waters. Reduction in volume and velocity results in deposition of bedload materials and of coarser suspended sediments. The terminal-segment channels are generally depositional features. In this segment of ephemeral streams, flow velocity is greatly reduced. Lower parts of terminal segments may receive flood water from less than half of the storms that result in flow in the upper segments. Species composition of vegetation along the terminal segments is largely the same as the surrounding area.

The desert riparian vegetation associated with ephemeral streams provides habitat for many animal species that are absent from the upland areas. Ephemeral stream channels are linear features that are characterized on a different spatial scale from the surrounding area. The patchiness of desert riparian vegetation probably reflects differences in water and nutrient contents of the sediments (Atchley *et al.*, 1999; Killingbeck and Whitford, 2000).

2.4 BASINS AND FLATLANDS

The most obvious factors affecting run-off, run-on, and type of organic matter transported by overland flow are slope and length of slope affected by flow. Areas with
slopes of less than 1% experience sheet flow or overland flow during some large
rainfalls. In arid regions with low slopes and with fine textured soils, vegetation may
develop in banded patterns (Mabbutt and Fanning, 1987). Banded vegetation patterns have been described in Africa, Australia, the Middle East and North America
(White, 1971). The most intensively studied banded vegetation is that of patches of
closely spaced trees (primarily *Acacia* spp.) interspersed in a matrix of much sparser
vegetation. In Africa, this type of vegetation is called 'brousse tigree' (tiger stripes)
and other authors refer to patterned vegetation as vegetation arcs. Vegetation in
banded patterns is not restricted to *Acacia* spp. woodlands. It has been described in
chenopod shrublands (Dunkerley and Brown, 1995), desert grassland (Worrall,
1959) and in tabosa grass (*Hilaria mutica*), tarbush (*Flourensia cernua*), mesquite
(*Prosopis glandulosa*) mosaics (Montana, 1992). Patterned vegetation divides a
slope into narrow contours of vegetation, representing run-on areas or sinks, separated by barren or low vegetation cover, run-off slopes. The effect of this spatial
arrangement of vegetation is to concentrate surface water into a small surface area,
thereby multiplying the effective moisture by a factor determined by the infiltration
rate on the run-off slope and the area of the run-off slope.

Banded vegetation develops on slopes as small as 0.5° (Dunkerley and Brown,
1995). In chenopod shrublands, the topography is a stepped microrelief of approximately 10 cm. The low benches are occupied by dense vegetation. The upslope
sides of the vegetation bands intercept water and transported debris from overland
flow (interception zone). The dynamics of banded vegetation described in a conceptual model by Tongway and Ludwig (1990) has been verified by the empirical
studies of Montana (1992). The model proposed by Tongway and Ludwig (1990)
relating the grove–intergrove run-off, run-on relationships is summarized in Fig. 2.9.

According to this model the downslope side of the grove and the intergrove
have low infiltration rates and excess water flows across the surface transporting
leaf fragments, kangaroo and sheep dung, seeds and other organic matter to the
upslope side of the next grove. The accumulation of organic matter on the upslope

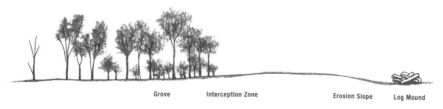

Grove   Interception Zone   Erosion Slope   Log Mound

Figure 2.9   A schematic of the spatial relationships of mulga (*Acacia aneura*) groves and
the intergrove barren patches or erosion slopes (Based on Tongway and Ludwig 1990)

side of the groves, increases the fertility of the soils in addition to the increased water storage which enhances establishment of seedling mulga trees. Thus, over long periods of time, *Acacia* die out on the downslope side of the groves, establish on the upslope side and the grove slowly moves up slope. In northern New South Wales grove–intergrove mulga woodlands are interspersed as plains with broad drainages with dense mulga groves at the landscape scale of tens of kilometers (Fig. 2.10). Infrequent very large rain events (e.g. 50–75 mm in 2–4 h) are thought to generate overland flows of sufficient magnitude to fill the depressions of the ancient drainage channels that developed during wetter periods in the past. These flows probably transport organic matter from the grove–intergrove subsystem into the drainage basin thus acting somewhat like a scouring event.

Sand plains and floodplains are relatively flat features occupying basins in basin and mountain range topography and are common in shield-platform deserts. There is evidence that the geological history of these seemingly stable landscapes is important as a determinant of soil characteristics and the vegetation that develops on those soils (Friedel *et al.*, 1993). In central Australia, the alluvial sand plains that had been deposited < 700 years BP and older landscape units such as red earth alluvial fans deposited in floods 1500–2000 years BP had vegetation dominated by long-lived trees and shrubs (*Acacia aneura, Ventilago viminalis,*

Figure 2.10   An example of banded vegetation shown from the perspective of the erosion slope looking toward the mulga (*Acacia aneura*) grove. Note the *A. aneura* skeleton on the erosion slope that forms a 'log mound' on the otherwise flat surface of the erosion slope. Photo from the Lake Mere station in northwestern New South Wales, Australia.

Figure 2.11   An example of a desert flatland, the Nullabor Plain, South Australia. The shrub dominant on this portion of the Nullabor Plain is bluebush, *Maireana* spp.

and *Atalaya hemiglauca*). The less stable surfaces are erosional features. The more stable surfaces exhibit erosion of only a few centimeters of topsoil. Shrubs with life spans of a few decades populate these surfaces. The erosional surfaces that are less stable have exposed subsoil and are populated by grasses and herbaceous plants (Friedel *et al.*, 1993). There are unstable depositional areas that are mobile sands that are moved by present day flooding and older depositional features that represent infrequent (episodic) flood deposition. The vegetation patterns on this landscape respond to similar soil-forming processes that McAuliffe (1994) documented for Sonoran Desert alluvial fans. Thus, in sand plain or flood plain landscapes, the characteristics of the landscape units vary as a function of return time of disturbance (either aeolian or fluvial) and the parent material from which the soil was formed (Fig. 2.11).

Intermediate slopes (> 0.5%–5%) are characterized by rills or runnels which are small channels that do not necessarily terminate in a larger channel. These conditions are met on the lower slopes of piedmonts or bajadas. Because of the steeper slopes, run-off tends to be channeled into rills or runnels. Sheet flow from small areas (tens of square meters) transports leaf and stem fragments, small feces, and fine sediments into the rill channels. The volume and velocity of the rills are relatively low, hence small obstructions or a reduction in slope angle result in spreading of the water and organic matter in relatively narrow bands

(1–5m in width) across an area of generally less than 10 m in length. This creates an interrupted dendritic pattern of water flow and organic matter accumulation behind debris dams. The vegetation in the accumulation zones has generally the same species composition as the run-off zone areas and differs primarily in density and/or size of individual plants.

2.5 DESERT PAVEMENTS AND STONY SURFACES

Stone strewn surfaces are common features in arid environments. They range from rock or boulder strewn surfaces to plains of gravels. Using the Arabic terms for such landscapes, areas covered with large rocks are called *hamadas* whereas those with a covering of smaller stones and pebbles are *regs*. The accepted explanation for the origin of such features is loss of fine materials by deflation until the stony surface protects the finer material below from further deflation. However, on sloping terrain stone pavement may form from below by differential erosion. In areas where stony tablelands have an underlying layer of clay stone or shale, stony gilgai relief may form. Gilgai, from the Australian aboriginal world for small waterhole, is thought to form from subsurface swelling by wet subsoils. These are characteristic landforms in the south and west of the Australian arid zone. Soils in the depressions are sandier than surrounding clays. This microrelief may be important for plant growth on such stony places. Other desert stony surface features develop from hard crusts that resist erosion, e.g. calcrete (cemented calcareous materials), gypcrete (cemented gypsum sulfates) and silcrete (silica cemented sand and gravels). These durable surfaces are of interest geologically but support little if any vegetation (Fig. 2.12).

Figure 2.12   An example of desert pavements and sand dunes. Foreground, stony pavements; background sand dunes–sand sea, Gobabeb, Namibia.

Batanouny (1973) provided an example of vegetation distribution in a stony or gravel landscape. He described the gravel surface as 'covered by closely packed gravels providing an impenetrable surface to root penetration, so it is usually sterile.' This surface is dissected by runnels (small channels) that capture wind- and/or water-borne materials. The type of vegetation that is supported in these channels is dependent on the depth of soil materials accumulated in the channel. The fine branches of the drainages support only ephemeral plants whereas the larger channels that result from the merging of small channels have deep soil and support perennial shrubs such as *Artemisia monosperma* and grasses such as *Panicum turgidum*.

2.6 RIVERS AND FLOODPLAINS

Rivers in desert regions may be conveniently divided into those that originate in wet mountainous areas and traverse a desert, e.g. Nile (Africa), Rio Grande and Colorado (North America), and Indus (Asia) and those that originate in semiarid or arid regions such as the Lake Eyre drainages in Australia. The former are generally permanent streams whereas the latter are largely ephemeral. Perennial rivers and their flood plains, because of their high water tables are considered to be mesic ecosystems. Although there are many interesting features of such systems, they are beyond the scope of this book.

Ephemeral streams in deserts tend to be wide and often trench-like. Many desert streams tend to become braided with individual channels splitting and recombining. Channels tend to have sand or fine gravel beds that are readily transported during floods as 'bedload'. Depending on the nature of the watershed, desert streams may also produce clayey flood plains as the discharge volumes vary depending upon storm characteristics in the source area. The variation in discharge and stream volume affects the degree of *aggredation* or *degradation* of various parts of stream channels. These processes in turn affect the characteristics of vegetation along stream margins and in the areas between channels.

2.7 SAND DUNES AND SAND FEATURES

The features that most people associate with the term desert are sand dunes. Sand dunes are important features of the driest deserts but a less common feature of semiarid deserts. Aeolian landforms result from the interactions of wind and the land surface and are, therefore, sensitive to changes in prevailing direction and intensity of wind. The formation of aeolian deposits and landforms require '(1) sources of sand and dust, (2) sufficient wind energy to transport these materials, and (3) conditions favoring accumulation of material in depositional areas' (Lancaster, 1994). The location and characteristics of present-day sand dunes, sand seas, and other aeolian features are products of geological history modified in some instances by recent land use practices. The importance of geological history in determining the characteristics of aeolian features as habitat for plant assemblages is similar to that described for alluvial piedmonts. For example, the Kelso sand dunes in California (USA) include active surface sands and stabilized dunes. The stabilized dunes are protected from deflation and sand movement either by vegetation or by gravel lag colluvium (Lancaster, 1994). In this dune field, the active dunes are composed of finer materials than the stabilized dunes. The most important variable affecting characteristics of these aeolian features was the spatial relationship to sources of sediment. A Pleistocene river and lake system that gradually dried provided the source of most of the material in the dune system (Lancaster, 1994).

The morphology of aeolian features is the basis of the general classification of such features. Crescent-shaped or barkhan dunes (a form of transverse dune) form by wind-blown sand grains that form mounds with slip faces on the lee side of the mound. These dunes move slowly in a downwind direction and tend to be mobile. In areas with large quantities of easily moved sand, barkhan dunes may coalesce into linkage groups of transverse dunes. Transverse dunes are the most active and tend to be migratory (Thomas, 1992). In areas with active transverse dunes, most of the vegetation is limited to the inter-dune slacks. However, because of the high infiltration rates characteristic of sand dunes, the flanks of transverse dunes may support dense populations of ephemeral plants in addition to scattered perennials.

Longitudinal (linear) dunes are elongated in the same direction as the sand transporting winds. These dunes generally have broad flanks and narrow crests. In Australia and the Kalahari, the mobile sand is largely limited to the crest and vegetation stabilizes the flank slopes. There are several theories for the formation of longitudinal dunes, such as extension of barkhans, wind sift and extension downwind from points of sand supply (Mabbutt, 1977). The linear dunes of shield-platform deserts are important landscape features that have a long history of development. This history is reflected in the complex vegetation communities that develop on the dunes. The dune soils differ in terms of lithologic origin and weathering history. The soils of the Kalahari can be divided into red, pink, and white sands with more than 90% sand (0.02–2.0 mm diameter) and fine soils that are sandy loams, sandy silts, and sandy clays (Skarpe, 1986). In the western Kalahari Desert, Skarpe (1986) described 16 vegetation types. The primary factor separating these vegetation types is the textural and lithologic origins of the sands and spatial characteristics of the dunes (height, breadth, and spacing). Sand-covered undulating flats and low dunes support open shrub savanna assemblages and/or scattered *Acacia* spp. and grasses. In areas with taller dunes, there are differences in species composition of the grasses on the dune crests (mostly *Stipagrostis amabilis*) and the dune slacks where *Stipagrostis obtusa* is dominant. The patterns of vegetation on individual linear dunes reflect differences in water and nutrient availability. The broader geographical differentiation of communities reflects differences in general soil types and in some areas the effects of recent anthropogenic disturbance.

There are similarities between the linear dune areas of southern Africa and central Australia. The landforms are the result of 'paleo-winds' (pevailing winds of past geological eras such as peak of Pleistocene glaciation) and there are distinctive soil differences among the crests, flanks and slacks or swales. The dominant vegetation of the Australian dune field is open spinifex hummock grassland (Buckley, 1981). Spinefex (or porcupine grass) is a common name applied to a group of hummock-forming grasses with tightly curled leaves that are spine-like in appearance and effect. The main species are *Triodia basedowii, Triodia pungens*, and *Plectrachne schinzii* (Buckley, 1981). The linear dunes of central Australia are 5–25 m high and 0.2–1.5 km apart. There is a consistent soil catena on the sand ridges and a consistent floristic gradient from dune crest to swale (Buckley 1982). Most plant species are restricted to a single topographic position. Species occupying the dune crests require loose sandy soil. The species on the lower dune flanks occupy sites with clayey sand. There is a strong dune to swale gradient in total soil nitrogen. Thus even in sand dune environments, subtle differences in soil texture and nutrient status resulting from the geomorphic history of the landform must be considered in order to understand the ecosystem processes of this type of landscape.

Dunes also form where vegetation traps moving sand. This is important in semiarid regions and especially important in areas where vegetation cover and composition is changing by direct or indirect effects of human use and management (Gibbens *et al.*, 1983). Trees and shrubs may serve as the foci for dune formation such as the coppice dunes formed around mesquite, *Prosopis glandulosa* in the southwestern USA and northern Mexico. In areas where mesquite cover increased with a concomitant decrease in grass cover, wind erosion accounted for between 3.4 cm and 65 cm (Gibbens *et al.*, 1983). Sand movement continues in coppice dune dominated areas resulting in deflation from interdune slacks and deposition within the coppiced shrubs.

Finer materials, aeolian dust, made up of soil particles less than 0.05 mm may be deposited in areas downwind from the source and form layers. This fine material or loess may be locally important as in the Negev Desert, Israel where loess has accumulated between steps on the limestone hills and has accumulated up to 3 m thick in the valleys. In Negev Desert sand dune areas, virtually all of the intercepted water infiltrates into the sand. The areas with highest shrub biomass and largest shrubs are the vegetated sand dunes. The adjacent loess-covered hills and valleys support much smaller shrubs. This vegetation pattern probably develops in this winter rainfall desert where water availability in the early part of the growing season is most important for productivity. Water stored in the fine-textured loess soils is less available (due to lower energy gradient between soil water and root water potentials). Water storage in dunes is also a function of the underlying geology. If the dunes form on a calcrete or duricrete base, water stored in the dune sand will be available to plants. If dunes form on a permeable substrate, infiltrated water may move to deep groundwater storage.

Dune soils tend to have very low nutrient levels because of the rapid percolation of water through dune profiles (Abrams *et al.*, 1997). Even the sandy sediments of ephemeral stream channels may have higher nutrient levels than dune soils. In the Namib Desert, most soil nutrients are linked to landscape position and are much less related to the plant communities.

2.8 LANDFORM STABILITY

As can be seen from this brief overview, the more extensive desert landforms interact with precipitation and wind to establish the particle-size gradients. Particle-size gradients interact with climate and vegetation to form the soils. Landform also affects velocity and spatial distribution of run-off water (water inputs exceeding infiltration) across the landscape. Landform with its unique pattern of soils is a product of the geological history that has shaped the landscape. This spatial–temporal context is essential for understanding patch dynamics and resource transport among patches and landscape units.

REFERENCES

Abrams, M. M., Jacobson, P. J., Jacobson, K. M., and Seely, M. K. (1997). Survey of soil chemical properties across a landscape in the Namib Desert. *J. Arid Environ.* **35**, 29–38.

Atchley, M. C., deSoyza, A.G., and Whitford, W. G. (1999). Arroyo water storage soil nutrients and their effects on gas exchange of shrub species Chihuahuan Desert. *J. Arid Environ.* **43**, 21–33.

Batanouny, K. H. (1973). Habitat features and vegetation of deserts and semi-deserts in Egypt. *Vegetatio* **27**, 181–189.

Buckley, R. (1981). Soils and vegetation of central Australian sandridges. I. Introduction. *Aust. J. Ecol.* **6**, 345–351.

Buckley, R. (1982). Soil requirements of central Australian sandridge plants in relation to the duneswale soil catena. *Aust. J. Ecol.* **7**, 309–313.

Dunkerley, D. L. and Brown, K. J. (1995). Runoff and runon areas in a patterned chenopod shrubland, arid western New South Wales, Australia: characteristics and origin. *J. Arid Environ.* **30**, 41–55.

Friedel, M. H., Pickup, G., and Nelson, D. J. (1993). The interpretation of vegetation change in a spatially and temporally diverse arid Australian landscape. *J. Arid Environ.* **24**, 241–260.

Graf, W. L. (1988). *Fluvial Processes in Dryland Rivers*. Springer-Verlag, New York.

Gibbens, R. P., Tromble, J. M., Hennessy, J. R., and Cardenas, M. (1983). Soil movement in mesquite dunelands and former grasslands of southern New Mexico from 1933 to 1980. *J. Range Mgmt.* **36**, 145–148.

Gile, L. H. and Grossman, R. B. (1979). *The Desert Project Soil Monograph: Soils and Landscapes of a Desert Region Astride the Rio Grande Valley near Las Cruces, New Mexico*. US Department of Agriculture, Soil Conservation Service.

Guerrero-Campo J., Alberto, F., Hodgson, J., Garcia-Ruiz, J. M., and Montserrat-Marti, G. (1999). Plant community patterns in a gypsum area of NE Spain: I. Interractions with topographic factors and soil erosion. *J. Arid Environ.* **41**, 401–410.

Killingbeck, K. T. and Whitford, W. G. (2001). Nutrient resorption in shrubs growing by design and by default in Chihuahuan Desert arroyos. *Oceologia* **128**, 351–359.

Lancaster, N. (1994). Controls on aeolian activity: some new perspectives from the Kelso Dunes, Mojave Desert, California. *J. Arid Environ.* **27**, 113–125.

Mabbutt, J. A. (1977). *Desert Landforms*. Australian National University Press, Canberra.

Mabbutt J. A., and Fanning, P. C. (1987). Vegetation banding in arid Western Australia. *J. Arid Environ.* **12**, 41–59.

McAuliffe, J. R. (1991). Demographic shifts and plant succession along a late Holocene soil chronosequence in the Sonoran Desert of Baja California. *J. Arid Environ.* **20**, 165–178.

McAuliffe, J. R. (1994). Landscape evolution, soil formation, and ecological patterns and processes in Sonoran Desert bajadas. *Ecol. Monogr.* **64**, 111–148.

Montana, C. (1992). The colonization of bare areas in two-phase mosaics of an arid ecosystem. *J. Ecol.* **80**, 315–327.

Noy Meir, I. (1973). Desert ecosystems: environment and producers. *Annu. Rev. Ecol. Systemat.* **4**, 25–52.

Noy Meir, I. (1981). Spatial effects in modelling of arid ecosystems. In D. Goodall and H. Perry (eds), *Arid Land Ecosystems*, pp. 411–432. Cambridge University Press, Cambridge.

Parker, K. C. (1988). Environmental relationships and vegetation associates of columnar cacti in the northern Sonoran Desert. *Vegetatio* **78**, 125–140.

Parker, K. C. (1995). Effects of complex geomorphic history on soil and vegetation patterns on arid alluvial fans. *J. Arid Environ.* **30**, 19–39.

Phillips, D. L. and MacMahon, J. A. (1978). Gradient analysis of a Sonoran desert bajada. *Southwestern Naturalist* **23**, 669–680.

Schulz, E. and Whitney, J. W. (1986). Vegetation in north-central Saudi Arabia. *J. Arid Environ.* **10**, 175–186.

Sharma, K. D., Choudhari, J. S., and Vangani, N. S. (1984). Transmission losses and quality changes along a desert stream: the Luni Basin in N. W. India. *J. Arid Environ.* **7**, 255–262.

Skarpe, C. (1986). Plant community structure in relation to grazing and environmental changes along a north–south transect in the western Kalahari. *Vegetatio* **68**, 3–18.

Thomas, D. S. G. (1992). Desert dune activity: concepts and significance. *J. Arid Environ.* **22**, 31–38.

Tongway, D. J. and Ludwig, J. A. (1990) Vegetation and soil patterning in semi-arid mulga lands of Eastern Australia. *Aust. J. Ecol.* **15**, 23–34.

White, L. P. (1971). Vegetation stripes on sheet wash surfaces. *J. Ecol.* **59**, 615–622.

Wondzell, S. M., Cunningham, G. L., and Bachelet, D. (1987). A hierarchical classification of landforms: some implications for understanding local and regional vegetation dynamics. In E. F. Aldon, V. Gonzalrd, E. Carlos, and W. H. Moir (eds), *Strategies for Classification and Management of Native Vegetation for Food Production in Arid Zones*, pp. 15–23. General Technical Report RM-150. US Department of Agriculture, Forest Service, Rocky Mountain Forest and Range Experiment Station, Fort Collins, Colorado, USA.

Wondzell, S. M., Cunningham, G. L., and Bachelet, D. (1996). Relationship between landforms, geomorphic processes and plant communities on a watershed in the northern Chihuahuan Desert. *Landscape Ecol.* **11**, 351–362.

Worrall, G. A. (1959). The Butana grass patterns. *J. Soil Sci.* **10**, 34–53.

Chapter 3 | Characterization of Desert Climates

When we think of 'deserts' we think of hot dry climates. Certainly hot and dry are good adjectives to describe deserts in a very general way. Operationally there are many aspects of 'hot and dry' climates that must be considered when attempting to understand the responses of plants and animals and ecosystem processes to desert climate. Bioclimatic regimes are generally determined by temporal and spatial averaging of weather patterns (Neilson, 1986). Regional climate establishes the limit for local climatic variation and determines the structure of plant communities (Neilson, 1986). Local weather variation results from these broad geographic-scale phenomena. Local weather variables drive processes such as primary production and nutrient cycling. These variables include seasonality of rainfall with respect to seasonal temperatures, storm intensity–duration relationships, frequency of precipitation events, frequency–event size relationships, return times of particular rain events, wind – season and intensity, daily patterns of microclimates, rainfall–potential evapotranspiration relationships. Episodic events, such as extended periods of drought, have long-term effects on the structure and function of desert ecosystems. Droughts may result from changes in sea surface temperatures of the world's tropical oceans (Palmer, 1986).

Drought is the climatic feature of arid and semiarid regions that causes the greatest concern to human populations and to governments. Drought is 'a creeping phenomenon' (Glantz, 1987). The onset and termination of 'drought' are not easily identified because drought is viewed as rainfall that is insufficient to support primary production. Glantz (1987) quoted Ivan Tannehill of the US Weather Bureau who said 'The first rainless day in a spell of fine weather contributes as much to the drought as the last day . . . No one knows precisely how serious it will be until the last dry day has gone and the rains have come again.' Although drought may be difficult to define precisely, the periodic retreat of moist air masses from arid regions during the period when rainfall is expected has many ramifications for the native biota and for the human inhabitants and their domestic livestock.

It is now apparent to global climatologists that rainfall or the lack thereof in the mid-latitudes of the planet is greatly affected by the surface temperatures of the world's tropical oceans. El Niño has become a household word in the past twenty years because of the marked effect that the surface water warming of the central Pacific has on global climates. El Niño is the periodic incursion of warm surface water into the eastern equatorial Pacific off the coasts of Peru and Ecuador. El Niño is a pattern of sea surface temperatures that are part of the Southern Oscillation. The Southern Oscillation is 'the seesawing of mean pressure differences (at sea level) between the western and eastern equatorial Pacific' (Glantz, 1987). Southern Oscillation events have been correlated with droughts, floods, and other climate anomalies around the world.

El Niño warming appears to have a greater effect on weather patterns in southern Africa than in eastern Africa and the western Sahel. Droughts in northern Africa and the Sahel have been correlated with global scale patterns of sea surface temperatures in the Atlantic, Pacific and Indian Oceans (Palmer, 1986). General circulation models of sea surface temperatures and rainfall in Africa have shown that warm sea surface temperatures in the Atlantic and Pacific have the effect of reducing rainfall over the western Sahel but warming in the Indian Ocean produces a slight enhancement of rainfall in this region. The results of the general circulation models show that for the eastern Sahel, sea surface temperature of the Indian Ocean is the most important factor in reducing rainfall (Palmer, 1986). The best-known characteristics of the El Niño–Southern Oscillation (ENSO) is the coincidence of drought or 'below-average' rainfall in many areas of the world at the same time (Ropelewski and Halpert, 1987). For example, droughts in India, northern China, Australia, parts of Africa and the Americas tend to occur almost simultaneously (Nicholls, 1991). This phenomenon is explained by sea surface temperatures that affect high altitude air mass flow patterns (the 'jet streams'). Changes in jet stream flows from highly zonal (west–east) to meridonal (north–south) produce distinct types of variation in weather patterns. Changes in jet stream flow have been shown to be responsible for extreme drought and excessive rainfall in the arid southwestern US (Neilson, 1986). Recent concern with global warming has focused considerable attention on how that warming will affect the periodicity of surface sea warming in the tropical oceans and the impacts of surface sea water temperatures on global circulation patterns.

The ENSO has a number of important effects on rainfall on the Australian continent. For the Australian continent, there is a tendency for droughts to coincide with anomalous warm sea surface temperatures in the east equatorial Pacific and for extensive wet periods to accompany periods of cool sea surface temperatures in the east equatorial Pacific (Nicholls, 1991). Other ENSO characteristics of interannual fluctuations in Australian rainfall include the following: variability is very large, droughts and wet periods have time scales of about one year, spatial scale is very large (continental), phase-locked with the annual cycle, and often followed/preceeded by the opposite rainfall anomaly (Nicholls, 1991).

These effects of the ENSO have important implications for the vegetation of the Australian arid region. 'Climates with the same general level of aridity can offer very different mixtures of growth opportunities, because of the patterning of rainfall in time; accordingly different growth forms are found' (Westoby, 1980). The virtual absence of succulents from the Australian arid and semiarid regions is attributed to twelve-month droughts that such plants are incapable of surviving.

The few species of succulents found in Australia occur in locations where run-off collects even after small rain events (Nicholls, 1991). Australia has more trees at a given level of aridity than any of the other desert areas of the world. This has been attributed to the large rainfalls characteristic of anti-ENSO periods. Trees with deep roots are able to remove deep-water stores in the soil that result from large rainfall events. These heavy rainfall events frequently persist for about one year or longer which produces conditions favorable for germination and establishment of tree species (Nicholls, 1991). These examples serve to emphasize the necessity of careful documentation and long-term records of climate patterns linked with ENSO events in the arid and semiarid regions of the world. By examining long-term climate records and records of ENSO and anti-ENSO events for local regions, it may be possible to modify management strategies for arid and semiarid lands to avoid loss of vegetation cover and soil erosion.

Coarse annual climate statistics (e.g. long-term mean annual rainfall, long-term mean temperatures) provide a first-order constraint upon the potential structure of any ecosystem. Seasonality, interannual variability and other factors produce the secondary constraints on ecosystem structure (Allen and Starr, 1982; O'Neill *et al.*, 1986). These generalizations led to the development of a conceptual model for the distribution of biomes and vegetation communities (Neilson, 1987). That model states that the seasonal timing of rainfall is in 'resonance' with the life history characteristics of the organisms that populate the biome. In other words, the timing of temperature and rainfall events within each biome allows for the selection of specific life history types adapted to this general climatic pattern and selects against other life history types. If seasonality of weather is in synchrony with the phenology and life history characteristics of an organism, that organism will survive. If they are not in phase, the organism will not survive.

These biological features were incorporated into a general theory of regional control of gamma diversity patterns. If regional climate is not overly stressful (i.e. not at the survival threshold for many organisms), the diversity of habitats in the landscape will be small. If regional climate is marginal (i.e. normal range of variation approaches or exceeds the physiological thresholds for many species) minor differences in habitat structure can place some of them outside the range of physiological tolerance. In this situation the diversity of microhabitats in the landscape will be higher than in the non-stress climatic region (Neilson and Wullstein, 1983). An example of this is when organisms change their use of habitats during drought and periods of 'average' rainfall. During periods of average rainfall, the available habitats for two species of kangaroo rats overlapped and included the entire range of habitats available in the Jornada Basin. During a drought, the suitable habitats were reduced to a much smaller subset and there was little overlap in the spatial distribution of the two kangaroo rat species (Whitford, 1976).

All of the major biotic regions (biomes) and their ecotones appear to be controlled by major climate patterns. Decreasing rainfall on an east to west gradient in North America establishes the location of the transitions between short-grass prairie and Great Basin Desert scrub. Further south are the southern deserts with spring and autumn dry periods and mid-summer rains (the Chihuahuan and Sonoran Deserts). West along a southern transect the summer rainfall deserts are replaced by the winter rainfall Mojave Desert. On a global scale, north–south gradients are primarily temperature controlled whereas the latitudinal gradients are rainfall controlled. Ecotones between major biomes represent thermal and water requirement thresholds for a large number of species (Neilson, 1987).

The standard parameters used to delineate arid climates are the water balance parameters developed by Thornthwaite (1948). The basis for Thornthwaite's classification of climates is the moisture index, I, calculated as:

$$I = 100 \left[(P/PE) - 1\right]$$

where P is average annual precipitation and PE is average annual potential evapotranspiration.

If the moisture index I is negative, it indicates a net annual water deficit. Meigs (1953) elaborated on this scheme. He classified an extreme-arid climate as one with an average precipitation of less than 65 mm, absence of precipitation seasonality, occurrences of rainless periods of 12 consecutive months and I values greater than –57.

Oberlander (1979) proposed a supplementary water balance index that accounts for shortcomings in the Thornthwaite method. Oberlander (1979) points out that the ratio between water deficit and precipitation is not a good indicator of soil moisture, since very high summer deficits can occur in winter rainfall areas with long periods of soil moisture recharge in the cool months of the year. He proposed an aridity index I_b calculated as:

$$I_b = 100 \left[(ST_{MAX} - D_{MIN})/P\right]$$

where ST_{MAX} is the maximum soil moisture storage and D_{MIN} is the minimum monthly deficit for the quantity of water necessary to remove the deficit in the least dry month.

Oberlander's I_b index has the advantage of including two measures of soil moisture, a deficit term that by itself is not sensitive to storage and a maximum storage term. This index produces + or – values indicating those arid climates with a period of soil moisture storage (+ values) and those with moisture deficits all year (– values). Minimum water deficit and maximum soil moisture storage form a continuum because only when monthly D decreases to 0 does ST begin to appear. Areas having soil moisture storage always have months with no moisture deficit (months in which storage plus precipitation exceeds the potential evapotranspiration). Areas that have moisture storage deficit in all months always have zero soil moisture storage.

The I_b index clearly shows the extreme arid desert core in North Africa grading into the less arid fringes to the north and south where winter rainfall or seasonal summer rainfalls prevail (Table 3.1). The winter rainfall areas of desert regions bordering the Arabian peninsula and the Iranian Desert are identified

with positive I_b values. The south Asian deserts are identified with I_b values ranging from + to – depending upon location and seasonality of precipitation (Table
3.1). The Central Asian deserts tend to have I_b values that are positive. Deserts of
Southern Africa have I_b values that vary from marginally negative to extremely
negative in the central Namib region. The highest intensity of aridity based on the
I_b values are in the coastal deserts of Chile, Peru, and Namibia (all deserts with
cold offshore ocean currents) and the eastern margin of North Africa from Libya
to Somalia (Oberlander, 1979).

The Oberlander index (I_b) by accounting for water deficit and water storage
appears to be the most useful index of aridity devised to date and may potentially
be used as an index of primary productivity. The Oberlander index (I_b) is far more
meaningful for examining the production potential of landscape units that receive
run-off than are any of the other indices. However, this index has not been applied
to the limited data sets on primary production in desert regions.

Table 3.1
Potential Evaporation Values and I_b Values for a Variety of Arid Regions

	PE	I_b
North Africa		
Fort Lamy, Chad	1746	+17.9
Aswan, Egypt	1505	−667
Luxor, Egypt	1394	−1700
El Azizia, Libya	1115	+32.6
Ghat, Libya	1404	−46.2
Timbuktu, Mali	1726	−23.6
Marrakech, Morocco	1002	+14.1
Niamy, Niger	1849	+13.6
Khartoum, Sudan	1828	−38.7
Gabes, Tunisia	992	+5.2
Western Sahara	1106	−77.8
Southwest Asia		
Kabul, Afghanistan	730	+58.3
Kandahar, Afghanistan	1077	+45.0
Tehran, Iran	960	+22.6
Baghdad, Iraq	1271	+39.9
Beersheba, Israel	999	+35.7
Amman, Jordan	889	+58.0
Dead Sea North, Jordan	1459	−1.0
Jidda, Saudi Arabia	1637	−55.7
Aden, Yemen	1788	−192
South Asia		
Bikaner, India	1639	−3.0
Jodhpur, India	1518	−2.3
Hyderabad, Pakistan	1563	−10.8
Karachi, Pakistan	1419	−11.9

Table 3.1 – *continued*

	PE	I_b
South America		
Anofagasta, Chile	793	–292
Iquique, Chile	830	–2350
Coquimbo, Chile	719	+1.7
Lima, Peru	884	–106
San Juan, Argentina	868	–15.4
Southern Africa		
Benguela, Angola	1402	–20.8
Khomo, Botswana	1107	–6.0
Upington, South Africa	1059	–8.3
Swakopmund, Namibia	732	–292
Australia		
Bourke, New South Wales	1085	+3.5
Broken Hill, New South Wales	938	+4.6
Alice Springs, Northern Territory	1162	–3.3
Tennant Creek, Northern Territory	1443	–7.7
Cloncurry, Queensland	1486	–5.8
Port Augusta, South Australia	1010	–1.3
Broome, Western Australia	1544	–1.2
Kalgoorlie, Western Australia	989	–0.4

Data from Oberlander, 1979.

3.1 SEASONALITY

Most of the desert areas of the world are characterized by a wet season (Table 3.2). Most of the world's deserts are seasonal, that is, there is a predictable 'season' of rainfall but not a predictable amount of rainfall in that season. The beginning and end of the wet season may vary by more than a month but significant rainfall that is sufficient for growth of some parts of the flora often has a probability of close to 1.0.

There are profound implications of seasonality of precipitation for desert organisms. For many species of animals, breeding season must anticipate the availability of resources needed to support the brood or young. There are a few species that are opportunistic with respect to reproduction but for most species the timing of reproduction is seasonal. For example, North American desert rodents in the Mojave desert, which is primarily a winter rainfall desert, breed in the late autumn and winter with the young recruited into the population coincident with the availability of fruits of spring annuals. In the Chihuahuan Desert peak recruitment of young rodents occurs in July at the beginning of the 'predictable' wet season (Whitford, 1976). Wallwork *et al.* (1984, 1986) in experimental studies of soil acari

Table 3.2
Seasonality of Selected Desert Areas of the World
(N, number of months of the wet season; Months, beginning and ending months of the wet season; % of total, percent of the mean annual rainfall that falls during the wet season; Annual, mean annual precipitation in mm; S – PET, potential evapotranspiration during the wet season; A – PET, mean annual potential evapotranspiration.)

Desert location	N	Months	% of total	Annual	S – PET	A – PET
Sahara						
Libya	6	Nov-Feb	95	267	350	870
North Morocco	6	Oct-Mar	91	239	510	1760
North Algeria	6	Oct-Mar	78	79	61	2000
South Algeria	2	Oct-Nov	100	8	410	3000
South Libya	2	Dec-Jan	100	9	180	2900
Chad	2	Aug-Sep	100	22	260	3410
Sudan	4	Jun-Sep	94	236	221	3570
Senegal	3	Jun-Aug	93	23	1006	3470
Niger	3	Jul-Sep	92	20	1006	3470
Sahel						
Senegal	5	Jun-Oct	98	620	890	2890
Niger	5	May-Sep	95	529	930	2890
Chad	5	May-Sep	93	945	690	2890
Senegal	5	Jun-Oct	95	557	930	2680
Karoo						
Puffuder	6	Nov-Apr	76	98	1100	1560
Sutherland	6	Mar-Aug	64	145	440	1100
Kalahari						
Botswana	6	Nov-Apr	84	293	1160	1970
Upington, South Africa	3	Feb-Apr	56	155	580	1970
Namib						
Windhoek	6	Nov-Mar	94	364	910	1990
Arabian						
Baghdad	6	Nov-Apr	95	145	640	2690
Basra	6	Nov-Apr	95	190	750	2160
Ammon	5	Nov-Mar	91	278	270	1420
Jidda	2	Dec-Jan	82	43	250	1900
Tehran	6	Nov-Apr	90	292	330	1560
Damascus	6	Sep-Feb	90	221	550	1680
Afghanistan	4	Dec-Mar	89	163	410	2110
Australian						
Alice Springs	5	Nov-Mar	70	251	1120	2000
Port Augusta	6	May-Oct	59	239	470	1410
Kalgoorlie	7	Feb-Aug	73	245	510	1410

Table 3.2 – *continued*

Desert location	N	Months	% of total	Annual	S – PET	A – PET
Indian						
Jodphur	4	Jun-Sep	89	320	780	2490
Hyderabad	4	Jun-Sep	85	178	830	2550
Cihuahuan (North America)						
El Paso, TX	3	Jul-Sep	54	221	560	1690
Sonoran (North America)						
Phoenix, AZ	8	Nov-Mar Jul-Sep	87	191	1410	2330
Yuma, AZ	5	Dec-Feb Aug-Sep	63	93	870	2410
Monte (South America)						
Mendoza	6	Oct-Mar	72	197	880	1320
Atacama (South America)						
Antofagosto, Chile	3	Jun-Aug	92	10	130	660

Data were taken from the British Air Ministry Meteorological Office Publication – *Tables of Temperature, Relative Humidity, and Precipitation for the World.* London (1958).

in the Chihuahuan Desert found that most taxa timed their reproduction such that the immatures entered the population in August and September, coincident with the highest probability of moist soils and microbial activity. This pattern persisted despite addition of sufficient moisture via irrigation in May–July to duplicate the late summer conditions. They also found that providing reliable moisture in winter favored an increase of a Mojave desert species that occurred at very low density in the Chihuahuan desert.

Several investigators studying the ephemeral flora have emphasized the importance of rainfall seasonality. For example, similarity in abundance of spring annuals between years was correlated with differences in September–October precipitation of the previous fall in the Mojave Desert (Bowers, 1987). In this winter rainfall desert, it has been suggested that phenological events of a number of species are triggered by rainfalls of >25 mm in the early part of the rain season (September–December (Beatley, 1974)). If the heaviest rainfall in this period is < 25 mm, but >15 mm, only scattered plants are physiologically active the following spring. If no autumn rain event approaches 25 mm, essentially all perennial plants are dormant in the following spring and ephemeral plants are absent (Beatley, 1974). These generalizations were confirmed by Bowers (1987) who concluded that the more similar the amount of precipitation in the autumn months, the more similar the relative abundance of annuals. However, the differences in species composition of the flora contributing to this relative abundance was not predictable based on seasonality and size of rainfall events. Species compositional dynamics appear to be tied to other factors that affect germination of species (such as soil nutrients).

Seasonality of precipitation has some important effects on vegetation. There are almost no C4 summer annuals in the Negev where rainfall seasonality is marked, i.e. probability of rainfall between May and September = 0. The phenologies of desert perennials are frequently seasonal and not opportunistically linked to rainfall. An appraisal of the importance of seasonality is available from the results of rainfall supplementation studies. Rainfall supplementation by irrigation did not change the timing of flowering and fruit production in creosotebush, *Larrea tridentata* (Fisher *et al.*, 1988). In studies of mesquite, *Prosopis glandulosa*, in several landscape units in the Chihuahuan Desert, we found that leaf set, flowering, and fruiting were seasonal and that variation in the initiation of these phenological events was a function of landscape position (i.e. relationship to cold air drainage). Mesquite on the piedmont slopes initiated leaf set and flowering two to four weeks earlier than mesquite on the basin floor. However, the phenology of mesquite was seasonal and the timing of events did not vary among drought years and wet years.

The data in Table 3.2 show that rainfall in some deserts is extremely seasonal, e.g. the extreme desert areas of Algeria and Libya where rainfall is recorded only in late autumn to early winter. The northern Sahara exhibits strong winter rainfall seasonality. The southern Sahara and Sahelian regions have predominantly summer seasonal rainfall. The Middle Eastern desert areas are also strongly seasonal. with winter precipitation accounting for more than 80% of the total annual precipitation. Winter precipitation is also characteristic of the higher rainfall areas of the Atacama desert in South America.

Precipitation in the arid and semiarid regions of southern Africa is not strongly seasonal. Climatically the Karoo and Kalahari are similar to the Chihuahuan Desert with more than half the precipitation occurring in the warm summer months, but with some probability of rainfall during any month of the year. The North American and South American Deserts are not strongly seasonal. The absence of strong seasonality has implications for the species composition of the ephemeral plant communities. In deserts that are not strongly seasonal, precipitation outside the wet season may be sufficiently frequent to support a distinct flora that does not develop in seasonal deserts. For example, in the Chihuahuan Desert, the C4 ephemeral flora develops virtually every year with the summer rains (except during droughts). In this desert sufficient winter rainfall to result in development of winter–spring C3 ephemerals occurs at a frequency of about one year in four. In the Sonoran Desert near Tucson, Arizona, where rainfall is distinctly biseasonal, there is a fairly predictable winter–spring ephemeral flora and a fairly predictable summer ephemeral flora.

In strongly seasonal deserts, a useful measure of relative aridity is the ratio of potential evapotranspiration (PET) of the wet season to the mean rainfall of the wet season. This ratio provides an index of moisture stress during the period when most of the primary production is likely to occur. If that ratio is less than 3.0, there is potential for high productivity during the wet season. In regions where this ratio is 10.0 or greater, productivity is severely limited due to lack of water.

3.2 SPATIAL EFFECTS

Although temporal variability of rainfall is emphasized by investigators working in desert environments, spatial variability is equally important. Spatial patterns of rainfall in arid environments are governed by topography and type of rainfall (frontal or convectional). Even frontal rainfall in desert areas may occur with cellular patterns. In Tunisia, a typical cell size is about 6–7 km^2 (Berndtsson, 1987). In the frontal storm, winter rainfall areas of the Negev Desert of Israel, cell sizes exceeded 20 km in diameter (Nativ and Mazor, 1987). Convectional storm cells vary greatly in areal extent (Fig. 3.1). Many convectional cells in the Chihuahuan Desert of southern New Mexico, USA, range in size from approximately 1 km in diameter to >10 km in diameter. Not only does storm depth vary spatially, storm intensity also varies. In a detailed analysis of storm intensities in northern Tunisia, Berndtsson and Niemczynowicz (1986) found that the spatial correlation of intense rainfall was less than 1 km. They concluded that the storm center reduction factor for reduction in intensity was 50% for an area of 20 km^2. Thus, areas experiencing the same storm will receive different amounts of rainfall. These

Figure 3.1 An example of a convectional rainstorm that affects a small spatial area. Photo from the Santa Rita Experimental Range south of Tucson, Arizona.

examples demonstrate the importance of spatial variability in rainfall as the driving variable for ecosystem processes in arid environments.

The primary factor affecting storm depth and intensity and local climate is topography. Desert mountains influence rainfall by orographic effects (rapid cooling of rising air masses) which results in increasing rainfall with increasing elevation. In Tunisia, the average annual precipitation in the Atlas mountains in the northwestern part of the country is about 1580 mm. In the flat terrain of southern Tunisia, the average annual rainfall is less than 50 mm (Berndtsson, 1987). Orographic effects are not all that dramatic but are significant. The average annual rainfall on the east side of the Organ Mountains (maximum elevation 2800 m) near Las Cruces, New Mexico is approximately 625 mm. In Las Cruces (elevation 1250 m) the average annual rainfall is 230 mm. Topography not only affects precipitation, in basin and range topography, cold air drainage from mountain slopes into the valley bottoms can have dramatic effects on the length of the growing season (Beatley, 1975). The cold air drainage and resulting reduction in growing season is a major determinant of the species composition of the vegetation. Spatial patterns in vegetation can interact with climate to affect water distribution following storms. In the Great Basin desert, where most of the precipitation is in the form of snow, the taller vegetation (*Atriplex confertifolia*) trapped more snow than mixed stands or pure stands of shorter *Ceratoides lanata* (West and Caldwell, 1983). The geographic location of mountain ranges is the most important factor causing mid-latitude aridity. The extensive dry regions of Asia (Turkestan east to the Gobi Desert of China) and the Great Basin Desert of North America are the result of orographic effect on the flow of air masses (Manabe and Broccoli, 1990). The mechanisms of mid-latitude aridity are not limited to 'rain shadow' effects of mountains but are largely the effects of mountains on the polar jet stream. The interaction of mountain ranges with the polar jet determines regions of frequent (or infrequent) passage of extratropical disturbances (Manabe and Broccoli, 1990).

Spatial variability in precipitation can have dramatic effects on the larger landscape scale. One large rainfall event in northwestern New South Wales covered an area of approximately 100 km in diameter. The flush of growth resulting from that event, attracted kangaroos, birds, and other animals from surrounding areas that experience little productivity during that growing season (personal observation).

3.3 PREDICTABILITY

The most frequent statement about rainfall in deserts is that it is unpredictable. It is necessary to examine that unpredictability seasonally in a 'seasonal' desert. It is also necessary to examine the characteristics of the rainfall in terms of effectiveness for various ecosystem processes and groups of biota. One common analysis that has been done is to calculate the coefficient of variation (standard deviation of annual rainfall ÷ the mean annual rainfall (×100). Analysis of data for a series of 26 desert areas showed that there was little relationship between the coefficient

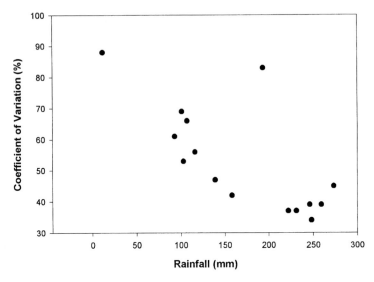

Figure 3.2   Coefficients of variation of rainfall as a function of average annual rainfall for desert areas of the world. (Data from Low 1979.)

of variation for rainfall and the mean annual rainfall of these regions (Fig. 3.2). Regression analysis of the data in Low (1979) yielded a significant negative relationship between mean annual rainfall and the coefficient of variation ($p < 0.002$). However, the r^2 was only 0.33 indicating that only 33% of the variation was attributable to mean annual rainfall (Fig. 3.2). Thus, although the coefficient of variation in annual rainfall may be higher in arid and semiarid regions than in wetter regions, that variability is not unique to desert regions. One finding is that the coefficient of variation is generally lower in forested and mesic grassland environments (Wagner, 1981). Conley *et al.* (1974) addressed the question of variability of rainfall by comparing precipitation records from Lansing, Michigan and Deming, New Mexico. They found that the interannual variation was larger in Michigan than in New Mexico, but suggested that variation approaching the physiological limits of the biota was the critical issue. This is indeed a crucial issue because it focuses directly on questions of management and sustainability. In ecosystems where precipitation inputs periodically fall below the physiological threshold limits for some proportion of the biota, loss of species can change many if not most of the ecosystem processes. If variability in rainfall periodically drops below the physiological limit thresholds, maintenance of a diverse flora and fauna is dependent on patch heterogeneity at the landscape scale (i.e. patches that provide refuges for such species). There is indirect evidence for the physiological threshold concept and its relationship to patch refugia in data from long-term studies of desert small mammal and lizard populations (Whitford, 1976; Whitford and Creusere, 1977). Several species of small mammals and lizards appeared in the trappable population in creosotebush shrubland and mixed grass–shrub mosaic habitats during a period with above average rainfall. These species disappeared when annual rainfall dropped to average or below average. Therefore, for desert regions the rainfall variability statistic of concern is the frequency of rainfall amounts below the physiological threshold for the dominant species.

3.4 INTENSITY, DURATION, FREQUENCY, AND RETURN TIME

The rainfall parameters that determine the effectiveness and the landscape effects of precipitation are intensity, duration, return-time and frequency of precipitation events. David Tongway of the CSIRO Rangelands Research Centre in Australia used a rainfall classification method suggested by Jackson (1958) to examine the distribution of rain events by event size. Jackson (1958) divided rain events into the categories ranging from very large (events > 25 mm) to very small (events < 3 mm). He documented that small rainfall events can account for a large proportion of the total rainfall in an arid region. Tongway used those categories to classify and compare the long-term rainfall at Yudnapinna Station in South Australia with that at Cobar in northwestern New South Wales. Yudnapinna is in a seasonal rainfall area of Australia where winter rainfall predominates. Cobar is in the area between the summer rainfall to the north and winter rainfall to the south. However, in both of these areas, rainfall events that were too small to result in water storage in the soil for sufficient time for plant growth tended to dominate the precipitation during some years (Table 3.3).

The relative effectiveness of short-duration rainfall varies greatly with season and geographical location. In semiarid regions at high latitudes, the cooler prevailing

Table 3.3

Comparison of Storm Distributions from 1939 to 1955 for the Yudnapinna Station, South Australia (mean annual precipitation, 203.4 mm yr^{-1}; vegetation, chenopod shrubland) and the Cobar Region of Northwestern New South Wales (mean annual precipitation 379.8 mm yr^{-1}; vegetation, eucalypt–mulga (*Eucalyptus – Acacia*) woodland)

	VH	H	M	S	L	VL
Yudnappinna Station (1939–1955)						
No. days cumulative	5	35	32	34	180	658
Mean no. days year^{-1}	0.25	2.1	1.9	2.0	10.1	38.7
Cobar Region (1963–1982)						
No. days cumulative	16	82	73	66	235	896
Mean no. days year^{-1}	0.8	4.1	3.7	3.3	11.8	44.8

Storm type classification from Jackson (1958): VH, very heavy, > 50 mm in 2 days; H, heavy, 23–50 mm in 2 days; M, moderate, 13–23 mm in 2 days; S, significant, 7.8–13 mm in 1 day; L, light, 2.8–7.8 mm in 1 day; VL, very light, < 2.8 mm in 1 day.

temperatures allow small rainfall amounts to percolate into the soil and provide effective moisture for plant growth (Sala and Lauenroth, 1982, 1985; Dougherty *et al.*, 1996). In short-grass steppe, a platyopuntia cactus (*Opuntia polyacantha*) is reported to respond to rain events as small as 2.5 mm 'although a 10 mm event has considerably more value than a 2.5 mm or 5 mm event' (Dougherty *et al.*, 1996).

Button and Ben-Asher (1983) summarized data from Avdat in the southern Negev desert in Israel using duration categories. Their summaries indicate the importance of small events, i.e. < 4 mm, in both seasonal (Negev and Yudnapinna) and nonseasonal (Cobar) areas. In the Negev a rainfall as little as 12 mm produced run-off. Rainfall intensities generally ranged between 1.5 and 30 mm min^{-1} for events lasting between 10 min and 720 min (Button and Ben-Asher, 1983). In the Chihuahuan Desert, we conducted a number of studies using simulated rainfall (sprinkler irrigation of experimental plots). These studies examined processes such as plant growth, decomposition rates, nitrogen mineralization and composition of the soil biotic communities when water was applied in a series of small events (irrigation with 6 mm of water at weekly intervals and irrigation with 25 mm of water every fourth week) (Fisher *et al.*, 1988; Whitford *et al.*, 1988a,b). When these studies were conducted during the hot months of June through September, there were no measurable differences in the soil water potential in unirrigated and irrigated plots 8 hours after irrigation. Although small rain events are effective in stimulating the activities of soil microbiota, they are relatively ineffective as a source of water for plant growth. There is considerable variation in the proportion of the total annual precipitation for a year that is composed of storms smaller than 6 mm (Table 3.4). Data on daily rainfall from a series of rain guages at Etosha National Park, Namibia and growth of grasses showed that between 35.5% and 57.6% of the total annual rainfall was effective (Du Plessis, 1999). It is not possible to establish a minimum threshold for effective rainfall that is applicable to all arid regions. The minimum storm depth that is effective for

Table. 3.4

Distribution of Rainfalls at a Recording Station on the Chihuahuan Desert Rangeland Research Center North of Las Cruces, New Mexico, USA, in the Northern Chihuahuan Desert

Year	Total (mm)	Total < 6 mm	Percent < 6 mm	Number of rainfall events > 25 mm	12–25 mm	6–12 mm	3–6 mm	< 3 mm
1979	228.6	33.8	15.0%	1	1	3	9	16
1980	227.8	25.6	11.2%	1	2	6	8	9
1981[a]	179.1	24.6	13.7%	0	2	5	6	10
1982	248.2	108.3	43.6%	0	2	6	14	35
1983	230.5	105.0	45.6%	0	1	6	17	19
1984	381.5	88.1	23.1%	2	2	11	20	18

[a]Data available for June through December only.

plant growth is dependent on soil characteristics, topography, cover and the species composition of the vegetation. However, minimum effective storm depth must be emperically determined for a particular landscape if rainfall is to be used to predict productivity.

In arid and semiarid regions that receive a significant part of the total precipitation from convectional storms, rainfalls are generally of short duration and limited in areal extent. The intensity–duration relationships of rain events are critical variables determining run-off and sediment transport. In a *Larrea tridentata* shrub dominated site in the northern Chihuahuan Desert, Bolin and Ward (1987), reported that run-off and sediment yields varied as a function of storm energy not vegetative cover at storm intensities less than 50 mm h^{-1} for 10 minutes. Most rain events fall below that threshold as seen in Fig. 3.3. Low-intensity storms tend to be of longer duration than high-intensity storms. Data from a single rain gauge while presenting the same general pattern of intensity–duration does not reflect the general magnitude of storm intensities for a large area of arid landscape. An intense–long duration precipitation event with a low probability of occurrence may have a marked effect on ecosystem processes for a number of years following such an episodic event. Because rare high intensity–long duration events may change geomorphology, i.e. backcutting of washes, channelizing rills, and redistributing large quantities of sediment, the storms may actually change the functional relationships and linkages among ecosystems of a landscape.

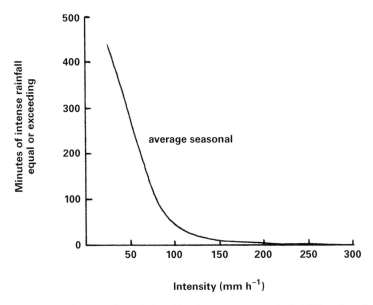

Figure 3.3 Intensity–duration relationship of average seasonal rainfall based on data from the Walnut Creek Hydrological Laboratory, Tombstone, Arizona. (Data from Osborn (1983), reproduced/modified by permission of American Geophysical Union).

Timing and intensity may be more important than total amount of rainfall. This was suggested by Gunster (1993) in a study of season of flowering in desert plants. Flowering in some (but not all) late-season flowering plants was controlled by timing and intensity of rain events rather than by the total amount of rain. Timing of rain events affects the phenologies of many but not all species of desert plants. Some species may respond to rainfall depth rather than to seasonal timing but these species are not the dominant species in desert plant communities.

3.5 MICROCLIMATE

The thermal–hydric condition of the atmosphere is a determinant of the degree of stress experienced by an organism and may determine the level of activity of the organism, e.g. stomates open or closed, animal foraging, cryptobiosis. Most desert organisms avoid conditions that are stressful by selecting environments with benign microclimates. Even in extreme ambient conditions, there are locations where conditions are moderate.

The most important determinant of climatic stress in subtropical deserts in mid-summer is the intense solar radiation. The generally low humidity and lack of closed canopy vegetation cover results in little light scatter and back reflection absorption in the atmosphere. Consequently exposed surfaces receive almost all of the radiative energy that reaches the top of the atmosphere without attenuation. Such an intense radiative environment increases the importance of structural features that produce moderate microclimates.

In arid environments moving from the surface of the soil, which is exposed to direct and intense solar radiation, to 15 cm below the surface, eliminates both thermal and hydric stress. Where soil surface temperatures reach or exceed 60°C at mid-day, temperatures at 15 cm and 45 cm below the surface fluctuate only 1–2 degrees around 30°C . As the air relative humidity at the soil surface or under a litter layer drops from 40–50% shortly after dawn to less than 10% at mid-day, the relative humidity of the air in soil interstices at 15 cm to 45 cm depth is consistently between 90% and 100%. These changes in water content of the atmosphere are also reflected in the water content of leaf litter on the soil surface under a shrub canopy. Litter loses water rapidly as the air temperature increases (Fig. 3.4). The intense radiation in deserts produces marked differences in the daily and seasonal maximum and minimum soil surface temperatures (Fig. 3.5). At 15 cm depth the difference between daily maximum temperatures and minimum temperatures ranges between 3°C and 7°C whereas at the soil atmosphere the daily difference between maximum and minimum temperatures may vary between 25°C and 50°C (Fig. 3.5).

Because the atmosphere over arid regions contains little moisture, solar radiation is intense and there is little 'greenhouse' effect to keep long-wave re-radiation at or near the soil surface. As a result, thermal loads are high and daily thermal amplitudes are large. Temperatures of objects on the soil surface rise rapidly after sunrise and drop quickly after sunset. For many organisms the temperature at the

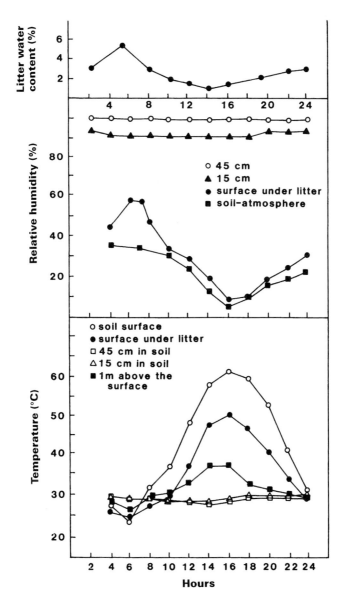

Figure 3.4   Variations in daily patterns of litter water content, relative humidity at the soil surface and in the soil, and ambient temperatures at various heights and depths in reference to the soil surface (data from Whitford *et al.*, 1981, 1982 and unpublished results).

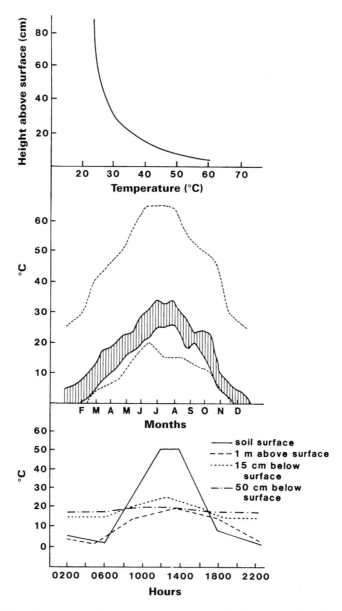

Figure 3.5   Variation in ambient temperature with height above the soil surface, maximum and minimum air temperatures and maximum and minimum soil temperatures (vertical lines), and diurnal fluctuations in surface temperatures and below ground temperatures in mid-summer in the northern Chihuahuan Desert. (Data from Whitford and Ludwig 1973, Whitford *et al.* 1973; Whitford and Ettershank 1975.)

soil surface is the most important thermal feature of the environment. Soil surface temperatures quickly rise to 60°C or above in mid-summer (Fig. 3.5) and even in mid-winter may rise to nearly 30°C (Heatwole and Muir, 1979; Whitford and Ettershank, 1975). In deserts at higher latitudes or at elevations considerably above sea level, mid-winter temperatures may fall below 0°C at the soil surface. However, at relatively short distances above and below the soil surface the thermal environment is considerably less stressful (Figs 3.4 and 3.5). Organisms that must move across the soil surface thus limit their activity to the early morning hours and to the short period preceding sunset.

The high temperatures that characterize summers in deserts exacerbate the general dryness of the atmosphere over a desert. As temperatures increase, the water-holding capacity of the air increases exponentially (Fig. 3.4). Thus at high temperatures the gradient between respiratory surfaces, internal spaces within leaves, etc. and the atmosphere is extremely steep. Rates of water movement from such surfaces to the atmosphere vary directly with the steepness of the diffusion gradient. Organisms that are immobile are forced to reduce or eliminate contact with the atmosphere in order to maintain water balance.

Moving into the soil to depths $\geq$ 15 cm is not the only effective means of avoiding thermal stress and the attendant hydric stress. The temperature of a shaded environment 1 m above the soil surface may be less than 40°C when soil surface temperatures are > 60°C. The temperatures within a shrub canopy are not attenuated as much as temperatures in the soil at depths > 15 cm but the environment at 0.5 m above the soil surface is considerably more benign than that closer to the soil surface. Moving away from the soil surface with reduced exposure to solar radiation reduces both the radiative and convective loads. The decrease in temperature with increasing height above the soil surface occurs rapidly because air mixing breaks up the boundary layer at the soil surface.

REFERENCES

Allen, T. F. J. and Starr, T. B. (1982). *Hierarchy: Perspectives for Ecological Complexity*. University of Chicago Press, Chicago.

Beatley, J. C. (1974). Phenological events and their environmental triggers in Mojave Desert ecosystems. *Ecology* **55,** 856–863.

Beatley, J. C. (1975). Climates and vegetation pattern across the Mojave/Great Basin Desert transition of southern Nevada. *Am. Mid. Nat.* **93,** 53–70.

Berndtsson, R. (1987). Spatial and temporal variability of rainfall and potential evaporation in Tunisia. In *The Influence of Climate Change and Climatic Variability on the Hydrologic Regime and Water Resources*, Proceedings of the Vancouver Symposium. IAHS Publication Number 168.

Berndtsson, B. and Niemczynowicz, J. (1986). Spatial and temporal characteristics of high-intensive rainfall in northern Tunisia. *J. Hydrol.* **87,** 285–298.

Bolin, S. B. and Ward, T. J. (1987). An analysis of runoff and sediment yield from natural rainfall plots in the Chihuahuan Desert. In USDA Forest Service. *Strategies for Classification and Management of Native Vegetation for Food Production in Arid Zones,* pp. 196–200. Rocky Mountain Forest and Range Experiment Station. General Technical Report RM-150.

Bowers, M. A. (1987). Precipitation and the relative abundances of desert winter annuals: a 6-year study in the northern Mohave Desert. *J. Arid Environ.* **12**, 141–149.

Button, B. J. and Ben-Asher, J. (1983). Intensity duration relationships of desert precipitation at Avdat, Israel. *J. Arid Environ.* **6**, 1–12.

Conley, W., Nichols, J. D., and Tipton, A. R. (1974). Reproductive strategies in desert rodents. In R. H. Wauer and D. H. Riskind (eds), *Transactions of the Symposium on the Biological Resources of the Chihuahuan Desert Region of the United States and Mexico*, pp. 193–215. National Park Service Transactions and Proceedings No. 3.

Dougherty, R. L., Laurenroth, W. K., and Singh, J. S. (1996). Response of a grassland cactus to frequency and size of rainfall events in a North American shortgrass steppe. *J. Ecol.* **84**, 177–183.

Du Plessis, W. P. (1999). Linear regression relationships between NDVL, vegetation and rainfall in Etosha National Park, Namibia. *J. Arid Environ.* **42**, 235–260.

Fisher, F. M., Zak, J. C., Cunningham, G. L., and Whitford, W. G. (1988). Water and nitrogen effects on growth and allocation patterns of creosotebush in the northern Chihuahuan Desert. *J. Range Mgmt.* **41**, 387–391.

Glantz, M. H. (1987). Drought in Africa. *Sci. Am.* **256**, 34–40.

Gunster, A. (1993). Does the timing and intensity of rain events affect resource allocation in serotinous desert plants? *Acta Oecol.* **14**, 153–159.

Heatwole, H. and Muir, R. (1979). Thermal microclimates in the pre-Saharan steppe of Tunisia. *J. Arid Environ.* **2**, 119–136.

Jackson, E A. (1958). *A Study of the Soils and Some Aspects of the Hydrology of Yudnapinna Station, South Australia.* CSIRO Division of Soils, Soils and Land Use Series No. 24.

Low, B. S. (1979). The predictability of rain and the foraging patterns of the red kangaroo (*Megaleia rufa*) in central Australia. *J. Arid Environ.* **2**, 61–72.

Manabe, S. and Broccoli, A. J. (1990). Mountains and arid climates of middle latitudes. *Science* **247**, 192–195.

Meigs, P. (1953). World distribution of arid and semi-arid holoclimes. *Arid Zone Hydrology*, pp. 203–210. UNESCO, Paris.

Nativ, R. and Mazor, E. (1987). Rain events in an arid environment – their distribution and ionic and isotopic composition patterns: Makhtesh Ramon Basin, Israel. *J. Hydrol.* **89**, 205–237.

Neilson, R. P. (1986). High-resolution climatic analysis and southwest biogeography. *Science* **232**, 27–34.

Neilson, R. P. (1987). Biotic regionalization and climatic controls in western North America. *Vegetatio* **70**, 135–147.

Neilson, R. P. and Wullstein, L. H. (1983). Biogeography of two southwest American oaks in relation to atmospheric dynamics. *J. Biogeog.* **10**, 275–297.

Nicholls, N. (1991). The El Nino/Southern Oscillation and Australian vegetation. *Vegetatio* **91**, 23–36.

Oberlander, T. M. (1979) Characterization of arid elements according to combined water balance parameters. *J. Arid Environ.* **2**, 219–241.

O'Neill, R. V., DeAngelis, D. L., Waide, J. B., and Allen, T. F. H. (1986). *A hierarchical concept of ecosystems.* In R. M. May (ed.), Monographs in Population Biology. Princeton University Press, Princeton, NJ.

Osborn, H. B. (1983). Timing and duration of high rainfall rates in the Southwestern United States. *Water Resources Res.* **19**, 1036–1042.

Palmer, T. N. (1986). Influence of the Atlantic, Pacific and Indian Oceans on Sahel rainfall. *Nature* **322**, 251–253.

Ropelewski, C. F. and Halpert, M. S. (1987). Global and regional scale precipitation patterns associated with the El Nino/Southern Oscillation. *Monthly Weather Rev.* **115**, 1606–1626.

Sala, O. E. and Lauenroth, W. K. (1982). Small rainfall events: an ecological role in semiarid regions. *Oecologia* **53**, 301–304.

Sala, O. E. and Lauenroth, W. K. (1985) Root profiles and the ecological effect of light rainshowers in arid and semi-arid regions. *Am. Midl. Nat.* **114**, 406–408.

Thornthwaite, C. W. (1948). An approach toward a rational classification of climate. *Geog. Rev.* **38,** 55–94.

Wagner, F. H. (1981). Population dynamics. In D. W. Goodall and R. A. Perry (eds), *Arid-land Ecosystems: Structure, Functioning and Management*, vol. 2, pp. 125–168. Cambridge University Press, Cambridge.

Wallwork, J. A., Kamill, B. W., and Whitford, W. G. (1984). Life styles of desert litter-dwelling microarthropods: a reappraisal based on reproductive behavior of Cryptostigmatid mites. *Suid-Afrikaanse Tydskrif vir Wetenskap* **80,** 163–169.

Wallwork, J. A., MacQuitty, M., Silva, S., and Whitford, W. G. (1986). Seasonality of some Chihuahuan Desert soil oribatid mites (Acari: Cryptostigmata). *J. Zool. Lond. (A)* **208,** 403–416.

West, N. E. and Caldwell, M. M. (1983). Snow as a factor in salt desert shrub vegetation patterns in Curlew Valley, Utah. *Am. Midl. Nat.* **109,** 376–379.

Westoby, M. (1980). Elements of a theory of vegetation dynamics in arid rangelands. *Isr. J. Bot.* **28,** 169–194.

Whitford, W. G. (1976). Temporal fluctuations in density and diversity of desert rodent populations. *J. Mammal.* **57,** 351–369.

Whitford, W. G. and Ettershank, G. (1975). Factors affecting foraging activity in Chihuahuan Desert harvester ants. *Environ. Entomol.* **4,** 689–696.

Whitford, W. G. and Ludwig, J. A. (1973). *Air temperatures, air relative humidity and solar radiation studies at the Jornada site*. US/IBP Desert Biome Publication. Utah State University, Logan, UT.

Whitford, W. G., Ludwig, J. A., and O'Laughlin, T. (1973). *Soil temperatures at 10 cm and 50 cm for the Jornada site*. US/IBP Desert Biome Publication. Utah Univeristy, Logan, UT.

Whitford, W. G., Stinnett, K., and Anderson, J. (1988a). Decomposition of roots in a Chihuahuan Desert ecosystem. *Oecologia* **75,** 8–11.

Whitford, W. G., Stinnett, K., and Steinberger Y. (1988b). Effects of rainfall supplementation on microarthropods on decomposing roots in the Chihuahuan Desert. *Pedobiologia* **31,** 147–155.

Chapter 4 | Wind and Water Processes

Visitors to arid regions are often impressed by the funnel clouds of 'dust-devils' with their vacuum cleaner power. Dust devils or dust whirlwinds are produced by thermals that suck up soil and plant fragments transporting these materials for various distances across the landscape. However dust-devils are only an interesting manifestation of an important and pervasive abiotic process. Wind is a characteristic environmental feature in deserts. Wind erosion is primarily a feature of arid regions and the probability of wind erosion affecting soil loss decreases exponentially as rainfall increases. Most arid lands have characteristic windy seasons during which high velocity wind 'storms' persist for hours and recur at high frequency. Some undetermined portion of the dust comes from uninhabited desert, but most desert dust storms probably originate in marginally arable semiarid areas due to erosion of surfaces no longer protected by vegetation, lag gravels, or stones.

4.1 WIND EROSION

Most of the available information on wind transport focuses on soil erosion and the environmental consequences of dust storms. Dust storms result in a substantial degree of deflation and erosion, especially in the world's arid zones (Goudie, 1978). Wind erosion occurs where there are long fetches (distance across unprotected soil perpendicular to prevailing wind or downslope with respect to water flow) of unprotected soil. Tilled agricultural fields are especially prone to soil loss by wind erosion. Most wind erosion research has focused on variables that affect wind transport of soil particles from agricultural fields. There are a number of variables that affect wind erosion and the generation of 'dust storms'. These include areal extent of bare soil, soil moisture, extent and type of vegetative cover, and wind speed. Most research has focused on wind tunnel studies of soil moisture, soil aggregate structure, and soil texture as factors affecting the threshold velocity for entrainment of particles in a wind stream.

Skidmore (1986) provided a general functional relationship for wind erosion based on numerous studies of factors affecting the entrainment of soil particles in moving air streams. That functional relationship is expressed as:

$$E = f(I, K, C, L, V)$$

where E is the potential average annual soil loss per unit area, I is a soil erodibility index, K is a soil ridge roughness factor, C is a climatic factor, L is the unsheltered median travel distance of wind across an unvegetated patch of soil (fetch), and V is an equivalent quantity of vegetative cover

Skidmore (1986) suggested that wind erosion occurs when the shear stress exerted on the surface by the wind exceeds the ability of the surface material to resist detachment and transport. Dryness increases the susceptibility of the surface to erosion as does surface roughness, and surface cover. Soil moisture affects the adhesive properties of the soil surface resisting detachment and surface roughness and cover effectively reduce the velocity of the wind. An important feature of wind erosivity is that erosivity varies with the cube of excess velocity over the threshold velocity necessary to initiate particle movement (Breed and McCauley, 1986). On barren or sparsely vegetated areas, the fine particles are winnowed from the coarser materials which move across the surface by saltation (saltatory movement involves particles bouncing across the surface). Because some of the energy of bouncing soil particles is transferred to surface particles, the threshold velocity is effectively reduced once the saltation process has started. This accounts for erosivity varying with the cube of velocity in excess of the threshold velocity. One problem with the threshold velocity concept is that there are temporal variations in saltation activity resulting from the unsteady behavior of wind. Natural wind erosion events consist of intermittent bursts of blowing soil interspersed with periods of inactivity. The level of intermittency is governed by whether typical wind fluctuations span the gap between mean wind speed and threshold wind speed (Stout and Zobeck, 1997).

The increase in flux of soil mass with distance downwind (the fetch effect) involves three mechanisms: (1) the 'avalanching' mechanism which occurs when one particle moving downwind dislodges one or more particles when it impacts with the surface; (2) an 'aerodynamic feedback' effect in which the aerodynamic roughness height is increased by saltation of particles thereby increasing the momentum flux of soil particles. These increases produce a positive feedback loop with respect to distance downwind; (3) 'soil resistance' which is primarily an expression of the change in threshold velocity with distance. Changes in threshold velocity result from inhomogeneous patchiness of soil, or progressive destruction of aggregates and crusts in the downwind direction of the fetch. Empirical studies in the Owens Dry Lake, California, support this three mechanism model (Gillette et al., 1996).

One difficulty encountered in modeling wind erosion is the determination of the threshold wind speed or friction velocity at which soil movement is initiated. Winds that are sufficient to initiate soil erosion are generally turbulent and gusty. When wind speeds are averaged over longer time periods (i.e. 30–60 s), the apparent threshold is considerably lower than the true wind speed at which saltation (sand grain movement) is initiated (Stout, 1998). High-frequency sampling of wind speed and saltation activity are critical to accurately determine the true threshold of a wind-eroding surface.

The coarser materials that move across an unvegetated surface are the texture of dune sand. The lighter clay and silt-sized particles are transported aloft as the 'dust'. Saltating particles break interparticulate bonds in soil aggregates and in crusts releasing additional particles for entrainment. Studies of the rates of erosion in relation to the flux of saltating grains for crusted soils of different strengths (measured by flat-ended penetrometer) found that the rate of dislodgment of surface particles decreased with increasing crust strength. Dust is release from craters formed by single impacts of saltating sand particles. The volume of material removed from impact craters is a linear function of the kinetic energy of the saltating particles in an unaggregated soil (Rice *et al.*, 1996).

The sand moved by saltation is frequently deposited in dunes or in dune-like structures around plants or other physical objects. In undisturbed deserts, even strong winds in excess of 30–50 km h^{-1} will not usually generate airborne dust because of the vegetative cover (Musick *et al.*, 1996). Even sparse vegetation reduces wind erosion by its effect as a roughness factor that reduces velocity by generating turbulence. In a laboratory wind tunnel study using wood dowels to simulate plant stems, Van de Ven *et al.* (1989) developed an equation to predict soil loss as a function of wind velocity, soil type, and vegetation characteristics. They found that if any of the vegetation parameters, density (*N*), height (*H*) and diameter (*D*), were increased, soil loss decreased by the square root of that factor. Their generalized equation was of the form:

$$SL = K\,(Uh - Ut)/\sqrt{(NDH)}$$

where *SL* is soil loss, *K* is an empirically derived constant, *Uh – Ut* is a velocity parameter specific for a soil's resistance to erosion. The more resistant the soil is to erosion, the higher *Ut* will become and the lower *Uh – Ut* becomes for a particular free-stream wind velocity. *N* is the number of sticks (stems) per unit area, *D* is the diameter of the sticks, and *H* is the height of the sticks above the soil surface.

The effectiveness of plants in natural ecosystems in reducing the transport capacity of wind has been the subject of very few studies. Studies of growing crop plants in wind tunnel experiments showed that soil loss was highly correlated with a plant area index and canopy cover. These plant variables reduced the transport capacity of wind. Canopy cover of 4% reduced the transport capacity of a 57 km h^{-1} wind by 50% (Ambrust and Bilbro, 1997). A study by Van de Ven *et al.* (1989), emphasized the importance of even dead annual plants in stabilizing soils in arid regions. Grazing and browsing by native or domestic herbivores that reduces height and/or density of grass clumps or sparse shrubs, will increase soil loss by wind erosion in arid regions. A variable not accounted for in this model is the porosity of the vegetation cover. Sparse multistemmed mesquite plants provide less wind resistance and less protection than a dense shrub of the same height and ground cover value. Compact desert shrubs and tussocks of grass probably provide more wind resistance and soil protection than large multistemmed shrubs that have very open canopies. In the Kalahari, removal of vegetation by fire, grazing, or drought resulted in a dramatic increase in near-surface wind velocity (up to 200%) and a decrease in shear stress velocity necessary to initiate sand movement on dune surfaces. Removal of vegetation by burning reduced areodynamic roughness by two orders of magnitude and reduced mean shear velocity by about 60% (Wiggs *et al.*, 1994). These authors concluded that the relictual nature of the Kalahari dunes was attributable to the extensive cover by grasses and other vegetation. Movement of Kalahari dune sands could only occur immediately after devegation by fire, drought, or overgrazing (Wiggs *et al.*, 1994).

The cohesive resistance of soil to wind erosion can be affected by variables other than water. In areas with widely spaced shrubs, the presence of annual plants, even if dead, contributes to the cohesive resistance of the soil because of the fine roots and fungi that grow on the dead roots. The dead roots contribute the source of carbon required by soil microflora for their growth and reproduction. Soil microbes interact with the soil particles to form aggregates. The cohesive resistance to wind erosion also varies directly with the percentage of the soil volume that is made up of aggregates (Leys and Raupach, 1991). Chepil (1951) indicated that as the fraction of nonerodible aggregates increased, erodibility decreased. The clay fraction of the soil contributes directly to the formation of soil aggregates (Tisdall and Oades, 1982). Erodibility of soils is greatest in soils with less than 10% clay content, minimal erodibility in soils with a clay content of 20–30% and increasing erodibility in soils with clay content >30% (Chepil, 1953).

Biological crusts stabilize many undisturbed desert soils. Biological soil crusts consist of cyanobacteria, green algae, lichens, and mosses. Threshold wind friction velocities may be considerably higher than the wind forces experienced at a site. Foot, vehicle and livestock traffic break up biological crusts. Areas with biological disturbed crusts are frequently exposed to wind speeds that exceed the stability thresholds of the damaged crusts (Belnap and Gillette, 1998). Because biological crust organisms are concentrated in the top 3 mm of soils, sand-blasting by wind can strip these crusts from the surface. Loss of biological crusts opens sites to wind and water erosion and subsequent loss of productivity.

Algal mats have been shown to stabilize sand dune areas (Pluis and Winder, 1989) and to resist destruction by saltating sand if the blow-out areas were sufficiently small. However, on large blow-out (barren) areas, algal crust development was not sufficiently fast to resist growth of the blow-outs. Disturbance that breaks up root-held soil or algal mats would therefore contribute to wind erosion. Even rain crusts which form from splash erosion (fine particles that are held together by physical forces to form a 'skin' on the soil surface) have the effect of increasing the threshold velocity. Soil piles originating from digging by animals, deposits of subsoil around ant nest entrances, and termite foraging galleries on the surfaces of plants disturb the surface crusts or place soil in locations from which it is easily detached by wind. Soil ejected by burrowing and digging animals dries rapidly and loses aggregate stability, which contributes to entrainment of this soil at low wind velocities. There are no published quantitative measurements of the effects of soil disturbance by animals on wind erosion and aeolian transport of soil. In arid and semiarid regions the relatively large volumes of soil moved by vertebrates and invertebrates (Whitford and Kay, 1999; Whitford, 2000) may significantly affect soil movement.

The most important variable affecting wind erosion in noncultivated desert and semi-desert areas is the fetch length of unvegetated patches. The existing wind erosion models do not effectively address this variable. Herrick (unpublished data) conducted empirical studies of threshold velocities for detachment of dyed soil particles. His data are for sand particle detachment, which provides information on the threshold fetch length for saltation. Data were collected on windy days in several landscape units at the Jornada Experimental Range in the northern Chihuahuan Desert. In a bunch grass – black grama (*Bouteloua eriopoda*) grassland and a burrograss (*Scleropogon brevifolia*) – tobosa grass (*Hilaria mutica*) association on fine textured soil, minimum fetch length for detachment was approximately 50 cm and threshold velocities for fetch lengths between 50 cm and 300 cm ranged between 25 and 52 km h^{-1} (Fig. 4.1) In mesquite coppice dune with a high degree of surface roughness due to varying heights of coppices and dunes, the minimum fetch length for saltation was approximately 500 cm. Threshold velocities ranged between 18 and 35 km h^{-1} (Fig. 4.1). In these

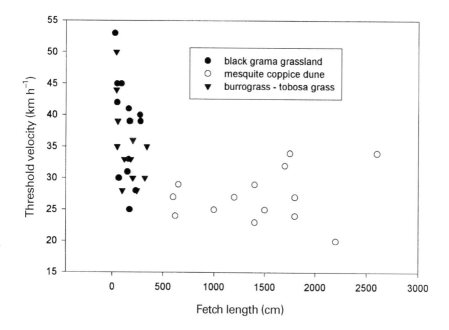

Figure 4.1   Relationship between fetch length (diameter of unvegetated patches) and threshold detachment velocity for sand particles. (Herrick unpublished data.)

ecosystems with different vegetation cover and height characteristics, the relationship between fetch length and threshold velocity for saltation was not apparent above some minimum value for fetch length.

Reduction in canopy height and cover of vegetation and disruption of the soil surface by grazers, especially during periods of reduced rainfall, can create small erosion cells (modified from the usage of Pickup, 1985), i.e. unvegetated patches of sufficient fetch length for wind to exceed threshold velocities for a particular soil. Continued movement of soil particles across an erosion cell will bury or abraid the vegetation on the downwind side of the barren patch. Barren patches in desert landscapes are also subject to water erosion, which exacerbates the growth of the cell. Wind and water erosion combine to increase the size of the erosion cells. Eventually cells will coalesce into large barren areas unless the barren patches are colonized by vegetation. When the coalesced erosion cells pass some size threshold, they will behave as an unvegetated cultivated field with respect to wind erosion. Many of the present desert landscapes have undoubtedly developed as a result of wind erosion on erosion cells that rapidly increased in size and coalesced as a result of grazing during periods of drought.

In terms of ecosystem function, aeolian transport of organic matter may be of greater importance than transport of soil. Soil losses resulting from disturbance are reduced over time as stabilization by vegetation or as lag gravels, and stones form a desert pavement. However, dead plant materials are susceptible to redistribution by wind and water. The redistribution of organic materials contributes to the development of the well-known 'islands of fertility' around desert shrubs (deSoyza *et al.*, 1997). Organic matter accumulations under shrubs are related to the morphologies of the shrubs, shrub densities and productivity of the surrounding areas. Because of the small size and mass, much of the organic debris, the threshold velocity to dislodge and entrain leaves, stems and fruits is quite low. Applying aerodynamic theory to the shape and mass of creosotebush, *Larrea tridentata* leaves, stems and fruits, deSoyza *et al.* (1997) calculated threshold velocities of 1.1 m s^{-1} (3.96 km h^{-1}) for leaves, 1.72 m s^{-1} (6.2 km h^{-1}) for fruits, and for a 25 mm stem segment 1.1 m s^{-1} (3.96 km h^{-1}). Once entrained in a wind stream, leaves, small stems, etc. are carried until turbulence and eddies reduce the wind velocity and the material is deposited. Turbulence develops as a result of resistance to flow by objects in the air stream path. In sparse shrubland, plants of different morphologies differ greatly in their effectiveness as litter 'precipitators' (Fig. 4.2).

Shrubs with a hemispherical shape accumulate more wind-borne litter than shrubs shaped like inverted cones. As shown in Fig. 4.2, based on aerodynamics theory, the streamline flow over and around a hemispherical shrub will cause a small local turbulence under the canopy, which will result in fragment deposition but not entrainment of that material. When that air flow pattern is compared to a conical-shaped shrub, the downflow will accelerate the flow of air around the base of the cone entraining litter in that flow. Although this analysis was based on idealized leaf, stem, and fruit shapes and shrubs reduced to simple geometrical

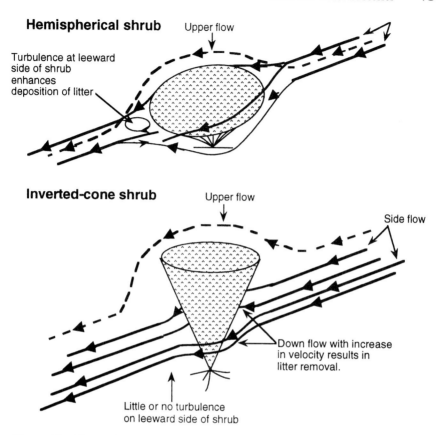

Hemispherical shrub Upper flow

Turbulence at leeward
side of shrub
enhances
deposition of litter

Inverted-cone shrub Upper flow

Side flow

Down flow with increase
in velocity results in
litter removal.

Little or no turbulence
on leeward side of shrub

Figure 4.2 Flow paths of wind encountering shrubs with cone-shaped canopies. Upper panel, example of a shrub with a hemispherical crown. Turbulence on the leeward side of hemispherical crowns causes deposition of litter. Lower panel, high wind velocity due to downflow at the base of conical shrubs causes turbulence and litter to be entrained in the wind stream. (From DeSoyza *et al.*, 1997.) Reprinted with permission by The American Midland Naturalist, University of Notre Dame Press.

shapes that were perfectly symmetrical, the results provide an explanation for the observed patterns of litter accumulation and soil mounding under *L. tridentata* shrubs in the Chihuahuan Desert of New Mexico and are probably generally applicable to other shrub-dominated desert ecosystems.

4.2 REDISTRIBUTION OF RAINFALL

Although wind is an important force in the structuring of patches in the landscape and in the concentration of organic matter in some patch locations, water may be an equally important or more important force affecting patch structure and dynamics. Because water is essential biological activity, those variables that affect redistribution of rainfall across a landscape are critical variables. The fate of rain falling on vegetation is very different from that falling on bare ground or rock. Rainfall can be partitioned into different units depending upon the pathway taken by the rain on its way to the soil or back to the atmosphere (Fig. 4.3).

Gross rainfall is the amount of water measured in the open or above the canopy of the vegetation. *Interception water* is the amount of rainfall per event retained by the canopy or by litter and which is evaporated without adding moisture to the soil. *Throughfall* is that portion of the rainfall directly reaching the soil surface or litter through spaces in the canopy and as drip from leaves, twigs and stems. *Stemflow* is water intercepted by the canopy and directed down the stems to the mineral soil or to the root crown. *Infiltration* is the movement or passage of water into the soil. *Percolation* is the movement of water through a soil column. *Run-off* is that portion of intercepted water that does not infiltrate into the soil.

A simplified water balance for a landscape can be expressed by: Precipitation = Evapotranspiration + Runoff + ΔStorage. In many landscapes there may be

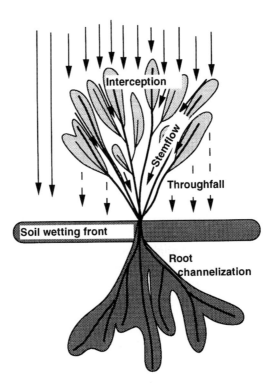

Figure 4.3 Partitioning of rainfall and the fate of rainfall on patches in a desert ecosystem.

additional inflows and outflows but the principle of balance still applies. The hydrological landscape can be envisioned as a series of layers where different types of hydrologic flows and storages occur (Ferguson, 1991–92). The atmospheric layer includes fluxes of water vapor across, and to and from the land surface. The surface layer includes overland flows and surface water channels and bodies. The vadose layer includes soil moisture stored in soil voids, or in transit toward groundwater, plant roots, or back to the surface. Vadose water is characterized by capillary tension (soil water potential ψ). Vadose water supports plant growth and provides the hydraulic head for maintaining stream base flow. The phreatic layer includes water in any subsurface saturated zone either being stored or in transit toward deep aquifers (Ferguson, 1991–92). The hydrologic structure of arid and semiarid landscapes varies considerably from place to place. Water infiltrating landscape units are formed of permeable materials with or without a significant soil mantle. Water-infiltrating landscapes are formed of sandstone, carbonates or unconsolidated material. Water spreading landscapes are formed of essentially impermeable materials including impermeable bedrock with little overlying soil. Lithology includes shale, slate, and granites. Drainage is mostly by surface run-off. Infiltration, subsurface storage, and subsurface flows are small. Many landforms are intermediate with impermeable bedrock overlain by a deep layer of permeable soil. These are the hydrologic units that characterize the basin and range topography of many desert regions.

The growth form of the vegetation of a given landscape unit affects the distribution patterns of water and nutrients within that unit as well as the nature and quantity of nutrients and water exported to adjacent units. Factors affected by the growth form of the vegetation include (1) canopy interception of water, stem flow, leaf drip, and evaporation from leaf surfaces, (2) infiltration and run-off, (3) organic matter transport, (4) water movement in the soil and (5) soil and organic matter accretion or loss.

4.2.1 Interception and Stem Flow

Many arid zone plants are morphologically well suited to promote stem flow (Fig. 4.4). It has been reported that up to 40% of the incident rainfall on an area in Australia covered by mulga (*Acacia* spp.) canopy was channeled down the main stems and entered the soil with minimal outwash at the bole (Pressland, 1976). Mallee (*Eucalyptus* spp.) has the inverted cone morphology and redirects 30% of the water in a rain event down the stems. That stemflow water is redirected along live root channels to depths as great as 28 m (Nulsen *et al.*, 1986).

Martinez-Meza and Whitford (1996) documented the importance of stem angle (with respect to the horizontal) and stem length as variables affecting stemflow in desert shrubs. In the three shrubs studied, stemflow yields were higher in shrubs with upward oriented branches than in shrubs with horizontally oriented branches. In mesquite (*Prosopis glandulosa*) maximum stemflow amounts were recorded on stems with stem angles of 70–75°. In tarbush (*Flourensia cernua*)

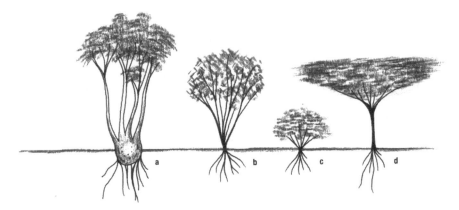

Figure 4.4 Morphologies of some desert shrubs and trees that facilitate stemflow and redistribution of stemflow to deep storage via live root channelization. (a) Mallee eucalyptus, *Eucalyptus* spp.; (b) mulga, *Acacia aneura*; (c) creosotebush, *Larrea tridentata*; (d) *Acacia* spp.

there was some stemflow recorded from branches with stem angles less than 40° but maximum stemflow volumes were recorded from branches with stem angles greater than 45° and stem lengths greater than 70 cm. In tarbush, stem angle and stem length accounted for 50% of the variability of the collected stemflow. In creosotebush, *L. tridentata*, stem angle and stem length accounted for 41% of the variability in stemflow. Stemflow volume was maximum at stem angles equal to or greater than 65° and there was little stem flow from branches with stem angles < 50°. This study re-emphasizes the importance of morphology of the dominant vegetation as a factor affecting water redistribution. The importance of shrub morphology and stemflow water was reinforced by deSoyza *et al.* (1997). They found that *L. tridentata* populations from the driest part of its range (Death Valley, California) were composed of plants with exterior stem angles >45°. Also, a population of young creosotebushes was almost entirely plants with exterior stem angles > 45°. Populations of creosotebushes in the Chihuahuan Desert were composed of plants varying from inverted cone shaped to hemispherical shaped plants.

Water that moves down the stems of shrubs and trees may follow live root channels and be stored at depths in the soil far below the rooting zone of ephemeral plants or perennial herbs and grasses that grow beneath the canopies of the shrubs. Nulsen *et al.* (1986) examined wetting and drying profiles and movement of dye. They found that stemflow water followed live root pathways and there was no evidence of that water moving along voided root channels, cracks in the soil or channels occupied by dead roots. Although the soils in the area of their study had low hydraulic conductivity, the soil at 4.5 m under mallee gained water rapidly after a 50.5 mm rainfall. The hydraulic conductivity of the soils was not high enough to account for the apparent rate of moisture transmission even if the soil water flow was assumed to be saturated. The study by Nulsen *et al.* (1986) provides reasonable evidence that live roots enhance the movement of water into the deep soil profiles.

Martinez-Meza and Whitford (1996) used a flourescent dye on the stems of the shrubs to trace the fate of stemflow water resulting from simulated precipitation. Stemflow water moved to 40 cm when the soil wetting front was only 6 cm (Fig. 4.5). In *L. tridentata*, there is also a dense network of short-fine roots, originating from the root crown, in the upper 10 cm of soil. This root distribution tends to maximize the water availability to this set of shallow roots at the stem base of the plant thus providing a relatively large quantity of water concentrated in the small area around the bole of the plant. The flourescent dye technique did not provide information on the quantities of stemflow retained in the shallow soil compared with the quantity transferred to deeper soil by root channelization.

In trees such as mallee, mulga, other acacias and shrubs such as creosotebush, water redirected to the root crown by stemflow may follow the channels provided by the living roots to depths in the soil greater than possible by infiltration. Relatively large amounts of creosotebush roots have been recovered from below indurated calcium carbonate layers. In excavations that cut through indurated layers roots were found growing through the layers and small channels that may be the result of termites excavating access tunnels traversing the indurated layers.

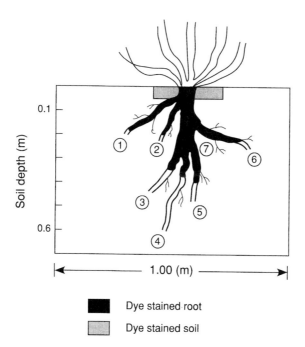

■ Dye stained root

▨ Dye stained soil

Figure 4.5 Rhodamine dyed roots and soil under creosotebush (*Larrea tridentata*) from dyed stems on shrubs subjected to simulated rainfall of 40 mm. (From Martinez-Meza and Whitford, 1996.)

Water channeled to deep soil layers via living roots is probably the resource that allows deep-rooted shrubs like creosotebush and mesquite to produce leaves and flower at the beginning of the growing season even in years when the upper 50 cm or more of soil is completely dry. That such a process may occur was reported by Allison and Hughes (1983) who traced water movement in the soil in a mallee-vegetated area and an adjacent area which had been cleared of mallee. On the mallee area recent water (that is post-1960 water) had penetrated to at least 12 mm depth but only to 2.5 mm on the cleared area. Since there would be a reduction in evapotranspiration and more water storage in an area cleared of vegetation (Schlesinger *et al.*, 1986) the difference in hydrostatic head should have resulted in water movement to greater depth on the cleared area. Channelized flow of water from the mallee stem along the roots to the deep soil is a plausible explanation for the observed data of Allison and Hughes (1983).

Other small trees and shrubs in arid regions are characterized by the inverted cone growth form that undoubtedly channels water down the stems into the soil at the root crown. Not all shrubs or small trees develop the inverted cone morphology. Many like the chenopods (*Atriplex* spp.), snakeweed (*Gutierrezia*), and burr sage (*Franseria* spp.) are spheroid in shape. It is doubtful that stemflow water confers any advantage to such shrubs since more than half of the stems in the canopy are at stem angles < 45°. A small fraction of the rainfall that falls on the leaves or stems of plants is retained on the plant surfaces and evaporated directly back to the atmosphere. If the quantity of water retained by shrubs is approximately 4–5% of the total as suggested by the studies by West and Gifford (1976), then losses by evaporation from leaf and stem surfaces can be considered insignificant.

4.2.2 Throughfall

The fraction of rainfall reaching the canopy of a plant that falls through the foliage or spaces in the canopy varies with storm depth. The percentage of the gross precipitation at the canopy surface that was measured as throughfall varied from 1% to 50–70% at storm depths between 1 mm and 5 mm (Martinez-Meza and Whitford, 1996). At the lowest storm depths, the smaller fraction of gross rainfall that appears as throughfall is due to water retained on the leaf and stem surfaces and which is lost by evaporation. The throughfall asymptote was at 5 mm storm depth. At the asymptote, the throughfall percentage varied between 30% and 70% in creosotebush, 60% and 80% in mesquite and 45% and 70% in tarbush.

Throughfall drop energies are lower than the kinetic energies of raindrops that fall on surfaces not below vegetation cover (Brandt, 1989). Since raindrop energy is a function of drop size, rain drops intercepted by canopy leaves or stems are broken up and fall through the canopy as finer droplets. The drop size of throughfall water is independent of canopy characteristics and of rainfall intensity (Brandt, 1989). The reduced energy of throughfall water enhances the infiltration of the throughfall fraction.

Small spheroid shrubs with dense foliage like many chenopods and tussock grasses are structurally capable of breaking up the large, high energy drops of intense, short duration rain storms, characteristic of summer convectional storm events of many hot desert areas. The density of foliage and/or fine branches of such plants close to the soil surface insures that fragments of leaves etc. will be trapped under the canopy. Rainfall intercepted by such canopies will enter the soil as a result of either leaf drip or stemflow with little or no transmission loss (Van Elewijck, 1989).

4.3 SPLASH EROSION-KINETIC ENERGY OF RAINDROPS

A raindrop impacts a bare soil surface something like a metal ball striking the surface. The energy of the raindrop impact is transferred to the soil particles directly under the drop. If the kinetic energy of the drop is sufficient, soil particles are detached from the surface and fly out in all directions. This is raindrop splash. Raindrop splash is the first step in erosion of soil. The volume of soil material detached by raindrop splash is determined by drop diameter, median grain size of the soil, and surface slope (Poesen, 1985). For a constant kinetic energy, the volume of splashed soil is not influenced significantly by diameter of raindrops nor of fall height of the raindrops except for low energy rainfall. Therefore kinetic energy is used to express the erosivity of rainfall in splash erosion equations. The kinetic energy of rainfall is translated into detachment of soil particles. Detachment of soil particles varies with the kinetic energy of raindrops and resistance of the soil to detachment. Empirical studies have shown that bulk density (mass per unit volume) of soil is the most significant parameter of resistance to detachment. Poesen (1985) provides an equation that expresses the total volume of soil detached by rainfall as:

$$V_{tot} = R^{-1} \text{ (B.D.)}^{-1} \text{ (K.E.)}$$

where R is the resistance of the material to drop detachment (J kg^{-1}), B.D. is the bulk density of the soil (kg m^{-3}), and K.E. is the kinetic energy of the rainfall (J m^{-2} year^{-1}).

Other factors are variables that affect the detachability of loose soil materials (Poesen, 1985). Slope may or may not affect detachability. The occurrence of positive relationships between slope and detachability probably depends on other factors such as texture and structure of the surface soil. The relationship between silty and clayey soils and volume of soil detached is greater than for sandy soils. Another variable that affects detachability of sloping soil is the effect of wind and the oblique trajectory of the raindrops that impact the soil surface. The variables other than those in the general model for total volume of soil detached are minor contributors to the variability in soil detachment.

Once soil particles are detached by raindrop splash, their contribution to sediment transport is determined by the particle size of the detached grains of soil.

Variation in transport as a function of time is explained by changing surface conditions such as surface water content, liquefaction, and waterfilm development (Poesen and Savat, 1981).

The reduction in kinetic energy of rainfall that is intercepted or deflected by plant foliage is substantial. In rainfall simulation experiments Wainwright *et al.* (1999) found that there was a 10% reduction in the intensity of subcanopy rainfall. They reported that the kinetic energy of the subcanopy rainfall was reduced to 70% of that hitting unprotected soil. This reduction in kinetic energy contributes to the development of mounds under the canopies of trees and shrubs.

4.4 INFILTRATION AND RUN-OFF

Infiltration, which is defined as the rate at which water enters a soil column, is an important determinant of water availability. There are a large number of variables that affect infiltration rates: vegetation cover, soil surface cover, soil texture, soil bulk density, abundance and distribution of macropores and antecedent moisture. In arid and semiarid regions factors such as extent of bedrock outcrop, areal extent of soil cover, slope area and differences in bulk density between channels of rills and small drainages and interfluve areas all contribute to variability in infiltration and run-off (Berndtsson and Larson, 1987). Antecedent rainfall has a greater effect on infiltration and run-off than rain event magnitude, average intensity, maximum intensity and duration (Istok and Boersma, 1986). In deserts where low intensity rainfall and frequent small magnitude rainfalls are characteristic of wet seasons, antecedent rainfall is an important consideration. Canopy morphologies, and foliage densities affect infiltration rates by changing drop size and by the quantity of litter trapped below the canopy. Infiltration rates are higher under shrub or tree canopies than in inter-shrub or inter-tree spaces (Lyford and Qashu, 1969; Elkins *et al.*, 1986) (Fig. 4.6). Infiltration rates are highest near stems of plants even on coppice dune sites. Coppice dunes have exceedingly high infiltration capacities (sometimes three to four times greater than interdune spaces) (Berndtsson and Larson, 1987). If there is a visible litter layer under the canopy, this can contribute to higher rates of infiltration of throughfall when compared to similar-sized shrubs that have little or no litter layer. However, woody shrubs and small trees probably have the greatest effect on infiltration by breaking up large drops of water that produce splash erosion and by stemflow as discussed in the previous section. The soils under shrub or tree canopies generally have higher organic matter content and lower bulk densities than soils in intercanopy spaces. These characteristics favor infiltration of the throughfall water. They also affect the water holding capacity of the soil because of the larger total capillary space resulting from the higher soil organic matter. Stemflow and the higher below canopy organic matter therefore combine to produce the higher infiltration rates measured on plots with shrub cover.

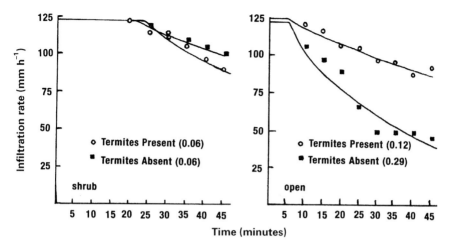

Figure 4.6 Comparison of infiltration rates from plots centered on shrubs and plots without shrubs. (Elkins *et al.*, 1986.) With permission from the Ecological Society of America.

Small grasses occurring at low density appear to have little if any effect on infiltration and run-off. However, larger tussocks have marked effects on infiltration. Large grass tussocks with standing dead foliage break up large drops and channel water along the leaf surface to the base of the clump (Van Elewijck, 1989). Tussock grasses have a dense network of relatively fine roots immediately below the vegetative clump. The presence of this root mass, the higher organic matter content of below clump grass soils and the reduction in drop size and energy by canopy interception combine to increase infiltration in comparison to unvegetated soil or soil with a sparse cover of small grasses and herbs.

Numerous methods have been used to measure infiltration rates and all have limitations with respect to the interpretation and use made of the data (Branson *et al.*, 1981). These methods include sprinkling infiltrometers, ponding infiltrometers, lysimeters, and run-off plots, which provide information about the infiltration characteristics of a small part of a landscape unit (see Fig. 4.7). At the other end of the scale are the run-off data collected from instrumented wiers on drainages from whole watersheds that often include several landscape units. Anyone who has studied in semiarid or arid regions will quickly point out that there appears to be little relationship between plot infiltration rates and water loss from a watershed. Comparison of run-off from grassland and shrubland habitats showed that run-off was initiated at lower thresholds of rainfall in shrublands than in grasslands (Schlesinger *et al.*, 2000). The runoff in shrublands averaged 18.6% of the rainfall over a seven-year period. Runoff from grassland plots varied between 5.0 and 6.3% over a five and one-half year period. Nutrient losses were higher from shrubland plots than from grassland plots (0.33 kg ha^{-1} yr^{-1} vs. 0.15 kg ha^{-1} yr^{-1}) because of the higher run-off volumes. However, in these Chihuahuan Desert

Figure 4.7 A run-off plot centered on a creosotebush, *Larrea tridentata*, shrub in the Chihuahuan Desert, New Mexico. The perimeter of the plot is a steel frame that is carefully driven into the soil. Water running off the plot passes over a sheet metal tray and flows into a slotted pipe. The pipe empties into a series of calibrated containers.

sites, there was a net accumulation of most nutrients due to inputs from atmospheric deposition (Schlesinger *et al.*, 2000). Comparison of run-off and sediment yield from two desert grassland types on the same soil type and geomorphic surface found that water run-off was three times greater from burro-grass areas (a low-growing rhizomatous grass) than from tobosa grass areas (a bunch grass with larger bare patches between plants) (Devine *et al.*, 1998).

Tree and shrub canopies affect infiltration indirectly by their effect on certain soil properties, i.e. bulk density and organic matter content. Lyford and Qashu (1969) measured infiltration rates at several distances from the plant centers of palo verdi trees (*Cercidium microphyllum*) and creosotebush (*L. tridentata*). They found infiltration rates nearly three times greater under the plants than in the intercanopy spaces. Since soil characteristics are important determinants of the type of vegetation that will develop in a given climate, it is difficult to assess the importance of soil characteristics *per se* on infiltration. Branson *et al.* (1981) summarize infiltration data for a variety of soils with predominantly grass cover. On sand or sandy soils with 65% vegetative cover, infiltration rates ranged between 42.4 and 79.5 mm h^{-1} compared with 36.8 mm h^{-1} on clayey soils. On dense clay, infiltration rate was only 12.7 mm h^{-1} but vegetative cover was only 40%. The only generalization that can be made is that sand or coarse soils will have higher infiltration rates than fine-textured clayey or silty soils. Another soil parameter that affects infiltration is soil depth. Shallow soils, that is, soils overlying a relatively impervious argyllic (clay) layer or indurated calcium carbonate hardpan will have reduced infiltration because the capillary space and pore space in the overlying soil quickly fills and percolation into and through the relatively impervious layer is slow. Thus shallow sand overlying caliche or clay will produce ponding very quickly in an intense storm whereas some storm intensity on that same land in a dune 1–2 m deep will not produce ponding.

In Australian chenopod shrublands vegetative cover by shrubs and herbaceous plants is around 50% and the remaining soil surface has a cover of lichen crusts. Final steady-state infiltration rates with an intact lichen crust were < 10 mm h^{-1} and with the crust removed averaged 46 mm h^{-1} (Graetz and Tongway, 1986). This relationship held even at low rates of applied rainfall (28 mm h^{-1}) when the infiltration rate with intact lichen crust was 7 mm h^{-1} and with the crust removed was 14 mm h^{-1}. However, the relationship between lichen or cryptogamic crusts and infiltration is not simple. Some studies have shown that the presence of cryptogamic crusts improves infiltration (Fletcher and Martin, 1948) or has no effect (Loope and Gifford, 1972; Eldridge *et al.*, 1997). Eldridge *et al.* (1997) found that removal of crusts affected infiltration rates only on plots where the soil surface had been disturbed by livestock. They concluded that on coarse-textured soils, crusts are poorly developed and dominated by physical (raindrop impact) crusts mixed with free-living fungi and cyanobacteria. Cyanobacterial sheaths and fungal hyphae may reduce infiltration in poorly developed soil crusts if they exclude water by occupying capillary pores (Greene and Tongway, 1989). Cyanobacterial sheaths and fungal hyphae may increase infiltration where they help to maintain continuous pores by their growth habits in the soil. The hydrophobic nature of microphytic crusts has been reported from sand dune soils in Australia and in the Negev Desert in Israel. In these areas removal of microphytic crusts greatly enhanced infiltration and reduced or eliminated run-off (Eldridge *et al.*, 1997). Removal of cyanobacterial crusts on sand dune soils and loess-covered hillslope in the Negev, resulted in a three- to fivefold increase in steady-state infiltration. However, removal of cyanobacterial crusts on loess floodplain had no effect on ponded infiltration (Eldridge *et al.*, 2000). The lack of cyanobacterial crust effects on infiltration on the loess floodplain was attributed to the exposure of surface silts to water and subsequent clogging of matrix pores and sealing of the surface. Comparisons of these studies are difficult because of differences in infiltrometers used and basic soil differences. If we are to generalize about the importance of cryptogamic crusts with respect to infiltration, we will need a more extensive set of data.

Another factor that affects infiltration into bare soils is the development of surface crusts (Tarchitzky *et al.*, 1984; Moore and Singer, 1990). These crusts are impervious in comparison to uncrusted soil. They apparently form as a result of raindrop impact disrupting soil aggregates. This results in fine particles aligning to form a 'skin', 0.1 mm thick and a 2–3 mm layer immediately below the 'skin' with higher bulk density in which the soil aggregates have been destroyed

(Tarchitzky *et al.*, 1984). This has been documented in sandy soils by Valentin (1991) who demonstrated that the kinetic energy of rainfall was the major crusting factor for the sandy soil. Waterlogging and subsequent slaking of aggregates led to structural and depositional crusts in a clay loam soil. Infiltration was significantly reduced by the surface crusting (Valentin, 1991). The result of crust formation is a marked reduction in infiltration.

Considering factors that affect water infiltration and influence the redistribution of water across a unit of landscape, the importance of the activity of animals must not be overlooked. The influence of large grazing herbivores either native or domestic have large effects on infiltration and run-off plus sediment losses and has been the subject of considerable research effort in arid regions (Graetz and Tongway, 1986). Obvious effects of grazing herbivores on infiltration are reduction in vegetative cover or changes in vegetative composition that affect infiltration and run-off as previously described. Somewhat less obvious are the effects that trampling by hoofed animals have on destruction of surface crusts and/or destruction of cryptogamic crusts. Disruption of the 'sealed' crust resulting from raindrop impact can result in greater infiltration and less run-off. Disruption of cryptogamic crusts may increase or decrease infiltration as previously discussed.

Probably the most important factor affecting water infiltration is the abundance of soil macropores (1–5 mm holes (Anderson *et al.*, 1990)). It is only in the past 20–25 years that soil scientists have rediscovered the importance of soil macropores in the spatial distribution of water in soil and the effects of these pores on water infiltration (Phillips *et al.*, 1989). Continuous pores or tubes transport flowing water to the deeper parts of the soil profile faster than predicted by theory. Macropores are widespread in many if not most ecosystems and are extremely important for infiltration in desert soils. Since the numbers of macropores per unit area need not be large to affect water infiltration, burrow/tunnel-constructing invertebrates are the most important contributors to macroporosity of desert soils.

The best-known sources of macropores are the burrows of earthworms (Lee and Foster, 1991). In arid and semiarid ecosystems, termites probably replace earthworms as the most important producers of macropores. Other invertebrates that produce pores in the soil include larvae of insects like beetles and cicadas, burrowing spiders, and ants. In studies in the Sahel comparing plots from which termites were excluded with plots with termites present, Mando and Miedema (1997) reported an average of 88 ± 25 large voids (0.8–1.2 cm diameter) per square meter. No voids were detected in the termite excluded plots. They reported that termites accounted for more than 60% of the macropores in this region. Mando (1998) in a study of water infiltration on plots with and without termites found significant increases in infiltration on plots with termites (Fig. 4.8) In his study an average of 51.3% of the precipitation infiltrated plots with termites but only 36.3% of the annual precipitation infiltrated plots from which termites were excluded. Elkins *et al.* (1986) reported on studies that utilized plots from which termites had been eliminated by a long-term, selective insecticide, chlordane. Infiltration rates on low-cover plots (plots with scattered small clump grass) were

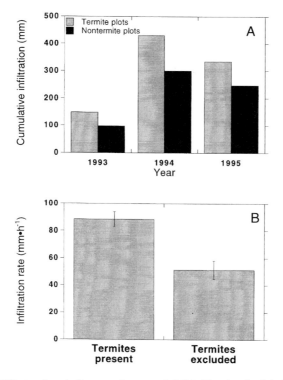

Figure 4.8 Effects of excluding termites on rainfall infiltration in Sahelian soils. (From data in Mando 1998 and Elkins *et al.*, 1986.)

51.3 ± 6.8 mm h^{-1} on plots with no termites and 88.4 ± 5.6 on plots with termites removed. Termite removal had affected soil bulk density: 1.99 g · cm^{-3} compared to 1.70 g cm^{-3} and porosity 24.9% vs. 35.8% on the termite free plots in comparison to plots with termites present. They concluded that the activity of subterranean termites in Chihuahuan desert soils reduced bulk density and maintained porosity. Additional evidence for the role of macropores produced by the feeding galleries of subterranean termites on water infiltration was provided by Eldridge *et al.* (1997). They found that these macropores were responsible for the deep levels to which infiltrating water percolated. When they compared saturated to unsaturated infiltration they found high values which has been reported to be a useful index of the macropore status of a soil. High values (1:6 or greater) suggest that most of the flow is through macropores. Ratios of 1:5 or less indicate that macropores are scarce and that capillary pores are the most important avenue for water movement into the soil. Data from other semi-desert systems in more tropical areas suggest that activities of termites reduce infiltration at least with respect to the nest structure itself (Tongway and Noble, personal communication). The cemented caps of termite mounds shed water.

The contributions of ant nests to infiltration in desert environments has shown that ant nest macropores contribute to water infiltration (Eldridge, 1993; Lobry de Bruyn and Conacher, 1994) but only when soils are saturated and water is ponding on the surface. Ant nest macropores behave differently from other types of macropores. Water movement in macropores does not follow capillary flow theory because the size of the pore allows water to flow as a film along the wall of a pore that is not completely filled with water (Bevan and German, 1982). In this way macropores contribute to rapid movement of water into the soil in the pores with little or no water flow through the soil matrix. Transitory ponding allows water movement into macropores open to the surface even when the soil is not saturated (Bevan and German, 1982) The ponding necessary to establish bulk flow into the soil via ant-produced biopores appears not to be transitory but to occur only when soils are saturated.

Many desert regions have large areas of surface covered with rock fragments or stones. Rock cover decreases the erosion potential compared to that of unvegetated soil with little or no stone cover. Run-off volume is controlled by stone cover and stone size (Abrahams and Parsons, 1991). Empirical studies have shown that surface cover of rock and vegetation provided better protection (less sediment yield) than either 100% vegetation cover or 100% rock cover (Benkobi et al., 1993). Stony soils with large surface area stones provide microhabitat for soil animals. Termites build galleries on the undersurfaces of rocks that are embedded in the soil and some species of ants build nests under rocks. In stony soils with scattered large rocks, run-off from the rock surfaces may not contribute to run-off because that water infiltrates rapidly at the buried margins of the rock because of the macropores produced by invertebrates. Poesen and Ingelmo-Sanchez (1992) examined the effect of rock and macropores on interill run-off and sediment yield. They found that the time at which the macropores ate the soil surface close is crucial as the determinant of when the effect of rock cover switches from reduction of run-off to an increase in run-off. Thus rock-covered soils in deserts may increase run-off and sediment transport under certain conditions.

The size distribution of rocks on the soil surface is important for the functional relationship of rock cover, infiltration, run-off, and sediment yield. Empirical studies in a desert shrubland showed that infiltration rate was negatively related to gravel cover. Sediment production was also negatively related to gravel cover. This apparent dichotomy results from the obstruction value of the gravel fragments that slow overland flow and the resistance to infiltration by the gravel fragments on small patches.

4.5 EXCHANGES AMONG LANDSCAPE UNITS

The most obvious avenue of exchange of materials among units of a landscape is that of transport of dissolved and suspended materials in water from upslope to downslope positions on a catena or watershed. Fluvial transport is the most important agent that produces soil gradients along catenas. The number of soil units that characterize a catena are at least partially determined by the length of the catena, the slope angle and/or variety of slope angles along the catena, the nature of the parent material at the top of the catena and the infiltration characteristics of the soil series of the catena. In turn the catenas soil units support different vegetation assemblages. The growth form cover and productivity of the vegetation assemblages varies with soil type and catenary position and influences run-off, infiltration and accumulation of sediment and transported materials. Infiltration and run-off at the watershed scale are very different from the rates and volumes estimated by studies of small plots on parts of the watershed. Flows in ephemeral stream beds monitored by flumes, are considerably lower than the estimates based on summing the run-off for square meter units on the watershed. Although transmission losses to the sediments of the ephemeral channels account for some of the reduction in measured run-off, the within drainage soil and vegetation heterogeneity are more important variables (Murthy and Chandrasekharan, 1985). Within arid and semiarid regions, sparse vegetation and the microtopography related to that vegetation are important variables affecting run-off at fine spatial scales (Bergkamp, 1998). Increased infiltration near vegetation reduces overland flow. Higher infiltration under plants relative to bare soil is attributed to lower soil bulk density, greater soil aggregate stability and greater density of macropores. Microtopography affects run-off by concentrating or diverging run-off flows. These spatial characteristics must be considered together with discontinuity in run-off that is related to storm depth and duration.

In order to understand run-off production, it is necessary to understand the discontinuity in run-off related to both the short duration and small size of storms and the heterogeneity in surface properties (DeBoer and Campbell, 1989). Microtopographic effects were documented in a study of a landscape made up of small terraces centered on grasses and small shrubs (terracettes), and hummocks that included the deep roots of the small trees that formed the center of the hummocks (Bergkamp, 1998). Bergkamp (1998) found that there was flow from the terracette to the surrounding bare space even during short duration and relatively low-intensity storms. However, there was no flow connection between terracettes. The plants on the hummocks redistributed water to deep soil layers, which controlled water movement on parts of the slope. At the part slope scale (1000–2000 m²) and the slope scale (1 ha) even with longer duration, higher intensity rainfall, the scattered distribution of saturated overland flow patches and the sinks (hummocks) inhibits the continuity of flow at this scale. Bergkamp concluded that run-off from the catchment cannot be directly related to run-off generation on the slope. It is suggested that flow in channels normally results from run-off in valley bottoms near stream channels. The spatial distribution of vegetation and bare patches on the slopes is the important landscape structural element that causes discontinuity in flow between fine scales and broader scale (Bergkamp, 1998).

Noy Meir (1981) lists four examples of arid ecosystems in which water redistribution by run-off at small distances is important: (1) plains and alluvial fans with a dense network of small runnels and channels; (2) plains and gentle slopes with regular (or irregular) slight depressions, e.g. gilgais or mulga grove–intergrove system; (3) rocky slopes interspersed with soil pockets; and (4) areas with strong microvariation in surface soil characterisitics, in particular infiltrability. This list is certainly not exhaustive and the examples were included in Noy Meir's paper to demonstrate the need to incorporate spatial redistribution into models for plant production. If we are to develop models that incorporate spatial patterns of water, organic matter and nutrient redistribution as effectors of primary production, we need also to consider how these spatial patterns vary with rainfalls of different quantity and intensity and how the return time of particular 'threshold' events affect the system.

4.6 EXCHANGES WITHIN LANDSCAPE UNITS

As discussed earlier, the morphology of the dominant woody perennials has a marked effect on the water and nutrient status of the soils beneath the shrub or tree canopy. Canopy interception, and stemflow frequently eliminates run-off from the soils under the canopies of the established shrubs. As a consequence, leaf litter and plant parts accumulate under the canopies of such shrubs and trees. If the slope angle and soil infiltration characteristics produce run-off from most rain events, the shrubland develops insular topography. In such habitats, transport of organic matter by run-off is limited to that material in the intercanopy space. The transport of organic matter resulting in its redistribution on a landscape is a function of variables discussed in the previous section. Shrubs and tree canopies break up rain drops reducing the energy of the drops such that throughfall does not disrupt the litter layer. The litter layer and higher organic matter content of subcanopy soils insures that virtually all of the throughfall infiltrates. Thus, there is no surface movement of water to remove litter from under canopies. Because there is little or no erosion from subcanopy soils, shrubs are often growing on slight mounds with an organic matter layer surrounded by the bare or nearly bare intershrub soil. Organic matter that is moved by sheet flow is limited to the leaves, stems, fecal pellets, etc. located in the intercanopy space. The distance that this material is moved downslope depends on sheet flow velocity and spacing of impending vegetation downslope. The net result from most rainfall events is an internal redistribution within a unit or accumulation in a downslope unit.

Run-off in arid and semiarid regions frequently results in the formation of litter dams and microterraces. These structures are commonly observed especially immediately after large rain events. Litter dams form quickly when there is general sheet flow across a landscape unit. Sheetflow causes leaf litter, animal feces, stem segments, etc. to be transported as 'litter rafts' (Mitchell and Humphreys,

1987). Obstructions on the surface such as protruding rocks, tussocks of grass, stems of shrubs or large logs (such as decaying *Yucca* logs) capture the rafting litter and produce a dam (Fig. 4.9). When a dam is formed, sediment accumulates in the ponded water behind the dam. Occasionally breaches form in a dam wall, which will temporarily drain the pond and cause rilling in the sediment infill. Typically these breaches are closed in subsequent run-off events. Microterraces form from the coarser mineral sediments that accumulate behind the litter dams.

In most deserts the litter dams are rapidly colonized by foraging termites and other arthropods. These organisms reduce the structural integrity of the litter dams and may reduce the whole structure to a microterrace of sediment and fine particulates. Litter dams and microterraces are transient features of arid land-scapes that develop as a result of within-landscape unit transport. The factors important in litter dam and microterrace evolution were listed by Mitchell and Humphreys (1987). These include: relatively bare slopes, slope angles less than about 10°, supply of organic debris, surface soil conducive to run-off and sediment yield, availability of obstructions, seed sources, sources of soil algae, and presence of litter breakdown mechanisms and bioturbating organisms. Despite the lack of quantitative data on litter dams and microterraces in arid and semiarid regions, it is obvious that these are important features of these environments and contribute significantly to the patch dynamics of these landscapes.

Figure 4.9   A debris–litter dam formed by a *Yucca elata* log in the Chihuahuan Desert, New Mexico.

A different kind of within-unit exchange occurs on steep hillslopes. Here run-off and sediment transported from the upper slopes has no chance of reaching the slope base during most rain events. Steep hillslopes with sedimentary rock lithography tend to be structured with narrow flat or concave benches between the steep-sided exposed rock surfaces. In the Negev, water moves from the top of the limestone hillslopes and ponds in the bench areas. Run-off and sediment yield measurements on these hillslopes clearly show that the bulk of the run-off and sediment transport is between upper and lower units of the hillslope (Yair *et al.*, 1980). On volcanic and granitic slopes, boulders and ridge structures probably tend to keep the run-off and sediment transport local.

4.7 EPISODIC EVENTS

If there is one generalization that can be made about arid and semiarid environments, it is that episodic events have a greater impact on the ecosystems and landscapes than the cumulative effect of 25–30 years of events within the 'average range'. Episodic events may be a wind storm with wind velocities more than 2 standard deviations above the long-term mean wind velocities for a season and location or an extremely intense, large storm depth rainfall that produces flooding that reaches or exceeds the capacity of the large drainage channels. Flood frequency analysis is based on dimensionless curves that relate return period in years to floods of different magnitudes (Farquharson *et al.*, 1992). The problem of predicting return period for floods of a given magnitude is that there are short records for arid areas. Return periods are based on calculation of mean annual flood for a given geographic location. In their synthesis of flood frequencies, Farquharson *et al.* (1992) examined the relationship between catchment area and mean annual flood for arid regions. Regression of all the available data gave an r^2 of 0.55 and regression of regions, i.e. North Africa, Southern Africa, Iran, and Queensland and the Arabian peninsula, yielded r^2 values between 0.43 and 0.92. Obviously catchment area is an important variable affecting flood size (measured in m^3 s^{-1}). However, the probability of a low frequency, large volume, intense storm that produces extreme flooding is unpredictable for arid and semiarid regions. Studies of paleofloods in the drainage of a 1400 km^2 catchment in the hyperarid area of the Negev Desert found 28 floods that ranged from 200 to 1500 m^3 s^{-1} over the last 2000 years (Greenbaum *et al.*, 2000). This study demonstrates the low frequency of episodic floods that affect the geomorphology and structure of desert landscapes.

The results of a study of erosion and deposition plus effects on vegetation in a wadi in the Negev Desert of Israel document the importance of episodic events in restructuring ecosystems. The watershed of Wadi Zin experienced a 60 mm rainfall within a period of one hour (Naomiish-Shalom-Gordon and Gutterman, 1991). The resulting flood was rated at a 100-year recurrence interval. The flood not only damaged vegetation, it lowered the surface of the wadi, uncovered bedrock in several areas and deposited large quantities of sediment as the flood waters waned. In this flood, vegetation was washed away from a portion of the wadi confined between vertical rock walls. These authors describe a two-phase flood regime. The high-energy first phase produced large loss of soil from the channel and benches. The second phase resulted from low-energy flow that produced soil deposition especially in the wider parts of the wadi. The result of the erosion and deposition from a single storm was a complete restructuring of the wadi that will affect vegetation development for years into the future (see Fig. 4.10).

Storms with different rainfall characteristics produce very different types of flow events in ephemeral channels. White (1995) characterized storm intensities by the kinds of flow events that occur in drainage channels. High-intensity storms produce debris flows (transport of organic debris, small cobbles, and soil sediments) whereas low intensity storms produce fluvial flows (flows with low quantities of suspended sediments). He characterized channels based on channel morphology and characteristics of the uplands that drain into the channels as debris-type flow channels or as fluvial-type flow channels. The flood bore in debris-type flow frequently 'scours' the watershed sufficiently that a subsequent rain event of similar intensity and duration will not produce a debris flow in a channel that is a debris-type flow channel (White, 1995). Scouring events are episodic in most desert regions.

Figure 4.10 An episodic flash flood in ephemeral stream channels in the Chihuahuan Desert, New Mexico.

In more than 35 years of research on the Jornada Experimental Range in New Mexico, I have experienced only one very large intense storm that 'scoured' the watershed that we were studying at the time. Although we had no gauging stations on the ephemeral drainages of the watershed, the bank cutting, bank overflooding, and large debris deposits at the terminus of identifiable drainage channels, provided evidence that the volume of water moving across the uplands of the watershed reached or exceeded the channel capacities of the drainage channels. The scouring action of the large overland flow removed all debris dams, dead wood from intershrub spaces, leaf litter (except under mounded shrubs), and dung. Large quantities of this material were deposited at the periphery of the ephemeral lake at the base of the watershed. The ephemeral lake basin overflowed to several meters beyond the fine-textured lacustrine soils of the basin. This event was a small, episodic flood in comparison to a flood in the local area at the turn of the century that was reported to have produced all of the head cutting of the present ephemeral stream channels.

These anecdotal observations reinforce the need for 'long-term' measurements in desert ecosystems. Even with a long-term ecological research program in place, we were not prepared to make all of the measurements needed to document all of the changes in the ecosystems of the watershed that resulted from this single storm. Except for studies conducted in the ephemeral lake, all of the changes in distribution patterns of organic materials etc. were qualitative. Episodic events 're-set' spatial patterns of organic matter, 're-set' soil processes and may 're-set' many other ecosystem processes.

4.8 EPHEMERAL PONDS AND LAKES

Ephemeral lakes and ponds are basins that remain flooded for short periods of time during a year but may not hold water for several years if the rainfall regime is not suitable to produce flooding (Figs 4.11 and 4.12). Filling of ephemeral lakes is episodic. Water entering lake basins may originate as run-off from a relatively small area surrounding the basin or may represent water in excess of the transmission loss capacity of the ephemeral streams that drain a large watershed. Motts (1969) proposed a flooding ratio, Rn the ratio of the number of days a basin contains water to the number of days it is dry. He proposed that a value of 0.33 or less is indicative of an ephemeral lake. Ephemeral lakes vary in their origin, size, flood timing, flood frequency, areal extent of flooding, salinity, relation to ground water, vegetation, soils, ratio of basin size to watershed size, surface morphology, mineralogy, and stratigraphy (Van Vactor, 1989). These physical variables affect the biota that inhabit ephemeral lakes and also affect the ecosystem processes of a lake when flooded.

In the desert rangelands of North America, the filling of ephemeral lakes does not necessarily require that the ephemeral stream channels reach the edge of the lake. Indeed, the more typical geomorphology is for the ephemeral stream

Figure 4.11   A dry ephemeral lake basin in the Chihuahuan Desert, New Mexico.

Figure 4.12   The flooded lake basin in Fig. 4.11.

channels to terminate at some distance from the dry lake basin (Wondzell *et al.*, 1996). In the mulga woodlands of northwestern New South Wales, Australia, some ephemeral lake basins received water directly from ephemeral drainage channels whereas other dry lake basins did not (personal observation). The importance of discharge from ephemeral stream channels has been questioned because of the probability of greatly diminished discharge because of transmission losses in the ephemeral stream channels. Measured infiltration rates in ephemeral stream channels in the Chihuahuan Desert varied from 57.3 to 22.4 cm h^{-1} with an average rate of 39.5 ± 14.8 cm h^{-1}. Infiltration rates of 90–100 cm h^{-1} were measured in another Chihuahuan Desert ephemeral stream bed using a different technique (Van Vactor, 1989). These high infiltration rates result in large transmission losses into the stream bed and suggest that only the most intense storms produce sufficient run-off to produce discharge and overland flow of that discharge from the terminus of the ephemeral stream channel. Basically then, ephemeral lake flooding occurs when rainfall inputs to the lake watershed exceed losses by infiltration thereby producing run-off that is transported by a combination of overland flow and channel flow to the dry lake basin (Van Vactor, 1989).

When flooding occurs, the water in the lake basin contains suspended organic materials, and suspended soil particles. The velocity and depth of the run-off that enters the basin determines the quantity and characteristics of the suspended matter. After a particularly intense storm, the suspended matter in a dry lake that we studied included leaf fragments, small sticks, and rabbit dung in addition to the unidentifiable suspended organics and inorganics. Ephemeral lakes thus function as the primary sink for organic matter in desert landscapes. The soil organic matter (soil carbon) of an ephemeral lake bed at the base of an intensively studied Chihuahuan Desert watershed was 3.4% compared with values ranging between 0.4% and 1.2% for the soils of the upland portions of the watershed (Nash and Whitford, 1995). The higher organic matter content of ephemeral lake soils may also result in higher rates of nitrogen mineralization because of the larger quantities of organic nitrogen in comparison to the soils on the remainder of the watershed (Nash and Whitford, 1995).

When an ephemeral lake is flooded the allochthonous (origin outside the water body) and the autochthonous (origin *in situ*) material is rapidly decomposed by a variety of organisms. This material forms the basis for the complex food webs that develop in desert ephemeral lakes. If the ephemeral lake basin supports terrestrial vegetation, the grazing management determines the residual vegetation remaining on the basin when a flooding event occurs. Most ephemeral lake basins are devoid of terrestrial vegetation or nearly devoid of such vegetation. The characteristics of the lake basin with respect to presence or absence of terrestrial vegetation is the primary factor that determines the structure of the developing aquatic food web during a flooding cycle. With terrestrial vegetation present, the aquatic food web starts as a heterotrophic system, which gradually shifts to an autotrophic system.

REFERENCES

Abrahams, A. D. and Parsons, A. J. (1991). Relation between sediment yield and gradient on debris-covered hillslopes, Walnut Gulch, Arizona. *Geol. Soc. Am. Bull.* **103**, 1109–1113.

Anderson, S. H., Peyton, R. L., and Gantzer, C. J. (1990). Evaluation of constructed and natural soil macropores using x-ray computed tomography. *Geoderma* **13**, 13–29.

Ambrust, D. V. and Bilbro, J. D. (1997). Relating plant canopy characteristics to soil transport capacity by wind. *Agron. J.* **89**, 157–162.

Belnap, J. and Gillette, D. A. (1998). Vulnerability of desert biological soil crusts to wind erosion: the influences of crust development, soil texture, and disturbance. *J. Arid Environ.* **39**, 133–142.

Benkobi, L., Trlica, M. J., and Smith, J. L. (1993). Soil loss as affected by different combinations of surface litter and rock. *J. Environ. Qual.* **22**, 657–661.

Bergkamp, G. (1998). A hierarchical view of the interactions of runoff and infiltration with vegetation and microtopography in semiarid shrublands. *Catena* **33**, 201–220.

Berndtsson, R. and Larson, M. (1987). Spatial variability of infiltration in a semi-arid environment. *J. Hydrol.* **90**, 117–133.

Bevan, K. and German, P. (1982). Macropores and water flow in soils. *Water Resources Res.* **18**, 1311–1325.

Bonell, M. and Williams, J. (1986). The generation and redistribution of overland flow on a massive oxic soil in a eucalypt woodland within the semi-arid tropics of north Australia. *Hydrol. Proc.* **1**, 31–46.

Brandt, C. J. (1989). The size distribution of throughfall drops under vegetation canopies. *Catena* **16**, 507–524.

Branson, F. A., Gifford, F. F., Renard, K. G., and Hadley, R. F. (1981). *Rangeland Hydrology*. Range Science Series no. 1, 2nd edn. Society for Range Management, Kendall/Hunt Publishing, Dubuque, Iowa.

Breed, C. S. and McCauley, J. F. (1986). Use of dust storm observations on satellite images to identify areas vulnerable to severe wind erosion. *Climat Change* **9**, 243–258.

Chepil, W. S. (1951). Properties of soil, which influence wind erosion: IV. State of dry aggregate structure. *Soil Sci.* **72**, 387–401.

Chepil, W. S. (1953). Factors that influence clod structure and erodibility of soil by wind. I. Soil texture. *Soil Science* **75**, 473–483.

DeBoer, D. H. and Campbell, I. A. (1989). Spatial scale dependence of sediment dynamics in a semi-arid badland drainage basin. *Catena* **16**, 277–290.

DeSoyza, A. G., Whitford, W. G., Martinez-Meza, E., and Van Zee, J. W. (1997). Variation in creosotebush (*Larrea tridentata*) morphology in relation to habitat, soil fertility and associated annual plant communities. *Am. Midl. Nat.* **137**, 13–26

Devine, D. L., Wood, M. K., and Donart, G. B. (1998). Runoff and erosion from a mosaic tobosagrass and burrograss community in the northern Chihuahuan Desert grassland. *J. Arid Environ.* **39**, 11–19.

Eldridge, D. J. (1993). Effect of ants on sandy soils in semi-arid eastern Australia: local distribution of nest entrances and their effect on infiltration of water. *Aust. J. Soil Res.* **31**, 509–518.

Eldridge, D. J., Tozer, M. E., and Slangen, S. (1997). Soil hydrology is independent of microphytic crust cover: further evidence from a wooded semiarid Australian rangeland. *Arid Soil Res. Rehabil.* **11**, 113–126.

Eldridge, D. J., Zaady, E., and Shachak, M. (2000). Infiltration through three contrasting biological soil crusts in patterned landscapes in the Negev, Israel. *Catena* **40**, 323–336.

Elkins, N. Z., Sabol, G. V., Ward, T. J., and Whitford, W. G. (1986). The influence of subterranean termites on the hydrological characteristics of a Chihuahuan Desert ecosystem. *Oceologia* **68**, 521–528.

Farquharson, F. A. K., Meigh, J. R., and Sutcliffe, J. V. (1992). Regional flood frequency analysis in arid and semi-arid area. *J. Hydrol.* **138**, 487–501.

Ferguson, B. K. (1991–92). Landscape hydrology, a component of landscape ecology. *J. Environ. Syst.* **21**, 193–205.

Fletcher, J. E. and Martin, W. P. (1948). Some effects of algae and moulds in the rain crust of desert soils. *Ecology* **29**, 95–100.

Gillette, D. A., Herbert, G., Stockton, P. H., and Owen, P. R. (1996). Causes of the fetch effect in wind erosion. *Earth Surf. Proc. Landf.* **21**, 641–659.

Goudie, A. S. (1978). Dust storms and their geomorphological implications. *J. Arid Environ.* **1**, 291–310.

Graetz, R. E. and Tongway, D. J. (1986). Influence of grazing management on vegetation, soil structure and nutrient distribution and the infiltration of applied rainfall in a semiarid chenopod shrubland. *Aust. J. Ecol.* **11**, 347–360.

Greenbaum, N., Schick, A. P., and Baker, V. R. (2000). The palaeoflood record of a hyperarid catchment, Nahal Zin, Negev Desert, Israel. *Earth Sur. Proc. Landf.* **25**, 951–971.

Greene, R. S. B. and Tongway, D. J. (1989). The significance of (surface) physical and chemical properties in determining soil surface condition of red-earths in rangelands. *Aust. J. Soil Res.* **30**, 213–225.

Istok, J. D. and Boersma, L. (1986). Effect of antecedent rainfall on runoff during low-intensity rainfall. *J. Hydrol.* **88**, 329–342.

Lee, K. E. and Foster, R. C. (1991). Soil fauna and soil structure. *Aust. J. Soil Res.* **29**, 745–775.

Leys, J. F. and Raupach, M. R. (1991). Soil flux measurements using a portable wind erosion tunnel. *Aust. J. Soil Res.* **29**, 533–552.

Lobry deBruyn, L. A. and Conacher, A. J. (1994). The bioturbation activity of ants in agricultural and naturally vegetated habitats in semi-arid environments. *Aust. J. Soil Res.* **32**, 555–570.

Loope, W. L. and Gifford, G. F. (1972). Influence of a soil microfloral crust on selected properties of soil under Pinyon–Juniper in southeastern Utah. *J. Soil Water Conserv.* **27**, 164–167.

Lyford, F. and Qashu, H. (1969). Infiltration rates as affected by desert vegetation. *Water Resour. Res.* **5**, 1373–1376.

Mando, A. (1998). Soil-dwelling termites and mulches improve nutrient release and crop performance on Sahelian crusted soil. *Arid Soil Res. Rehabil.* **12**, 153–164.

Mando, A. and Miedema, R. (1997). Termite-induced change in soil structure after mulching degraded (crusted) soil in the Sahel. *Appl. Soil Ecol.* **6**, 26–263.

Martinez-Meza, E. and Whitford, W. G. (1996). Stemflow, throughfall and channelization of stemflow by roots in three Chihuahuan Desert shrubs. *J. Arid Environ.* **32**, 271–287.

Mitchell, P. B. and Humphreys, G. S. (1987). Litter dams and microterraces formed on hillslopes subject to rainwash in the Sydney Basin, Australia. *Geoderma* **39**, 331–357.

Moore, D. C. and Singer, M. J. (1990). Crust formation effects on soil erosion processes. *Soil Sci. Soc. Am. J.* **54**, 1117–1123.

Motts, W. S. (1969). *Geology and Hydrology of Selected Playas in Western United States.* Air Force Cambridge Research Laboratories Report No. AFCRL-69-0214. Bedford, MA.

Murthy, K. N. and Chandrasekharan, H. (1985). Spatial variability of infiltration in an arid watershed in India. *J. Arid Environ.* **8**, 189–198.

Musick, H. B., Trujillo, S. M., and Truman, C. R. (1996). Wind-tunnel modeling of the influence of vegetation structure on saltation threshold. *Earth Surf. Proc. Landf.* **21**, 589–605.

Naomiish-Shalom-Gordon and Gutterman, Y. (1991). Soil disturbance by a violent flood in Wadi Zin in the Negev Desert Highlands of Israel. *Arid Soil Res. Rehabil.* **5**, 253–263.

Nash, M. H. and Whitford, W. G. (1995). Subterranean termites: regulators of soil organic matter in the Chihuahuan Desert. *Biol. Fertil. Soils* **19**, 15–18.

Navar, J. and Bryan, R. (1990). Interception loss and rainfall redistribution by three semi-arid shrubs growing in northeastern Mexico. *J. Hydrol.* **115**, 51–63.

Noy Meir, I. (1981). Spatial effects in modelling arid ecosystems. In D. W. Goodall and R. A. Perry (eds), *Arid Land Ecosystems: Structure, Functioning and Management*, vol. 2, pp. 411–432. Cambridge University Press, Cambridge.

Nulsen, R. A., Bligh, K. J., Baxter, I. N., Solin, E. J., and Imrie, D. H. (1986). The fate of rainfall in a mallee and heath vegetated catchment in southern western Australia. *Aust. J. Ecol.* **11**, 361–371.

Phillips, R. E., Quisenberry, J. M., Zeleznik, J. M., and Dunn, G. H. (1989). Mechanisms of water entry into simulated macropores. *Soil Sci. Soc. Am. J.* **53,** 1629–1635.

Pickup, G. (1985). The erosion cell – a geomorphic approach to landscape classification in range assessment. *Aust. Rangeland J.* **7,** 114–121.

Poesen, J. (1985). An improved splash transport model. *Z. Geomorphol.* **29,** 193–211.

Poesen, J. and Savat, J. (1981). Detachment and transportation of loose sediments by raindrop splash. part II. Detachability and transportability measurements. *Catena* **8,** 19–41.

Poesen, J. and Ingelmo-Sanchez, F. (1992). Runoff and sediment yield from topsoils with different porosity as affected by rock fragment cover and position. *Catena* **19,** 451–474.

Pressland, A. J. (1976). Soil moisture redistribution as affected by throughfall and stemflow in an arid zone shrub community. *Aust. J. Bot.* **24,** 641–649.

Rice, M. A., Willetts, B. B., and McEwan, I. K. (1996). Wind erosion of crusted soil sediments. *Earth Surf. Proc. Landf.* **21,** 279–293.

Schlesinger, W. H., Fonteyn, P. J., and Reiners, W. A. (1989). Effects of overland flow on plant water relations, erosion, and soil water percolation on a Mojave Desert landscape. *Soil. Sci. Soc. Am. J.* **53,** 1567–1572.

Schlesinger, W. H., Ward, T. J., and Anderson, J. (2000). Nutrient losses in runoff from grassland and shrubland habitats in southern New Mexico: II. Field plots. *Biogeochemistry* **49,** 69–86.

Skidmore, E. L. (1986). Wind erosion control. *Climate Change* **9,** 209–218.

Stout, J. E. (1998). Effect of averaging time on the apparent threshold for aeolian transport. *J. Arid Environ.* **39,** 395–401.

Stout, J. E. and Zobeck, T. M. (1997). Intermittent saltation. *Sedimentology* **44,** 959–970.

Tarchitzky, J., Banin, A., Morin, J., and Chen, Y. (1984). Nature, formation and effects of soil crusts formed by water drop impact. *Geoderma* **33,** 135–155.

Tisdall, J. M. and Oades, J. M. (1982). Organic matter and water stable aggregates in soils. *J. Soil Sci.* **33,** 141–163.

Valentin, C. (1991). Surface crusting in two alluvial soils of northern Niger. *Geoderma* **48,** 201–222.

Van de Ven, T. A. M., Fryrear, D. W., and Spaan, W. P. (1989). Vegetation characteristics and soil loss by wind. *J. Soil Water Conserv.* **44,** 347–349.

Van Elewijck, L. (1989). Influence of leaf and branch slope on stemflow amount. *Catena* **16,** 525–533

Van Vactor, S. (1989). *Hydrologic and edaphic patterns and processes leading to playa flooding.* MS. Thesis. New Mexico State University, Las Cruces, NM.

Wainwright, J., Parsons, A. J., and Abrahams, A. D. (1999). Rainfall energy under creosotebush. *J. Arid Environ.* **43,** 111–120.

West, N.E. and Gifford, G. F. (1976). Rainfall interception by cool-desert shrubs. *J. Range Mgmt.* **29,** 171–172.

White, K. (1995). Field techniques for estimating downstream change in discharge of grave-bedded ephemeral streams: case study in southern Tunisia. *J. Arid Environ.* **30,** 283–294.

Whitford, W. G. and Kay, F. R. (1999). Bioperturbation by mammals in deserts: a review. *J. Arid Environ.* **41,** 203–230.

Whitford, W. G. (2000). Keystone arthropods as webmasters in desert ecosystems. In D. C. Coleman and P. F. Hendrix (eds), *Invertebrates as Webmasters in Ecosystems*, pp. 25–41. CABI Publishing, Wallingford, UK.

Wiggs, G. F. S., Livingstone, I., Thomas, D. S. G., and Bullard, J. E. (1994). Effect of vegetation removal on airflow patterns and dune dynamics in the southwest Kalahari Desert. *Land Degrad. Rehabil.* **5,** 13–24.

Wondzell, S. M., Cunningham, G. L., and Bachelet, D. (1996). Relationship between landforms, geo-morphic processes and plant communities on a watershed in the northern Chihuahuan Desert. *Landsc. Ecol.* **11,** 351–362.

Yair, A., Sharon, D., and Lavee, H. (1980). Trends in runoff and erosion processes over an arid lime-stone hillside, northern Negev, Israel. *Hydrol. Sci.* **25,** 243–255.

Chapter 5   Patch – Mosaic Dynamics

There are many different kinds of patches that are included in a single landscape unit. Patches differ in size, species composition and the way they function in the landscape (Ludwig and Tongway, 1995). Patches may be composed of a single plant (one species) or a group of plants (several species) or of soil modified by activities of animals. Mosaics are repeating units of several patch types across the landscape. Mosaics are characterized by clumped distributions of plant species created by endogenous factors such as the tendency of one species to create favorable or unfavorable microsites for other species or by exogenous factors such as the differential germination and survival of individuals in distinct microenvironments (Yeaton and Manzanares, 1986). The structure of a patch dominated by a single plant species or produced by an animal species may be a function of landscape position. Position on a landscape, soil characteristics and relationship to other patches are important variables determining the biological composition of patches. The structure of patches may also reflect competitive interactions. In deserts, where many of the organisms are living at or very near the threshold for surviving the climatic extremes, the availability of resources in patches is a critical variable. Competition for those limiting resources by other organisms is probably an important process determining the species composition and characteristics of a patch.

5.1 SEEDS: GERMINATION–ESTABLISHMENT SITES

The species composition of patches results largely from the effect of seeds trapped in favorable germination–establishment sites. 'The patchiness seems typical of arid plant communities where relatively rare safe sites are interspersed among more abundant, barren, unfavorable germination sites' (Henderson *et al.*, 1988). Seed distributions are patchy and related to obstructions and/or traps across the landscape. Seed accumulations occur in naturally occurring or animal-produced depressions or pits (Reichman, 1984; Whitford and Kay, 1999). Seeds and leaves are readily moved by low-velocity wind (DeSoyza *et al.*, 1997). Wind velocity that exceeds the detachment velocity for seeds and small organic matter (leaves and small stems) need not carry the material aloft in order to redistribute it. Winds of sufficient velocity to move seeds and leaves occur almost every day in desert regions. Maximum wind speeds > 10 km h^{-1} were recorded on an average of 302 days yr^{-1} at the Jornada Experimental Range in southern New Mexico. Wind transport of seeds accounts for the distribution patterns recorded by Reichman (1984) (Fig. 5.1).

In desert landscapes there are microsites with high seed densities within micro-habitats of low seed density. In the Sonoran Desert, seed densities were lowest in an ephemeral stream channel and in open areas between shrubs. Seed densities were not correlated with shrub volumes (Reichman, 1984). In the Mojave Desert, seeds accumulated around shrubs and the densities of seeds were correlated with the volumes of the shrubs (Nelson and Chew, 1977). Differences in seed densities in relation to shrub size in the Mojave and Sonoran Deserts are probably the result of differences in patch characteristics in these two deserts. In the Mojave Desert, rodent burrows are primarily associated with shrubs (deSoyza *et al.*, 1997; Soholt and Irwin, 1976). In the Sonoran Desert and in the Chihuahuan Desert surface pits produced by rodents are relatively common and exhibit clumped distribution (Steinberger and Whitford, 1983). Rodent-produced pits are important as seed

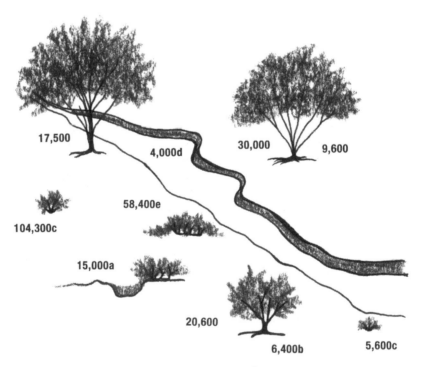

Figure 5.1   Estimated densities of seed (number m^{-2}) in various microsites in the Sonoran Desert. The prevailing wind is from right to left. a, rodent cache pit; b,windward side of small shrub; c, subshrub; d, rill channel; e, grass tussock. (Modified from Reichmann, 1984.) With permission of *Journal of Biogeography*, Blackwell Science Ltd.

traps and as favorable germination sites. In the Great Basin Desert virtually all of the patches of Indian rice-grass, *Oryzopsis hymenoides*, occurred as a result of seeds trapped with litter in rodent excavations (Longland, 1995). Heteromyid rodents enhance the germination of the *O. hymenoides* seeds buried in caches by removing the seed-investing structures (lemma and palea) on approximately one-third of the seeds placed in cache pits (McAdoo *et al.*, 1983).

In desert grassland, the spatial pattern of the vegetation was patchy but species associations within the patches were weak (Henderson *et al.*, 1988). The absence of strong positive or negative species association was considered to be consistent with the conclusions of Grime (1979) that competition and dominance are reduced in unpredictable desert environments. However, it is likely that this generalization may be restricted to arid grasslands and savannas and not to all desert environments. The high correlation between the seed bank and vegetation were hypothesized to be the consequence of frequent and unpredictable disturbance (especially variation in rainfall) and by the effects of seed consumers and the spatial patterns of their foraging behavior on seed bank turnover rates (Henderson *et al.*, 1988). Patchiness is typical in desert plant communities where safe sites for seeds, and germination are interspersed among more abundant unfavorable sites for seeds to escape seed feeding animals and to encounter conditions suitable for germination.

Subcanopy microsites may be safe and favorable germination–establishment sites for annual plants and for some perennial plants (Whitford *et al.*, 1995a; DeSoyza *et al.*, 1997). Shrub patch development in south Texas semiarid savanna occurred in part by colonization by mesquite (*Prosopis glandulosa*) followed by establishment of other shrubs beneath the mesquite. The presence of mesquite shrubs attracted birds that dispersed the seeds of the shrub, *Celtis pallida*. *C. pallida* exhibited increased germination and establishment under mesquite canopies where shade reduced soil temperatures to levels that favored germination and growth. The absence of *C. pallida* in the patch interspaces was attributed to lack of dispersal by birds and inhibition of germination by high temperatures (Franco-Pizana *et al.*, 1996). The patchy distribution of ephemeral plants has frequently been attributed to enhanced germination and establishment in the soils modified by shrubs and by the moderate subcanopy microclimate. The subcanopy soils and microclimate associated with shrubs has a positive effect on many, but not all, ephemeral plant species (Gutierrez and Whitford, 1987; Tielborger and Kadmon, 1995). In a demographic study of four desert annual plant species, three of the species exhibited higher emergence densities and higher seedling mortality under shrub canopies. One species (*Ifloga spicata*) had higher densities in open spaces. The subcanopy surviving plants produced more seeds than conspecifics growing in the openings. *Ifloga spicata* growing in open spaces produced more seeds than individuals of this species growing under shrub canopies. The emergence densities of plants germinating below shrub canopies also varied with the species of shrub (Tielborger and Kadmon, 1995). This provides support for the general observation that the structure of the perennial plant community affects the composition and abundance of the ephemeral plant community.

Animal-produced pits are collection points for seeds and organic matter. The decomposing organic matter provides nutrients and increases water-holding capacity of the microsite. There are several types of animal-produced pits that serve as collection areas for seeds, litter, and dung. These include cache pits and foraging pits produced by small granivorous mammals, bedding sites of large mammals like the Nubian Ibex, North American jackrabbits, and Australian kangaroos (Fig. 5.2) and pits dug by porcupines (Whitford and Kay, 1999). Animal-produced pits are more important as germination sites in some locations than in others. In inter-dune pans, areas with fine-textured soils, 96% of all seedlings found were growing in diggings. In the South African Karoo, several mammals (aardvark, *Orycteropus afer*, porcupine, *Hystrix austro-africanae*, bat-eared fox, *Otocyon megalotis* and Cape fox, *Vulpes chama*) excavated pits that ranged in size from 7 to 24 cm long, 3 to 12 cm wide. Three types of pits were identified; (1) shallow, basin-shaped scoops made by aardvarks while 'prospecting' for termites; (2) narrow, trench-like holes dug by foxes to extract beetle and other insect larvae; and (3) circular or conical holes dug by rodents to extract insect larvae or seeds. Pits in which seeds germinated were deeper (6.7 cm) than pits without seedlings (4.5 cm). Deeper pits collect more run-off water and accumulate more plant debris and litter which contributes to the differences in germination (Dean and Milton, 1991).

Figure 5.2   Hip holes dug by kangaroos, *Macropus* spp., in a mulga, *Acacia aneura*, grove in northwestern New South Wales, Australia.

Pits dug by some species are concentrated in areas where establishment of plants is essential to the functioning of a larger unit of landscape. For example, in the banded mulga system in Australia, pits produced by goannas (varanid lizards) accumulate plant debris, some fecal material and seeds of several species of plants that are important components of the interception zone of the mulga system (Whitford, 1998) (Fig. 5.3). Reichman (1984) emphasized the importance of surface disturbance as sites for seed accumulation. Although shrub canopies may generate seed accumulations, they are probably not as important as patch generators as the soil disturbances made by animals (Fig. 5.4).

Another factor that has to be considered in seed dispersal and location of safe germination sites is animal dispersal. Many desert shrubs and small trees produce relatively large fruits that are dependent on dispersal by animals. Birds are effective dispersers of fleshy-fruited plant species. In a study of vegetation patch dynamics in the Kalahari, plant species with fleshy fruits occurred in 8% of the treeless plots, 17% of the plots centered on *Acacia erioloba* saplings and 90% of the plots under large trees. The distribution of fleshy-fruited plants was attributed to roosting birds and mammals that utilized the shade of the large trees (Dean *et al.*, 1999). Plants selected as preferential perch sites by birds serve as focal species for development of patches composed of bird-dispersed seeds. In the Chihuahuan Desert, the cactus wren selects tall plants such as aborescent *Opuntia*

Figure 5.3   Pit dug by a goanna lizard, *Varanus gouldii*, in the excavation of a ground-nesting spider, northwestern New South Wales, Australia.

Figure 5.4   Recently excavated cache pits produced by kangaroo rats, *Dipodomys* spp., in the northern Chihuahuan Desert, New Mexico.

spp. and *Yucca elata* as nest sites. Cactus wren nests are constructed of grasses, many of which retain viable seeds in the seed heads. Abandoned nests fall to the ground providing a mulch and seed source in the patch where the nest was constructed (Milton *et al.*, 1998). This suggests that patches, which have as focal species, plants selected as nest or perch sites by birds, may be sites where grasses selected for nest materials may establish albeit late in patch development.

5.2 SINGLE SPECIES PATCHES

The significance of environmental modification by a single plant interfacing with the physical environment was initially described as the development of 'fertile islands' by Garcia-Moya and McKell (1970). Their studies showed that the desert shrubs modify the soil by accumulating nutrients in the subcanopy soils. Subsequent studies in warm deserts cold deserts and a variety of other arid and semiarid ecosystems corroborated these observations (Charley and West, 1975; Barth and Klemmedson, 1982; Crawford and Gosz, 1982; Virginia and Jarrell, 1983). However the 'fertile island' that develops in the vicinity of a single shrub species is not limited to nutrient concentration. Indeed nutrient concentrations are the product of a whole series of microenvironmental modifications resulting from the presence of a plant. As described above, a shrub, grass clump or tussock can serve as a seed and litter trap. The accumulation of organic material within or under the canopy enriches the soil carbon stores and, if sufficiently thick, provides a habitat for a variety of soil organisms (Santos *et al.*, 1978). The soil environment under a single grass plant is modified primarily by the addition of materials from root death and decomposition and from root exudates. Shading directly affects microclimate. The presence of a canopy reduces raindrop kinetic energy and the throughfall and stemflow water is retained within the patch.

5.3 INTRASPECIFIC MORPHOLOGICAL VARIATION

The characteristics of single-plant patches vary by species and by growth form of a single-species that exhibits morphological variation. A single-species patch may provide a significant obstruction to overland flow or no effective obstruction depending on growth form. Single-species patches may or may not be effective as wind breaks. Creosotebush (*L. tridentata*) exhibits considerable variation in morphology. Creosotebush shrubs with exterior stem angles >45° (conical-shaped shrubs) trap no litter beneath the canopy and there are essentially no differences in subcanopy soils and intershrub space soils in terms of texture, organic carbon content and concentrations of mineral nutrients. In contrast, shrubs of this species with a growth form approaching hemispherical, accumulate subcanopy litter layers. Subcanopy soils of these plants have nutrient concentrations higher than the intershrub soils forming islands of fertility. Conical shrubs transfer a larger proportion of the stemflow to the root crown than do hemispherical shrubs (De Soyza *et al.*, 1997) (Fig. 5.5). These differences in subcanopy environment are reflected in production differences and species richness of annual plant species growing in the subcanopy space (DeSoyza *et al.*, 1997). Some of the important desert legumes (*Prosopis* spp., *Acacia* spp.) exhibit considerable variation in growth form – single-stemmed small trees or multi-stemmed shrubs. The range of morphological variation in the dominant species of plants in desert communities has not been adequately documented. However, this example based on creosotebush, *L. tridentata*, from North American deserts is one example of how intraspecific morphological variation contributes to differences in patch characterisitics of single species patches.

The morphology of shrubs also affects the energy of drip-line drops that may be responsible for the splash erosion contribution of mounds under shrubs. The protection provided by shrubs to the soil surface beneath the canopy results in differential rain-splash. More sediment is splashed into the area under the canopy than is splashed outward. The surface materials of the mounds under shrubs have the same size distribution as the sediments. Some of the mounding is attributed to soil loss from the intershrub space. Desert pavement (surface covering of small stones and rock) in intershrub spaces is formed of lag materials remaining after off-site transport of soil by rain-splash erosion. The microtopography and

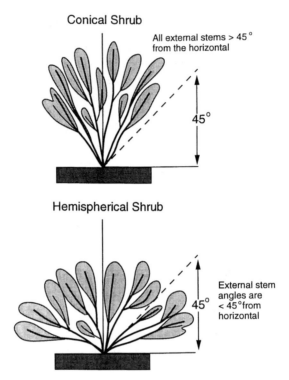

Conical Shrub

All external stems > 45°
from the horizontal

45°

Hemispherical Shrub

External stem
angles are
< 45° from
horizontal

45°

Figure 5.5 Diagram of two shrub morphologies and the effects of that mophology on stem-flow and throughfall.

distribution of soil surface materials of plant patches and bare spaces is thought to be the consequence of establishment of shrub vegetation during the past century (Parsons *et al.*, 1992)

5.4 MOSAICS AND MULTI-SPECIES PATCHES

The species composition and spatial relationships of the species are primary determinants of how the mosaic or patch functions in the landscape. Species composition may not be very predictable. For example, in a species-rich area of the southern Chihuahuan Desert dominated by a small tree, *Acacia schaffneri* and a cactus, *Opuntia streptacantha*, every combination of all common species within the association was found in the mosaics. However, the mosaics did have a common structural pattern: a perennial legume at the center, an erect (shrubby or aborescent) form of platyopuntia (flat cladodes) near or under the legume, and a decumbent form of platyopuntia at the periphery (Yeaton and Manzanares, 1986). It was hypothesized that these patches or mosaics developed following the establishment of a legume that modified the soil nutrient base. The colonization by aborescent platyopuntia occurs soon after the establishment of the legume. It was also suggested that *O. strepacantha* eventually replaces the aborescent legume, which accounts for some of the variability recorded in the structure of the mosaic. This study provides evidence that patch-mosaic structures are dynamic within the spatial and temporal constraints of the patch. Patch structure is not dependent on the temporal and spatial scales of the component species.

In contrast, a study in another part of the Chihuahuan Desert found a range of patches from species poor to species rich. All of the patches were composed of species distributed around focal or 'nurse' plant species. Single species patches of creosotebush shrubs, *L. tridentata*, were the most species poor. Creosotebush also occurred in two species patches with cacti (*Opuntia* spp.) and another shrub (*Jatropha dioica*). Some species, mostly cacti, tended to occur in the largest patches. Sixteen species associations were positive and nine were negative. All but one of the negative associations involved one or the other of the two dominant plants, i.e. *L. tridentata* and *Opuntia rastera*. Neither of these exhibited any significant positive association with any species in the complex (Silvertown and Wilson, 1994). This led to a conclusion similar to that of Yeaton and Manzanares (1986) that focal species modify the environment allowing colonization by other species. Some species, including shrubs, such as *Prosopis* and *Foquieria*, occurred more frequently than expected on medium-sized patches. Other species, especially cacti and leaf succulents such as *Yucca* spp. and *Agave* spp. occurred primarily on large patches. There was a significant spatial structure associated with six of ten species examined in this study (Silvertown and Wilson, 1994) all of which occurred in clumped patches more frequently than expected. These studies suggest that the vegetation associations of patches and mosaics depend on the presence of a large focal species or 'nurse' plant for initial arrival (seed capture), establishment (favorable microsite), and survival. The bare zones between vegetation patches are taken as evidence of this dependence. The presence of a focal plant therefore operates a switch that converts local bare desert into vegetation (Wilson and Agnew, 1992).

Desert vegetation is commonly dominated by plants that live for decades, centuries and even millenia (Vasek, 1980). Because of the longevity of dominant plant species, the underlying mechanisms of the dynamics of desert vegetation are difficult to study. McAuliffe (1988) provided data to support a Markovian Dynamics model of Sonoran Desert plant communities. He determined that establishment of the large shrub, *L. tridentata*, was predominately beneath the dense canopies of the small shrub *Ambrosia dumosa*. The *Ambrosia* shrubs eventually die and the patch is dominated by a large *Larrea*. *Ambrosia* is capable of colonizing bare (open) patches whereas *Larrea* requires *Ambrosia* as a nurse plant for its establishment. Creosotebush, *L. tridentata*, patches provide nurse plant environmental modification necessary for the establishment of large long-lived palo verdes (*Cercidium* spp.) and saguaros, *Carnegiea gigantea*. These large plants eventually dominate the patches in which they establish (Fig. 5.6). This results in

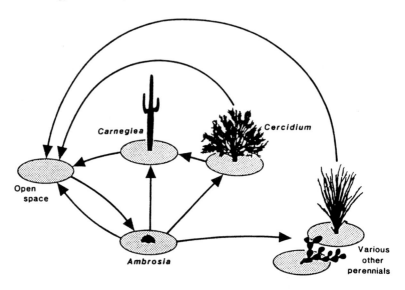

Figure 5.6   Diagram showing the principal transitions in the *Larrea–Ambrosia* communities in the southern Mojave Desert. (From McAuliffe, 1988.) With permission by the University of Chicago Press.

Sonoran Desert ecosystems consisting of four types of patches: *Ambrosia* shrubs, *Ambrosia* shrubs with small *Larrea*, dead *Ambrosia* shrubs with large *Larrea*, *Larrea* shrubs, and open, bare patches. In the Markovian model, McAuliffe assigns transition probabilities to each of the patch states. The transitions from one patch type to another may take decades or centuries (especially in the transition of the long-lived *Larrea* to bare, open patches).

5.5 COMPETITION VERSUS FACILITATION

Conventional wisdom, based on statements by influential desert biologists, questions the importance of and the existence of competition among plants in arid and semiarid regions (Shreve, 1931; Grime, 1979). Most of the early evidence for competition among desert plants was based on spatial distribution patterns (Fowler, 1986). The problem with spatial distribution as evidence for or against competition is that a regular distribution of plants may be the result of competition but the absence of a regular distribution is not evidence for the absence of competition. In a review of the literature on spacing in desert plants Fowler (1986) lists 33 studies of which nine found clumped distributions (contagious), 10 in which distributions were either clumped or random, and only two that were regular. Fowler (1986) also listed 16 studies in which there were differences in size of individuals that were related to distances between pairs of individuals (an indirect measure of competition). Studies of spacing of desert plants combined with information on size distributions of the plants can provide some insights into competition and its influence on patch development. In studies of vegetation of the Mojave and Sonoran Desert, small shrubs tended to be clumped, medium-size shrubs tended to be randomly distributed and large shrubs tended toward a regular dispersion pattern (Phillips and MacMahon, 1981). This pattern was interpreted as evidence of the increasing importance of competition as plants grow. Root systems were sufficiently extensive to overlap or abut in the interplant open spaces. The results were not dependent upon the position of a site on the large climatic gradient across the winter-rainfall Mojave to the bi-seasonal rainfall of the Sonoran Desert (Phillips and MacMahon, 1981).

Studies of direct effects of competition involve removing one or more species from a patch and monitoring the responses of the remaining species. Studies on the Santa Rita Experimental Range south of Tucson, Arizona, showed that removal of mesquite led to an increase in annual and perennial grasses (Cable and Tschirley, 1961). Removal of sagebrush, *Artemisia tridentata*, led to increased biomass of several species of native grasses (Robertson, 1947). Treatment of creosotebushes with a herbicide resulted in a rapid and marked increase in bush muhly, a grass that utilizes *L. tridentata*, as a nurse plant (Whitford *et al.*, 1978). These early experimental manipulations provided evidence for competition among desert plants but provided no insights into the mechanisms of that competition. These insights came from studies of two Mojave Desert shrubs (creosotebush (*L. tridentata*) and burr sage (*Ambrosia dumosa*)). Water-related interference competition occurred when water availability was low. However, competition was more intense between species than between individuals of the same species (Fonteyn and Mahall, 1981). This research suggested that regular distribution of plants may be due to past competition not to present competition. This generalization is particularly applicable to long-lived perennials such as creosotebush.

Not all close spatial associations of plants are negative. As suggested in the development of patches and mosaics, focal or nurse plants are frequently required for patches to develop. The nurse plant concept was developed to explain the establishment of saguaro cacti, which are largely limited to areas under canopies of trees or shrubs (Shreve, 1931; Steenberg and Lowe, 1977) (Figs 5.7 and 5.8). Nurse plants may protect young plants from excessive heat or cold or from mechanical injury and predation. Nurse plant protection is necessary not only for large columnar cacti, but it is important for a number of smaller cacti and also for some grasses (Franco and Nobel, 1989; Livingston *et al.*, 1997). Cactus species apparently differ in their need for nurse plants. About 89% of the saguaro seedlings were found in sheltered microhabitats under canopies of burr sage (*Ambrosia deltoidea*) and palo verde (*Cercidium microphyllum*). Nearly 30% of the seedlings of *Ferocactus acanthodes* were in unsheltered microhabitats where maximum soil surface temperatures reached 71°C. The sheltered seedlings of *F. acanthodes* were under canopies of a bunch grass, *Hilaria rigida* (Franco and

Figure 5.7 Saguaro, *Carnegia gigantea*, columnar cacti associated with creosotebush, *Larrea tridentata*, in the Sonoran Desert, Arizona.

Figure 5.8 Saguaro, *Carnegia gigantea*, columnar cactus associated with ironwood tree, *Olynea tesota*, in the Sonoran Desert, Arizona.

Nobel, 1989). This kind of commensal symbiosis can shift to a negative interaction (competition) when the nursed plants achieve a certain size. The classical nurse plant relationship between saguaros and palo verde changes into a relationship in which the palo verde is negatively affected by the presence of the cactus. Both stem dieback and mortality occurred in nurse trees. Stem dieback accelerated the local loss of individual nurse trees (McAuliffe, 1984a). We have observed more dead stems of creosotebush with large understory bush muhly (*Mulhenbergia porterii*) grass plants than in shrubs with no understory perennial grass plants. The response of this grass to the removal of creosotebush suggests interspecific competition, most probably for water (Whitford *et al.*, 1978).

Nurse plant–commensal relationships are not limited to large cacti. Other species of cacti such as *Echinocereus* spp. may receive protection and microclimate modification by other species of cacti, *Opuntia fulgida* (McAuliffe, 1984b). Nurse plant associations may also protect the 'protégé' species from browsing by herbivores. This has been suggested for *Mammalaria* spp. growing under various shrub species in Baja California (Cody, 1993). Protection from browsing or microclimate modification may account for the relationship between creosotebush, *L. tridentata*, and pencil cholla, *Opuntia leptocaulis*. In that relationship the pencil cholla eventually outcompetes the creosotebush which when it dies, exposes the cactus to the effects of soil erosion on the shallow root system and excavation and consumption of roots by rabbits and rodents (Yeaton, 1978). Intermediate size cacti such as the cylindropuntia or chollas of the Mojave Desert have also been shown to be associated with nurse plants (Cody, 1993). Young individuals of three cholla species were associated with nurse plants, especially the perennial grass, *Hilaria rigida*. Adult chollas also served as nurse plants for young chollas. However, no shrubs were found to serve as nurse plants for these clindropuntias.

The nurse plant association is not obligate for these species because the frequency of cholla growing in the open was approximately the same as those growing in association with other species or itself. As earlier studies documented, maturing cholla cacti compete with nurse plants for water and nutrients which frequently results in the demise of the nurse plant. Cacti and stem succulents are not the only plants that benefit from nurse plant associations. In the Sonoran Desert, ironwood tree, *Olneya tesota*, modify the habitat providing nurse plant benefits to columnar cacti, and large shrubs with berry type fruits (Suzan *et al.*, 1996).

Nurse plants may serve to trap seeds, modify the microclimate of germination sites and protect seedlings from predators (Cody, 1993). These associations contribute to the variation in species composition and structure in patches in hot deserts. The facilitation of growth of species associated with nurse plants may change to a competitive relationship over time. Nurse plant associations are reported for arid environments but appear to be rare or absent in more mesic environments. The functional explanation for this observation is attributed to differences in physiological responses of the species in the association. In arid environments, facilitation occurs when improvement of plant water relations under the canopy exceed the costs produced by lower light levels (Holmgren *et al.*, 1997). Sparse canopies of desert shrubs and the over-riding importance of water as a limiting factor tips the balance in favor of facilitation and nurse plant status. In more mesic environments, canopy densities reduce light sufficiently that competition rather than facilitation is observed.

5.6 ANIMAL-PRODUCED PATCHES

Patchiness in deserts is not limited to spatial patterns of plants in relationship to soils and topography. Patchiness is also a function of animal effects on soil properties and the spatial distribution of soil disturbances (biopedturbations) on the landscape. Patchiness caused by disturbance is important in the development and maintenance of spatial and temporal heterogeneity in arid and semiarid ecosystems (Whitford and Kay, 1999; Whitford, 2000). The agents of soil disturbance that create patches in arid environments include mammals, reptiles, and a variety of arthropods. Because soil materials are transported from deep soil layers to the surface, animal disturbance patches are areas of active pedogenesis. It has been suggested that, in deserts, digging by small mammals may be the most important mechanism for 'pumping' soluble nutrients to the soil surface and may be the only mechanism for bringing insoluble materials to the surface for weathering (Abaturov, 1972). However, many species of ants and termites construct burrow systems to greater depths in the soil than mammals, and many of these transport insoluble and soluble material from more than 200 cm deep in the soil profile. The combined effect of invertebrate and vertebrate burrowing activities can produce ejected materials that cover more than 20% of the soil surface area in a single season (Whitford and Kay, 1999).

The importance of animal activity in generating soil patches is probably best illustrated by the example from the banded mulga (*Acacia aneura*) landscapes in Australia. In these landscapes vegetation is concentrated in stripes that are perpendicular to the slope of the landscape. The large patches of virtually barren ground between stripes are erosion slopes that shed water during intense rains. Scattered across the erosion slopes are scattered logs of dead mulga trees. These logs are perched on perceptible mounds that are riddled with termite galleries and ant nest chambers. The mounds are probably formed from the foraging gallery sheets of termites. The mound material has the same textural composition as termite gallery sheeting and has lower bulk density and higher sand content than soils of the erosion slope. The macropores resulting from the galleries and chambers of the ants and termites are responsible for the higher infiltration rates recorded on log mounds in comparison with rates on adjacent unmodified soils (Tongway *et al.*, 1989). The activities of the arthropods also contribute to the higher nutrient levels in the soils of the log mounds. The log mounds represent scattered productive patches on the barren erosion slopes. In this system, termites and ants establish positive feedback loops where increased soil moisture produces more suitable conditions for the insects thereby reinforcing the water–nutrient enrichment of the patch.

Some of the most impressive patches of soil disturbance are produced by mammals or by the combined efforts of mammals and arthropods. Among the largest soil patches in desert landscapes are heuweltjies (pronounced haer–vulkies) which are large soil mounds found in southern Africa. These large mounds are produced by mole rats and termites and cover areas of up to 800 cm^2. The largest mounds are produced by hairy-nosed wombats (*Lasiorhinus latifrons*) (Fig. 5.9). These mounds can cover up to several hundred square meters and are visible on LANDSAT satellite imagery of Australia (Loffler and Margules, 1980). Other large soil disturbance features produced by mammals include the mounds of the now extinct burrowing bettong (*Bettongia lesuer*) in Australia (Noble, 1993) and mima like mounds in Argentina (Ojeda *et al.*, 1996).

Not all soil disturbances produced by mammals cover large areas. Soil disturbances vary from cache pits and foraging digs (10–15 cm^2) to mounds produced by fossorial mammals (30–50 cm^2) and the complex burrow systems of colonial rodents (50–2000 cm^2) (Whitford and Kay, 1999). Despite large variations in size, the pedturbations produced by mammals result in a number of changes in soil properties and vegetation responses (Table 5.1).

In a review of the literature on mammalian pedturbations, the probable importance of landscape position, soil properties, and geomorphological history as a

Figure 5.9   Mound system of a hairy-nosed wombat, *Lasiorhinus latifrons*, in South Australia.

Table 5.1

Effects of Mammalian Pedturbation on Soil Properties and on Vegetation in the Area or Disturbance and/or on the Mounds of Soil Ejected from the Burrows and Digs (modified from Whitford and Kay, 1999)

Soil property	Disturbance type	Vegetation response
Litter and seed accumulation	Cache pits/foraging pits	> Annual plants
	Nubian ibex bedding sites	> Annual plants
	Crested porcupine digs	> Density geophytes
Fecal pellet accumulation	Gemsbok resting form	> Density annuals
Subsoil to surface	Aardvark, bat-eared fox, Cape	> Density geophytes
	porcupine foraging digs	> Density annual forbs
		< Perennial grasses
< Bulk density	Burrow system, *Dipodomys*	> Biomass annuals
> N	*spectabilis*	< Cover perennials
		Δ Species annuals
	Burrows *Spermophilis* spp.	> Grass biomass,
		> *Artemisia tridentata*
		biomass
< Soil moisture	Burrow system – desert gerbils	
> Infiltration	Rodent burrows	< Native annual grass
		> Biomass two shrubs
	Abandoned *Gerbillus* colonies	> Grass production
> Total soluble salts, >CaCO₃	Burrow ejecta mounds	> Biomass production
> K, >P, >SOM	Ejecta mounds pocket gopher	> Biomass production
		Δ Species composition
<SOM, < N	Ejecta mounds, mole rats	Δ Species diversity
>N, >SOM	Ejecta mounds mole rats	Establish geophytes
No change soil properties	Ejecta mounds fossorial system	> Pioneer species,
		> exotic
		Mesembryanthemum
>SOM, >N	Mound *Dipodomys spectabilis*	> Biomass annuals
>N, K, Ca, Mg, Zn, Mn, B, Fe, Al	Heuweltjies	Δ Species composition
		> Productivity
	Giant kangaroo rat system	> Exotic annuals
	Fossorial system *Ctenomys*	> Biomass, > species
	australis	Richness
	Mima-like mounds, Argentina	> Exotic tree
	Gerbil colonies	< Grass production
	Red viscacha rat mounds	Succulent halophytes
	Bathyergid mole rats	< Plant biomass,
		Δ community structure

<, less than; >, greater than; Δ, change in; SOM, soil organic matter.

variables affecting the magnitude and direction of changes resulting from disturbance became more apparent. Most of the conflicting results reported in the literature probably resulted from the failure to account for these temporal and spatial variables. Despite these inconsistencies, it is obvious that animal-produced patchiness is an important feature of desert ecosystems.

Most studies of the effects of ants on soil nutrients have focused on single locations (Lobry deBruyn and Conacher, 1990). The published data indicate that some species of ants affect soil chemistry, whereas other species do not. Ants that transport fruits and seeds, plant parts, and dead insects to the nest concentrate organic matter in the vicinity of that nest. Ant species that have an effect on soil chemistry are characterized by long-lived colonies (>10 years) and by milling seeds or processing plant parts at the nest. Long-lived colonies of *Pogonomyrmex barbatus* and *P. rugosus* mill seeds within the nest and deposit chaff in midden piles that are usually located within a meter of the nest entrance (Wagner *et al.*, 1997). *P. rugosus* was reported to discard 60% of the total energy content of materials collected by foragers and returned to the nests. Soil nutrients that are concentrated in the soils of long-lived *Pogonomyrmex* spp. include N, P, and K (Whitford, 1988; Whitford and DiMarco, 1995; Wagner *et al.*, 1997). However, the effect of harverster ant nests on soils and vegetation cannot be generalized for an entire watershed or landscape. Large differences in soil nutrients were found around ant nests in mid-slope locations on a watershed but no significant differences in soil nutrients around nests and away from nests in locations at the base of the watershed (Whitford and DiMarco, 1995). It is likely that the relative effects of ants on soils vary with landscape position and the nutrient status of the soils on a particular landscape unit.

Arthropods indirectly affect increased plant production both by their activities as modifiers of soil fertility and water infiltration and by their effects as consumers. Ants and termites affect both the species composition and biomass production of soils that are modified by their activities (Table 5.2). Ants that enrich the soil in the immediate vicinity of their nests include seed harvesting ants, leaf-cutting (Attine) ants, and species that transport plant and animal material to the nests. Some species of ants create chaff–trash piles around the nest perimeter that increase the organic matter content of the soil under the ejected material (Wagner *et al.*, 1997). Nests of an ant, *Formica perpilosa*, that constructs nests under small leguminous trees (*P. glandulosa* and *Acacia constricta*), was found to enhance seed production by 1.9 times in relation to trees without nests. The increased seed production was attributed to deep wetting of soil as a result of bulk flow along nest gallery tunnels (Wagner, 1997).

A corona around the base of epigeic termitaria is generally characterized by higher nutrient concentrations and soil organic matter than soils not associated with mounds (Lee and Wood, 1971a, b; Coventry *et al.*, 1988; Lobry deBruyn and Conacher, 1995). Nutrient enrichment and modified water infiltration and storage affect both the species composition and the productivity (biomass production) of the vegetation in the vicinity of ant nests (Table 5.2). However, the limited data on the effects of termites and ants on plant biomass production and species composition must be interpreted with caution. The species of plants that increase in biomass and that are favored by the insect-modified soils are not necessarily the same for the same arthropod species on different landscape units within a region (Whitford and DiMarco, 1995). There are also inter-annual differences in the

Table 5.2

Effects of Ants and Termites on Soils in Arid and Semiarid Ecosystems

Species	Location	Effect on soil	Reference
Ants			
Camponotus acvapimensis	Humid savanna Africa	> Sand fraction, aggregates, clay, coarse sand, macroporosity, C_{total}	Levieux (1976)
Paltothyreus tarstus	Humid savanna Africa	> Clay, coarse sand, macroporosity, C_{total}; turnover 30 g m^{-2} yr^{-1}	Levieux (1976)
Aphaenogaster barbigula	Semiarid woodland, Australia	> Infiltration, soil turnover 3360	Eldridge and Pickard (1994)
Aphaenogaster spp.	Semiarid woodland, W. Australia	Turnover; 465 g m^{-2} yr^{-1} (gray sand loam); 223 g m^{-2} yr^{-1} (yellow sand) >clay	Lobry deBruyn and Conacher (1994)
Formica perpilosa	Southwestern USA	≫ Nitrogen mineralization rates ≫ NO_3, NH_4, P, H_2O	Wagner (1997)
Ant community	Chihuahuan Desert ecosystems	Soil turnover 21.3–85.8 kg h^{-1} yr^{-1}	Whitford *et al.* (1995b)
Pogonomyrmex rugosus	Chihuahuan Desert, USA	≫ Cover four species of annuals ≫ NO_3, N_{total}, < Ca, Mg	Whitford and DiMarco (1995)
Pogonomyrmex occidentalis, P. owyheei	Great Basin Desert, Oregon, USA	> Biomass of herbaceous plants	Sneva (1979)
Pogonomyrmex occidentalis	Great Basin Utah, USA	> Microbial biomass, and vesicular–arbuscular mycorrhizae	Friese and Allen (1993)

Pogonomyrmex rugosus, Veromessor pergandei	Sonoran Desert Arizona, USA	> Density and cover six species of annual plants	Rissing (1986)
Veromessor andrei	California, USA annual grassland	< Abundance of two dominant plants, > Abundance two alien annuals	Hobbs (1985)
Ant community	*Atriplex vesicaria* shrubland, Australia	Soil turnover: 350–420 kg ha^{-1} yr^{-1} > C_{total}, N_{total}, P_{total}	Briese (1982)
Pogonomyrmex rugosus	Burro-grass, Chihuahuan Desert	> N_{total}, Organic matter; biomass, average mass per plant (*Erodium texanum*)	Whitford (1988)
Termites			
Macrotermes subhyalinus, M. michaelsenii	Semiarid savanna, Kenya	Mounds: >CEC, $C_{organic}$, N, clay	Arshad (1981)
Subterranean termites	Savanna, Sahel Africa	≫ Porosity, hydraulic conductivity, cumulative infiltration, final infiltration rate, water storage ≫ > bulk density	Mando (1997) Mando *et al.* (1996)
Odontotermes spp.	Arid grassland Brushland, Kenya	Sheeting: rate of formation: 1059 kg ha^{-1} yr^{-1}, > clay, sand, C_{total}, Ca, K, < silt, pH, CEC, Mg, Na	Bagnine (1984)
Macrotermitinae and others	Semiarid *Acacia* brushland, Tanzania	> $C_{organic}$, N, K, < Ca, Na	Jones (1989)
Trinervitermes trinervoides	Semiarid grassland, South Africa	> Production of grasses (*Themeda triandra, Tagus koeleroides*), Mg, Ca, N, P, TEC < pH	Smith and Yeaton (1998)
Amitermes vitiosus, Drepanotermes perniger; Tumultitermes pastinator	Semiarid woodland, Australia	Turnover 300–400 kg ha^{-1} yr^{-1} > (2–7 times reference soil) EC, N_{total} $P_{soluble}$, $C_{organic}$, Ca, Mg, K, Na,	Coventry *et al.* (1988)

Table 5.2 – *continued*

Species	Location	Effect on soil	Reference
Termites			
Amitermes, Tumultitermes Drepanotermes, Coptotermes Nausititermes spp.	Arid and semiarid Australia	Mounds: >clay, silt, $C_{organic}$, N_{total}, Ca, Mg, K, TEC	Lee and Wood (1971b)
Amitermes vitiosus and other species	Semiarid woodland, Australia	0.025–0.50 mm yr^{-1} soil accumulation	Holt et al. (1980)
Drepanotermes tamminensis, Amitermes obeuntis, A. neogermanus	Semiarid woodland, Western Australia	Mounds: > clay, $C_{organic}$, < pH; Sheeting: > clay, $C_{organic}$, < pH	Lobry de Bruyn and Conacher (1995)
Amitermes spp., Nasutitermes magnus, Drepanitermes rubiceps	Semiarid woodland, Northeastern Australia	> Biomass production of annual grass and a legume	Okello-Otaga and Spain (1986)
Gnathamitermes tubiformans	Chihuahuan Desert, USA	Turnover rate of standing stock of nutrients in upper 10 cm, times per year: N-3.5, P-2.5, S-2.5	Schaefer and Whitford (1981)
Gnathamitermes tubiformans	Chihuahuan Desert, USA.	Gallery sheeting: production 1000–5650 kg ha^{-1} yr^{-1}	MacKay and Whitford (1988)
Heterotermes aureus, Gnathamitermes perplexus	Sonoran Desert, Arizona	Sheeting: 744 kg ha^{-1} yr^{-1}, > pH, Ca, Mg, K, Na, $C_{organic}$, N_{total}, > clay by 21 kg ha^{-1} yr^{-1}	Nutting et al. (1987)

CEC, cation exchange capacity; TEC, total exchangeable cations.

responses of vegetation in the vicinity of ant colonies. These differences are probably the result of time lags in soil processes such as nitrogen immobilization and mineralization (Fisher and Whitford, 1995).

The effect of termite mounds on vegetation varies with the age and status of the mounds. For example in semiarid grassland in South Africa, a grass species preferred by grazing animals (*Themeda triandra*) was dominant in the area around active mounds of *Trinervitermes trinervoides* (Smith and Yeaton, 1998). Chihuahuan Desert subterranean termites affect biomass production of the dominant shrub, creosotebush (*L. tridentata*) and the species composition and biomass production of annual plants (Gutierrez and Whitford, 1987). The primary effect of termites on annual plants was through their effect on soil organic matter, water infiltration and soil nutrients (especially nitrogen). The higher biomass production of shrubs on plots with termites present was attributed to both greater water and nutrient availability when compared to plots where termites had been eliminated (Gutierrez and Whitford, 1989).

Long-lived nests of termites are reported to have effects similar to those of ant nests on soil nutrients. For example, termite mounds frequently have nutrients concentrated in the soils adjacent to the base of the mounds similar to the pattern reported for harvester ant nests. The soils of such mounds are modified by the addition of excreta and/or saliva, so that the soils contain concentrations of plant nutrients relative to the nutrient content of undisturbed soil. These mound soils are enriched in organic matter, N, P, Ca, and Mg (Malaka, 1977; Arshad, 1982; Pomeroy, 1983; Wood *et al.*, 1983). Hence, soil eroded from nest mounds produces a nutrient rich area around the base of the mound. The relative importance of mound-building termites on the patch distribution of soil nutrients is likely as dependant upon landscape position and nutrient status of the matrix soil as are ant nests (Fig. 5.10).

5.7 TEMPORAL DYNAMICS AND FEEDBACKS

The structure and nature of patches is constantly changing. A plant that establishes in a small unvegetated patch may eventually have an adverse effect on its neighbors (Whitford *et al.*, 2001). Seedling creosotebushes that were planted in three areas on a Chihuahuan Desert watershed failed to survive in the black-grama grassland on the toe slope of the mountain, and in the creosotebush-dominated bajada location. More than half of the seedlings survived in a heavily grazed site lower on the bajada. These plants grew to nearly 1 m tall and 1 m canopy diameter in 15 years. The area below the canopies of these young creosotebushes was virtually devoid of perennial vegetation and there were few annuals that grew in the soil below the shrub canopy. This was very different from the patch stucture of mature creosotebushes which represented rich resource islands. These differences appear to be the result of the complex interactions between the shrubs and soil. Newly established creosotebush shrubs

Figure 5.10   Epigeic termite mound in northwestern New South Wales, Australia.

produce very different microenvironments as a result of differences in stemflow and throughfall, litter capture, etc. than older well established shrubs (DeSoyza *et al.*, 1997). There are temporal changes in composition, structure, and function of patches within the landscape that result from commensal to competition shifts, soil disturbances by animals, direct and indirect competition, and herbivory. Many of the temporal changes result from animal activity. The spread of some patches may be primarily due to dispersal of seed by herbivores and the generation of favorable germination–establishment sites by the same herbivore or other herbivores. An example of this feedback from animal activity is the spread of mesquite (*Prosopis glandulosa*) into desert grassland. Cattle feed on mesquite seed-pods and deposit the scarified seeds in their dung pats. Seedlings that germinate from dung pats that were deposited on bare soil have a higher probability of survival than seedlings from dung pats deposited on vegetation. Grazing reduces grass canopy cover and results in larger bare patches that are suitable establishment sites for mesquite seedlings (unpublished data). Animal feedbacks on patch structure and change in patch structure is not limited to seed dispersal. Selective and repeated browsing of some shrubs leads to the development of very dense (pruning effect) canopies. These shrubs provide a very different sub-canopy microclimate than unbrowsed shrubs and have a very different relationship to neighboring plants (Whitford, 1993). The temporal changes in patches and mosaics can be clearly understood in the dynamics of striped or banded vegetation. Interception zones (upslope margins of the band) are the locations where establishment is successful (Mauchamp *et al.*, 1993). The interception zones are also where animal activity is concentrated and where seed trap–germination sites, resulting from that activity, are concentrated (Whitford, 1998). The feedbacks from animal activity contribute to the temporal and spatial dynamics of patches.

REFERENCES

Abaturov, B. D. (1972). The role of burrowing animals in the transport of mineral substances in the soil. *Pedobiologia* **12,** 261–266.

Arshad, M. A. (1981). Physical and chemical properties of termite mounds of two species of *Macrotermes* (Isoptera, Termitidae) and the surrounding soils of the semiarid savanna of Kenya. *Soil Sci.* **132,** 161–174.

Arshad, M. A. (1982). Influence of the termite *Macrotermes michaelseni* (Sjost) on soil fertility and vegetation in a semi-arid savannah ecosystem. *Agro-Ecosystems* **8,** 47–58.

Bagnine, R. K. N. (1984). Soil translocation by termites of the genus Odontotermes (Holmgren) (Isoptera:Macrotermitinae) in an arid area of northern Kenya. *Oecologia* **64,** 263–266.

Barth, R. C. and Klemmedson, J. O. (1982). Amount and distribution of dry matter, nitrogen, and organic carbon in soil–plant systems of mesquite and palo verde. *J. Range Mgmt* **35,** 412–418.

Briese, D. T. (1982). The effect of ants on the soil of a semi-arid saltbush habitat. *Insectes Soc.* **29,** 375–382.

Cable, D. R. and Tschirley, F. H. (1961). Responses of native and introduced grasses following aerial spraying of velvet mesquite in southern Arizona. *J. Range Mgmt* **14,** 155–159.

Charley, J. L. and West, N. E. (1975). Pant-induced soil chemical patterns in some shrub-dominated semi-desert ecosystems of Utah. *J. Ecol.* **63,** 945–964.

Cody, M. L. (1993). Do cholla cacti (*Opuntia* spp., subgenus *cylindropuntia*) use or need nurse plants in the Mojave Desert. *J. Arid Environ.* **24,** 139–154.

Coventry, R. J., Holt, J. A., and Sinclair, D. F. (1988). Nutrient cycling by mound-building termites in low fertility soils of semi-arid tropical Australia. *Aust. J. Soil Res.* **26,** 375–390.

Crawford, C. S. and Gosz, J. R. (1982). Desert ecosystems: Their resources in space and time. *Environ Conserv* **9,** 181–195.

Dean, W. R. J. and Milton, S. J. (1991). Disturbances in semi-arid shrubland and arid grassland in the Karoo, South Africa: mammal diggings as germination sites. *Afr. J. Ecol.* **29,** 11–16.

Dean, W. R. J., Milton, S. J., and Jeltsch, F. (1999). Large trees, fertile islands, and birds in arid savanna. *J. Arid Environ.* **41,** 61–78.

DeSoyza, A. G., Whitford, W. G., Martinez-Meza, E., and Van Zee, J. W. (1997). Variation in creosotebush (*Larrea tridentata*) morphology in relation to habitat, soil fertility and associated annual plant communities. *Am. Midl. Nat.* **137,** 13–26.

Eldridge, D. J. and Pickard, J. (1994) Effects of ants on sandy soils in semi-arid eastern Australia: II. Relocation of nest entrances and consequences for bioturbation. *Aust. J. Soil Res.* **32,** 323–333.

Fisher, F. M. and Whitford, W. G. (1995). Field simulation of wet and dry years in the Chihuahuan Desert: soil moisture, N mineralization and ion-exchange resin bags. *Biol. Fertil. Soils* **20,** 137–146.

Fonteyn, P. J. and Mahall, B. E. (1981). An experimental analysis of structure in a desert plant community. *J. Ecol.* **69,** 883–896.

Fowler, N. (1986). The role of competition in plant communities in arid and semiarid regions. *Annu. Rev. Ecol. Syst.* **17,** 89–110.

Franco, A. C. and Nobel, P. S. (1989). Effect of nurse plants on the microhabitat and growth of cacti. *J. Ecol.* **77,** 870–886.

Franco-Pizana, J. G., Fulbright, T. E., Gardiner, D. T., and Tipton, A. R. (1996). Shrub emergence and seedling growth in microenvironments created by *Prosopis glandulosa*. *J. Veget. Sci.* **7,** 257–264.

Friese, C. F. and Allen, M. F. (1993). The interaction of harvester ants and vesicular–arbuscular mycorrhizal fungi in a patchy semi-arid environment: the effects of mound structure on fungal dispersion and establishment. *Funct. Ecol.* **7,** 13–20.

Garcia-Moya, E. and McKell, C. M. (1970). Contribution of shrubs to the nitrogen economy of a desert-wash plant community. *Ecology* **51,** 81–88.

Grime, J. P. (1979). *Plant Strategies and Vegetation Processes.* J. Wiley, Chichester.

Gutierrez, J. R. and Whitford, W. G. (1987) Chihuahuan Desert annuals: importance of water and nitrogen. *Ecology* **68,** 2032–2045.

Gutierrez, J. R. and Whitford, W. G. (1989). Effect of eliminating termites on the growth of creosotebush, *Larrea tridentata. Southwest. Nat.* **34,** 549–551.

Henderson, C. B., Petersen, K. E., and Redak, R. A. (1988). Spatial and temporal patterns in the seed bank and vegetation of a desert grassland community. *J. Ecol.* **76,** 717–728.

Hobbs, R. J. (1985). Harvester ant foraging and species distribution in annual grassland. *Oecologia* **67,** 519–523.

Holmgren, M., Scheffer, M., and Huston, M. A. (1997). The interplay of facilitation and competition in plant communities. *Ecology* **78,** 1966–1975.

Holt, J. A., Coventry, R. J., and Sinclair, D. F. (1980). Some aspects of the biology and pedological significance of mound-building termites in a red and yellow earth landscape near Charters Towers, North Queensland. *Aust. J. Soil Res.* **18,** 97–109.

Jones, J. A. (1989). Environmental influences on soil chemistry in centrial semiarid Tanzania. *Soil Sci. Soc. Am. J.* **53,** 1748–1758.

Lee, K. E. and Wood, T. G. (1971a). *Termites and Soils.* Academic Press, London.

Lee, K. E. and Wood, T. G. (1971b). Physical and chemical effects on soils of some Australian termites, and their pedological significance. *Pedobiologia* **11,** 376–409.

Levieux, J. (1976). Deux aspects de l'action des fourmis (Hymenoptera: Formicidae) sur le sol d'une savane preforestiere de Cote D'Ivoire. *Bull. Ecol.* **7,** 283–295.

Livingston, M., Roundy, B. A., and Smith, S. E. (1997). Association of overstory plant canopies and the native grasses in southern Arizona. *J. Arid Environ.* **35,** 441–449.

Lobry deBruyn, L. A. and Conacher, A. J. (1990). The role of termites and ants in soil modification. *Aust. J. Soil Res.* **28,** 55–93.

Lobry deBruyn, L. A. and Conacher, A. J. (1994). The bioturbation activity of ants in agricultural and naturally vegetated habitats in semi-arid environments. *Aust. J. Soil Res.* **32,** 555–570.

Lobry de Bruyn, L. A. and Conacher, A. J. (1995). Soil modification by termites in the central wheatbelt of Western Australia. *Aust. J. Soil Res.* **33,** 179–193.

Loffler, E. and Margules, C. (1980). Wombats detected from space. *Remote Sensing Environ.* **9,** 47–56.

Longland, W. S. (1995). Desert rodents in disturbed shrub communities and their effects on plant recruitment. In B. A. Roundy, E. D. McArthur, J. S. Haley and D. K. Mann (eds), *Proceedings: Wildland Shrub and Arid Land Restoration Symposium. General Technical Report INT-GTR 315*, pp. 209–214. US Department of Agriculture, Forest Service, Intermountain Research Station, Ogden, UT.

Ludwig, J. A. and Tongway, D. J. (1995). Spatial organisation of landscapes and its function in semi-arid woodlands, Australia. *Landscape Ecol.* **10,** 51–63.

MacKay, W. P. and Whitford, W. G. (1988). Spatial variability of termite gallery production in Chihuahuan Desert plant communities. *Sociobiology* **14,** 281–289.

Malaka, S. L. (1977). A study of the chemistry and hydraulic conductivity of mound materials and soils from different habitats of some Nigerian termites. *Aust. J. Soil Res.* **15,** 878–891.

Mando, A. (1997). Effect of termites and mulch on the physical rehabilitation of structurally crusted soil in the Sahel. Land *Degrad. Dev.* **8,** 269–278.

Mando, A., Stroosnijder, L., and Brussaard (1996). Effects of termites on infiltration into crusted soil. *Geoderma* **74,** 107–113.

Mauchamp, A., Montana, C., Lepart, J., and Rambal, S. (1993). Ecotone dependent recruitment of a desert shrub, *Flourensia cernua*, in vegetation stripes. *Oikos* **68,** 107–116.

McAdoo, J. K., Evans, C. C., Roundy, B. A., Young, J. A., and Evans, R. A. (1983). Influence of heteromyid rodents on *Oryzopsis hymenoides* germination. *J. Range Mgmt.* **36**, 61–64.

McAuliffe, J. R. (1984a). Sahuaro–nurse tree associations in the Sonoran desert: competitive effects of sahuaros. *Oecologia* **64**, 319–321.

McAuliffe, J. R. (1984b). Prey refugia and the distribution of two Sonoran Desert cacti. *Oecologia* **65**, 82–85.

McAuliffe, J. R. (1988). Markovian dynamics of simple and complex desert plant communities. *Am. Nat.* **131**, 459–490.

Milton, S. J., Dean, W. R. J., Kerley, G. I. H., Hoffman, M. T., and Whitford, W. G. (1998). Dispersal of seeds as nest material by the cactus wren. *Southwest. Nat.* **43**, 449–452.

Nelson, J. F. and Chew, R. M. (1977). Factors affecting seed reserves in the soil of a Mojave Desert ecosystem, Rock Valley, Nye County, Nevada. *Am. Midl. Nat.* **97**, 300–320.

Noble, J. C. (1993). Relict surface-soil features in semi-arid mulga (*Acacia aneura*) woodlands. *Rangeland J.* **15**, 48–70.

Nutting, W. L., Haverty M. I., and LaFage, J. P. (1987). Physical and chemical alteration of soil by two subterranean termite species in Sonoran Desert grassland. *J. Arid Environ.* **12**, 233–239.

Ojeda, R. A., Gonnet, J. M., Borghi, C. E., Giannoni, S. M., Campos, C. M., and Diaz, G. B. (1996). Ecological observations of the red viscacha rat, *Tympanoctomys barrarae*, in desert habitats of Argentina. *Mastozool. Neotrop.* **3**, 183–191.

Okello-Otago, T. and Spain, A. V. (1986). Comparative growth of two pasture plants from north-eastern Australia on the mound materials of grass and litter-feeding termites (Isoptera: Termitidae) and on their associated surface soils. *Rev. Ecol. Biol. Sol* **23**, 381–392.

Parsons, A. J., Abrahams, A. D., and Simanton, J. R. (1992). Microtopography and soil-surface materials on semi-arid piedmont hillslopes, southern Arizona. *J. Arid Environ.* **22**, 107–115.

Phillips, D. L. and MacMahon, J. A. (1981). Competition and spacing in desert shrubs. *J. Ecol.* **69**, 97–115.

Pomeroy, D. E. (1983). Some effects of mound-building termites on the soils of a semi-arid area of Kenya. *J. Soil Sci.* **34**, 555–570.

Reichman, O. J. (1984). Spatial and temporal variation of seed distributions in Sonoran Desert soils. *J. Biogeog.* **11**, 1–11.

Rissing, S. W. (1986). Indirect effects of granivory by harvester ants: plant species composition and reproductive increase near ant nests. *Oecologia* **68**, 231–234

Robertson, J. H. (1947). Responses of range grasses to different intensities of competition with sagebrush (*Artemisia tridentata* Nutt.). *Ecology* **28**, 1–16.

Santos, P. F., DePree, E., and Whitford, W. G. (1978). Spatial distribution of litter and microarthropods in a Chihuahuan Desert ecosystem. *J. Arid Environ.* **1**, 42–48.

Schaefer, D. A. and Whitford, W. G. (1981). Nutrient cycling by the subterranean termite, *Gnathamitermes tubiformans* in a Chihuahuan Desert ecosystem. *Oecologia* **48**, 277–283.

Shreve, F. (1931). *The Cactus and Its Home*. Williams and Wilkins, Baltimore, Maryland.

Silvertown, J. and Wilson, J. B. (1994). Community structure in a desert perennial community. *Ecology* **75**, 408–417.

Smith, F. R. and Yeaton, R. I. (1998). Disturbance by the mound-building termite, *Trinervitermes trinervoides*, and vegetation patch dynamics in a semi-arid, southern Africa grassland. *Plant Ecol.* **137**, 41–53.

Soholt, L., and Irwin, W. K. (1976). The influence of digging rodents on primary production in Rock Valley. *US/IBP Desert Biome Research Memorandum* **76-18**, 1–10.

Sneva, F. A. (1979). The western harvester ants: their density and hill size in relation to herbaceous productivity and big sagebrush cover. *J. Range Mgmt* **32**, 46–47.

Steenberg, W. V. and Lowe, C. H. (1977). *Ecology of the Saguaro: Reproduction, Germination, Establishment, Growth, and Survival of the Young Plant*. National Parks Service Scientific Monograph Series No. 8. U. S. Government Printing Office, Washington, DC, USA.

Steinberger, Y. and Whitford, W. G. (1983). The contribution of rodents to decomposition processes in a desert ecosystem. *J. Arid Environ.* **6**, 177–181.

Suzan, H., Nabhan, G. P., and Patten, D. T. (1996). The importance of *Olneya tesota* as a nurse plant in the Sonoran Desert. *J. Veget. Sci.* **7,** 635–644.

Tielborger, K. and Kadmon, R. (1995). Effect of shrubs on emergence, survival, and fecundity of 4 coexisting annual species in a sandy desert ecosystem. *Ecoscience* **2,** 141–147.

Tongway, D. J., Ludwig, J. A., and Whitford, W. G. (1989). Mulga log mounds: fertile patches in the semi-arid woodlands of eastern Australia. *Aust. J. Ecol.* **14,** 263–268.

Vasek, F. C. (1980). Creosote bush: long-lived clones in the Mojave Desert. *Am. J. Bot.* **67,** 246–255.

Virginia, R. A. and Jarrell, W. M. (1983). Soil properties in a mesquite-dominated Sonoran Desert ecosystem. *Soil Sci. Soc. Am. J.* **47,** 138–144.

Wagner, D. (1997). The influence of ant nests on seed production, herbivory, and soil nutrients. *J. Ecol.* **85,** 83–93.

Wagner, D., Brown, M. J. F., and Gordon, D. M. (1997). Harvester ant nests, soil biota, and soil chemistry. *Oecologia* **112,** 232–236.

Whitford, W. G. (1993). Animal feedbacks in desertification: an overview. *Rev. Chilena Hist. Nat.* **66,** 243–251.

Whitford, W. G. (1988). Effects of harvester ant, *Pogonomyrmex rugosus*, nests on soils and a spring annual, *Erodium texanum*. *Southwest. Nat.* **33,** 482–485.

Whitford, W. G. (1998). Contribution of pits dug by goannas (*Varanus gouldii*) to the dynamics of banded mulga landscapes in eastern Australia. *J. Arid Environ.* **40,** 453–457.

Whitford, W. G. (2000). Keystone arthropods as webmasters in desert ecosystems. In D. C. Coleman and P. F. Hendrix (eds), *Invertebrates as Webmasters in Ecosystems*, pp. 25–41. CABI Publishing, New York, NY.

Whitford, W. G. and DiMarco, R. (1995). Variability in soils and vegetation associated with harvester ant (*Pogonomyrmex rugosus*) nests on a Chihuahuan Desert watershed. *Biol. Fertil. Soils* **20,** 169–173.

Whitford, W. G. and Kay, F. R. (1999). Biopedturbation by mammals in deserts: a review. *J. Arid Environ.* **41,** 203–220.

Whitford, W. G., Dick-Peddie, S., Walters, D., and Ludwig, J. A. (1978). Responses of shrub defoliation on grass cover and rodent species in a Chihuahuan Desert ecosystem. *J. Arid Environ.* **1,** 237–242.

Whitford, W. G., Martinez-Turanzas, G., and Martinez-Meza, E. (1995a). Persistence of desertified ecosystems: explanations and implications. *Environ. Monitoring Assessment* **37,** 319–332.

Whitford, W. G., Forbes, G. S., and Kerley, G. I. H. (1995b). Diversity and spatial variability and functional roles of invertebrates in desert grassland ecosystems. In M. P. McClaran and T. R. Van Devender (eds), *The Desert Grassland*, pp. 152–195. University of Arizona Press, Tucson, AZ.

Whitford, W. G., Nielson, R., and DeSoyza, A. G. (2001). Establishment and effects of establishment of creosotebush, *Larrea tridentata*, on a Chihuahuan Desert watershed. *J. Arid Environ.* **47,** 1–10.

Wilson, J. B. and Agnew, A. D. Q. (1992). Positive-feedback switches in plant communities. *Adv. Ecol. Res.* **23,** 263–336.

Wood, T. G., Johnson, R. A., and Anderson, J. M. (1983). Modification of soils in Nigerian savanna by soil-feeding *Cubitermes* (Isoptera, Termitidae). *Soil Biol. Biochem.* **15,** 575–579.

Yeaton, R. I. (1978). A cyclical succession between *Larrea tridentata* and *Opuntia leptocaulis* in the northern Chihuahuan Desert. *J. Ecol.* **66,** 651–656.

Yeaton, R. I. and Manzanares, A. R. (1986). Organization of vegetation mosaics in the *Acacia schaffneri–Opuntia streptacantha* association, southern Chihuahuan Desert, Mexico. *J. Ecol.* **74,** 211–217.

Chapter 6 | Adaptations

It is important to point out that desert inhabiting organisms share most genetic and physiological traits with closely related species living in mesic environments. The suite of adaptive physiological traits of desert species are those that allow the organisms to remain within survivable thermal energy balance and water balance. It is important to understand that many species of desert plants and animals are living close to their limits of tolerance for one or more environmental variables.

Many desert animals are able to behaviorally seek out benign microclimates. For most, but not all organisms, tissue temperatures greater than 42°C are lethal, hence it is necessary to maintain tissue temperatures < 42°C. At tissue temperatures lower than the optimum for a given species, rates of reactions decrease by a factor of 2 for each drop of 10°C (the Q_{10} phenomenon). Animals that are capable of maintaining temperatures close to optimum for longer periods of time have an advantage in obtaining necessary resources. High temperatures and dry air increase water needs thus requiring behavioral and/or physiological mechanisms for reducing water loss.

There are general characteristics of desert plants that are considered to be adaptive in hot, arid environments. These include features like small leaf size, reduction in numbers of stomata per unit surface area of leaf, concentrations of stomata on underside of leaves, hairs and waxy surfaces on leaves. Adaptive specializations of desert plants include rooting patterns, stem photosynthesis, succulence, and different photosynthetic pathways.

There are several general categories of adaptations of both plants and animals that contribute to the success of many species in hot, arid, environments. These include: (1) avoidance of extremes, (2) physiological and morphological characteristics that reduce water losses, and thermal stress, (3) rapid response to availability of water and nutrients (4) reduced metabolic rates, and (5) mechanisms to conserve essential nutrients.

6.1 AVOIDANCE OF EXTREMES

The most successful adaptation for life in a desert or any extreme environment is for a species to avoid those conditions that exceed the limits of tolerance for life. The examples provided here are not an exhaustive listing of all avoidance mechanisms that have been documented. These examples were selected to emphasize the variety of avoidance behaviors and mechanisms that have been described in desert inhabiting species.

6.1.1 Annual Plants

Ephemeral or annual plants are found even in the extreme deserts. Seeds of many ephemeral plants may remain viable in the dry soil for several years. In most species of desert ephemeral plants there is germination inhibition in some fraction of the seeds of a given cohort. Because of the unpredictable nature of precipitation in arid ecosystems, precipitation at the beginning of the growing season will not always be followed by enough rain for plants to complete their life cycle and set seed (Tevis, 1958; Burk, 1982). However, in the Chihuahuan Desert, most species probably achieve some seed production even in years with scant rainfall (Kemp, 1989). Even species that produce some seeds under adverse conditions, exhibit delayed germination. The variable, delayed germination behavior of seeds is a genetic–physiological adaptation that allows desert ephemeral plants to avoid conditions when soils are too dry or temperatures too high for survival to reproduction.

Delayed germination may be simply dormancy during a cold or dry season or the population of seeds produced by an individual plant may exhibit extended dormancy for one or more years. Desert ephemerals can be divided into three groups: cool season, warm season, and aseasonal. In North American deserts, cool season ephemerals typically germinate in the autumn and persist through the winter season as leaf rosettes. With the advent of warmer temperatures in late winter or spring, the rosettes rapidly bolt and produce flowers and seeds. Summer ephemerals typically germinate after the first large rainfall in July or August (Mulroy and Rundel, 1977).

In seasonal species, seeds exhibit a dormant season and nondormant season. Delayed germination in desert ephemeral plants where germination is spread out for more than one year is referred to as 'bet-hedging', an adaptation to the unpredictability of desert environments (Phillipi, 1993). Although there are numerous theoretical papers on delayed germination in desert ephemerals, there are few rigorous emperical studies. In a study of seeds from several populations of the winter–spring ephemeral, *Lepidium lasiocarpum*, Philippi (1993) reported that the germination fraction of the seed population during the first year varied between 16% and 68%. He found some correlation between mean precipitation during the growing season and percentage germination. Heavier seeds were more likely to germinate in the first year. Lighter seeds were more likely to germinate after two years of dormancy. Philippi (1993) also documented that there is little dispersal of seeds away from the producing-parent location. Seed rain was very localized, with 3000 or more seeds from a large plant falling within a 10 cm radius of the plant. Although the seed rain from a single ephemeral may be fairly rapidly depleted by seed-harvesting ants or by other granivores, sufficient seeds escape to generate competition among germinants in subsequent years.

There are a number of models that have been proposed to explain dormancy polymorphism in annual plants (Westoby, 1981). Westoby (1981) argued that the group selectionist assumptions of such models are unnecessary to explain dormancy polymorphism. A common feature of seed dispersal patterns is for seeds to be concentrated around the mother hence selection acting on the mother's fitness would select for diversification in germination of offspring. Diversified germination behavior, where understood, is imposed by the seed coat, i.e. by variation in seed coat hardness or leachable inhibitors. Since the seed coat and appendages are maternal tissues, the genetic characteristics of the parent plant are reflected in the characteristics of the seed coat (Westoby, 1981).

Whatever the mechanism, inhibition of germination of some fraction of the seed crop of a given year insures that the species will persist even if conditions that stimulate germination are followed by conditions unsuitable for growth and seed production. It is not unusual for large numbers of germinants of desert annuals to die nor is it unusual for most of the annuals that germinated to establish after a wet period, to die before maturing and setting seed (Moorhead *et al.*, 1988). This phenomenon may be more common in a desert where winter rains are not reliable than in a winter rainfall desert.

6.1.2 Drought Deciduous Perennials

Shedding leaves during drought is an avoidance mechanism used by some plants. In drought deciduous species, leaves are shed when soil water potentials drop below some threshold value. Drought deciduous species can produce several cohorts of leaves during a single growing season. When the threshold soil water content value is reached, the leaves begin to senesce. In the senescing process, essential nutrients are resorbed and stored in stem or root tissues. Conventional wisdom holds that resorption efficiency should be extremely high in drought deciduous plants. This is because low rates of nutrient release from decomposing leaves would be insufficient to support the production of several cohorts of leaves. The loss of leaves greatly reduces water losses by transpiration allowing stem and root tissues to remain hydrated.

The North American ocotillo (*Foquieria splendens*) is a classic example of a drought deciduous desert shrub. Although ocotillo flowers and sets fruits at the beginning of the growing season, leaf production does not occur until there is sufficient rainfall. When the soils dry, ocotillo leaves senesce and are shed. Ocotillo may produce four or five non-overlapping cohorts of leaves during a single growing season. However, nutrient resorption in ocotillo is an enigma. There is considerable temporal variability in nutrient resorption efficiency. In one year, ocotillo resorbed essentially no nitrogen and the shed leaves had the same nitrogen content as newly formed leaves (Killingbeck, 1992). Three years later the same plants resorbed 72% of the leaf nitrogen which is close to the highest resorption proficiencies recorded (Killingbeck, 1993). Phosphorus resorption values were also low (Killingbeck, 1992). Studies of stem growth, nutrient resorption and reproduction in ocotillo have continued for more than 10 years and document the considerable temporal variability in this 'classic' drought deciduous shrub. The studies of ocotillo clearly show that conventional wisdom is frequently wrong. Generalizations based on life history characteristics or life form relationships must be considered with caution especially for species living in desert environments.

A modified drought deciduous strategy is exhibited by several Sonoran and Mojave Desert shrubs (*Fransaria dumosa*, *Isomeris aborea*, and *Encelia farinosa*) that not only produce new leaves in response to precipitation, but retain some of the leaves during part or all of the dry season. The brittlebush, *Encelia farinosa* produces a few, very small, densely pubescent leaves during hot, dry periods. Leaves of *E. farinosa* produced after large rains are abundant, large, and sparsely pubescent. The leaves of this species range in size from 50 cm^2 maximum under favorable moisture conditions to 0.5 cm^2 under hot, dry conditions (Cunningham and Strain, 1969). As soil moisture is depleted, the leaves on the higher nodes are smaller and have denser mesophyll. Further depletion of soil moisture causes the larger leaves with low density mesophyll to be lost to desiccation. The small leaves that are retained well into the dry period while having lower rates of photosynthesis, are able to continue to contribute to carbon gain during a period when the more efficient larger leaves would be lost to desiccation. If dry conditions persist for longer periods of time, more of the small leaves are lost until the shrub is completely leafless (Cunningham and Strain, 1969). This is an example of a plant with a variable avoidance mechanism, producing different size and density leaves following rains and gradually losing leaves with greater surface areas and higher transpiration losses as soil dries.

6.1.3 Amphibians

Because amphibians have thin, highly vascularized skin and are generally intolerant of ambient temperatures above 30°C, they are unlikely inhabitants of deserts (Shoemaker, 1988). The skin of amphibians provides no resistance to the movement of water from the animal to the atmosphere. Amphibian survival on dry land in moving air under moderate but constant temperature and humidity is very short (< 1 day). The usual nitrogen waste products of amphibians are ammonia and urea with urea being the predominant product in most species. Although urea is not very toxic, considerable water is required to excrete this soluble form of nitrogenous waste in animals that lack a concentrating kidney. Despite these limitations, there are amphibians in the world's hot deserts and these belong to several families of anurans: Pelobatidae (spadefoot toads), Bufonidae (Old World Toads), Leptodactylidae (frogs primarily of the southern hemisphere) and Hylidae (tree frogs).

The vast majority of amphibians in arid and semiarid environment avoid the desiccating environment of the surface by burrowing underground. Most desert dwelling amphibians share the characteristic of horny protuberances on the metatarsals of the rear feet, which are used to dig into the soil. The 'spade' is the structure, which gives the North American spadefoot toads (*Scaphiopus* spp.)

their name (Fig. 6.1). By digging deep into moist soil a frog can absorb moisture from the soil or at least not lose water to the soil as long as the water potential of the soil (the matric potential) is greater than that of the body fluids of the animal. Spadefoot toads (*Scaphiopus[hammondi] multiplicatus*) exhibit increases in the concentration of total solutes and urea in the plasma as the soils dry (Shoemaker, 1988) thus maintaining a favorable water potential gradient between the animal and the surrounding soil.

A more common means of reducing water loss to dry soils is the formation of a 'cocoon' that is composed of multiple layers of cornified skin (stratum corneum) separated by thinner intercellular layers. Cocoon formation has been observed in several species of several families of amphibians from South America, Africa, and Australia (Lee and Mercer, 1967; McClanahan *et al.*, 1976; Loveridge and Crayne, 1979). All of these species are fossorial and experience prolonged drought periods. The cocoon layers are produced at a rate of about one a day for the first few weeks that the frogs are in their burrow (Shoemaker, 1988). Species that form cocoons do not accumulate plasma solutes. These differences in avoidance adaptations probably evolved independently. Cocoon formers are all Gondwanaland species from regions with long geological histories of arid environments. The plasma solute concentrators are from areas of the world where arid environments are geologically more recent.

Two genera of aboreal frogs inhabit semiarid regions of South America and Africa. Frogs of the genera *Chiromantis* and *Hyperolius* from southern Africa and

Figure 6.1   A spadefoot toad, *Scaphiopus multiplicatus*, Chihuahuan Desert, New Mexico.

Phyllomedusa from the Chaco region of South America excrete nitrogen in the form of precipitated uric acid. *Phyllomedusa* spp. waterproof themselves with wax secreted from glands on the skin and spread this wax over their entire body. For these amphibians, their primary avenue for water intake is via the insects that they eat. During the day these frogs remain perched in trees and remain above ground even during the dry season (Shoemaker, 1988).

Other morphological characteristics shared by amphibians that inhabit arid areas are short, stout legs, a 'seat' patch of highly vascularized skin with enhanced permeability to water, a large urinary bladder and globose body shape. These morphological features are important for burrowing. Desert-inhabiting amphibians avoid dry conditions by burrowing in the soil and remaining in the soil for periods up to and exceeding one year in duration (Seymour and Lee, 1974). In the burrows, humidity is close to 100% and temperatures are moderate. For example, in June and July soil surface temperatures range between 20°C and > 60°C but at 42 cm, where spadefoot toads have been recovered from their burrows, temperatures ranged from 23.5 to 24.8°C (Ruibal *et al.*, 1969). Toads emerge with summer rains, breed in ephemeral ponds and feed heavily and rapidly during the brief periods of moist soils and high humidity. This is necessary in order to store sufficient energy and nutrients for the next dormant period. While below ground, their energy metabolism drops to approximately one-fifth that of surface-active toads, urea accumulates in the blood, and plasma electrolytes accumulate (Ruibal *et al.*, 1969).

The critical stages in the life cycle of desert amphibians are the larval or tadpole stages and recently transformed stage (toadlet). The water temperatures in ephemeral ponds must be between 16°C and 33°C for normal development of most desert-inhabiting frogs in North America (Zwiefel, 1968). The eggs of some *Scaphiopus* spp. continue normal development to 35°C. Desert ephemeral ponds often dry quickly and the shallow remnants reach lethal levels for tadpoles. Selective advantage accrues to tadpoles with higher growth rates and large size at metamorphosis. Differential growth rates are increased by some individuals that develop a horny beak in place of the filter feeding mouth parts characteristic of most tadpoles. Tadpoles with the horny beak mouth parts are cannibalistic on their filter-feeding brethren. The cannibals grow rapidly, metamorphose quickly and become large toadlets (Creusere and Whitford, 1977). The survival of toadlets in their burrows for the first year is a function of body mass and fat storage at the time the toadlets are forced to dig into the soil for the long dry hibernation period. Juvenile spadefoot toads tolerated loss of 16% of their initial body water and 40% of their total fat reserves. Average total body lipid content of the juvenile toads dropped from 3.2% to 2.0% in six months. Only 50% of the *Scaphiopus* (*hammondi*) *multiplicatus* survived for six months in their burrow (Meltzer and Whitford, 1976). These data show that the ability to burrow into the soil to avoid environmental extremes must be accompanied by sufficient water and fat stores.

One characteristic of desert anurans is their sudden appearance after the first heavy rain of summer. Dimmitt and Ruibal (1980) reported that toads remain in the 20–90 cm deep 'winter' burrows then move to 'summer burrows' 2–10 cm deep.

North American spadefoot toads emerge in small numbers on nights with light rain of insufficient quantity to generate run-off and flooding of the ephemeral breeding ponds. However, the toads in the shallow burrows are sensitive to sound and emerge in great numbers in response to heavy rainfalls (Dimmitt and Ruibal, 1980).

6.1.4 Reptiles

Snakes and lizards are common inhabitants of all of the worlds hot deserts. Most of these reptiles maintain body temperatures close to 38°C by behavioral thermoregulation. The most common thermoregulatory behavior of lizards and snakes is movement between shade and sun allowing heat gain or loss by radiation and conduction. Typical behavior of lizards immediately after emergence from their night-time retreat, is to select a location where the animal can maximize energy gain by radiation. Lizards select large rocks, branches or trunks of trees or open ground as basking sites. When their body temperatures reach a preferred temperature, these animals move between shade and sun to maintain body temperatures within 1–2°C of their preferred body temperature (Huey *et al.*, 1977). Desert lizards store fats in visceral fat bodies or in their tails (e.g. geckos). These fat stores thus do not retard heat flux by radiation, conduction, or convection in animals that are dependent on behavioral thermoregulation.

In sand dune environments where shade-giving plants are absent some species of lizards and snakes have skin flaps that cover the nares and lizards have fringes on the toes that allow them to 'swim' into the sand to avoid over-heating. Most lizards and snakes seek out burrows or rock crevices where the microclimate is moderate for part of each day and for extended stays during extremely dry or cold periods. The avoidance of 'stress' by reptiles in desert environments because of their efficient thermoregulatory behavior means that few if any reptiles experience thermal stress. However, the avoidance of stress of dehydration when prey become scarce is problematic. For example, during a prolonged hot, dry summer when virtually no surface foraging was observed in seed-harvesting ants (*Pogonomyrmex* spp.), we found several emaciated, and desiccated horned lizards (*Phrynosoma cornutum*) (Whitford, unpublished field notes). We concluded that the lizards died from a combination of inadequate energy intake and most importantly an inadequate supply of preformed water to replace water lost through evaporation.

6.1.5 Mammals

Most small desert mammals live in burrows of their own construction and most are nocturnal. The burrows may be simple tunnels at the base of shrubs or other plants or elaborate systems of tunnels such as the burrows of the banner-tailed kangaroo rat (Kay and Whitford, 1978). The air temperature in burrows and burrow systems is similar to but not equal to that of the surrounding soil. However, at depths >15 cm, the ambient temperatures in burrows remains at 35°C or less even when soil surface temperatures exceed 60°C. Air relative humidities in burrows are generally near saturation. In deep burrow systems there may be an accumulation of carbon dioxide in the burrow atmosphere. The amount of carbon dioxide accumulation is a function of soil moisture and temperature and the effectiveness of wind in burrow ventilation (Kay and Whitford, 1978).

Most but not all small desert mammals are nocturnal thereby avoiding the low humidities and high temperatures of daylight hours in summer. Diurnal species limit their activity to the early morning and late afternoon hours in mid-summer (crepuscular activity cycle). However, some diurnal species of desert mammals utilize behavioral means to extend their foraging period. In the Cape ground squirrel, *Xerus inauris*, the tail is raised and curled over the back to shade the body. The tail of this species is large, dorsoventrally flattened, and nearly as long as the body of the animal (Fig. 6.2). During the hotter part of the day the squirrels orient themselves to maximize the area of the back that is shaded by the tail. The shade provided by the tail reduces the surface temperature of the animal by more than 5°C which allows the squirrels to expand their foraging hours (Bennett *et al.*, 1984). The Cape ground squirrel is able to forage continuously on the surface for up to 7 hours but would be limited to approximately 3 hours of surface foraging without tail shading.

Intermediate size mammals such as rabbits, hares, jackals, foxes, etc. may construct burrows or enlarge burrows constructed by other species or may simply seek refuge in the shade of shrubs or small trees. Some species, such as jackrabbits, retreat to burrows only when air temperatures are extremely hot (> 42°C) and then remain in the burrows for short periods during the hottest part of the day (Costa *et al.*, 1976). Except in the hottest part of the range of the species, jackrabbits are nonburrowers. *Lepus* spp. generally use shallow forms at the base of shrubs as their only shelter. In the shaded form, jackrabbits lose

Figure 6.2   The use of a parasol tail by the Cape ground squirrel *Xerus inauris*. (From Bennett *et al.*, 1984.) With permission by The University of Chicago Press.

heat by radiation from the highly vascularized ears to the shaded shrub structures and by conduction to the soil of the form.

6.1.6 Birds

Behavioral avoidance of desiccating conditions and thermal loads is the primary adaptation of birds that inhabit deserts. Most species of breeding birds select sheltered nest sites or locations in tall vegetation where ambient temperatures $> 38°C$ are rarely encountered. Ground-nesting species select shaded sites. Foraging activities are generally confined to early morning hours or early evening when ambient temperatures are lower and air relative humidities higher than during mid-day. Most desert birds obtain sufficient moisture from their diet to meet or exceed water lost to evaporation, urine and feces, and thermoregulatory panting (Dawson, 1984). In extremely dry and unpredictable deserts birds may exhibit nomadism. A nomad is a bird that moves from place to place without regard to season or direction (Davies, 1984). Nomadic birds may have to move long distances once the resources needed to support life are depleted in an area. Although it is not known how nomadic birds find rich resource patches, nomadic species are successful in the dry center of Australia and in other extremely arid environments. Not all individuals of a species move at the same time, orient in the same direction or terminate in the same place. Nomadic populations obviously face high mortality rates but the advantages to that part of the population that locates new resource rich patches must outweigh these mortality losses. However, nomadism is a characteristic of only a fraction of the bird species inhabiting a desert area. Many other species obtain all the life support resources required from the same area (Davies, 1984)

6.1.7 Arthropods

Most desert arthropods exhibit some kind of avoidance behavior. Some like the desert cockroaches, *Arenivaga* spp. escape harsh surface conditions by burrowing 20–60 cm into the sand (Hawke and Farley, 1973). At night these insects move to the surface. The timing of that activity rhythm may be triggered by the level of the temperature inversion in the soil (Cloudsley-Thompson, 1975). In hot deserts the majority of arthropods, including many desert beetles are nocturnal. In cool coastal deserts most arthropod species are diurnal (Cloudsley-Thompson, 1991). Arthropods such as myriopods (millipedes) that lack effective morphological and physiological water conservation adaptations remain in burrows or other humid microsites until rains increase surface humidity. Even species of myriopods that have morphological and physiological adaptations exhibit thermoregulatory behavior (Wooten *et al.*, 1975). Wooten *et al.* (1975) reported that at mid-afternoon millipedes ceased feeding and moved to cooler microsites within shrubs or moved into mammal burrows at the base of the shrubs. In mid-summer, the inactivity period in mid-afternoon began when shrub temperatures first exceeded 34°C at around 1400 h. Millipedes resumed activity when the cool shrub microsites dropped to 32°C (Fig. 6.3). Peak feeding and activity on the surface occurred immediately after dawn and in the early evening when ambient temperatures were less than 30°C (Fig. 6.3). Desert centipedes, *Scolopendra* spp., which are common ground-dwelling predators in North American deserts are nocturnal and spend daylight hours in holes in the soil, under rocks or large debris. These animals also exhibit an endogenous activity rhythm (Cloudsley-Thompson, 1975).

Species of insects that live in or on vegetation move up and down in the canopy throughout the day to take advantage of moderate microclimates. This behavior has been reported in grasshoppers and probably occurs in many species of plant bugs, plant hoppers, etc. that are common inhabitants of desert perennial plants.

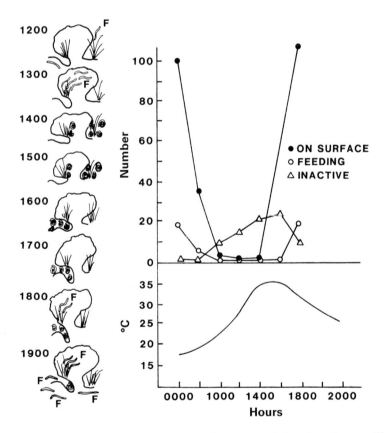

Figure 6.3 Diagram of the diel behavioral patterns of the giant desert millipede, *Orthoperus ornata*. Diagrams on the left depict the spatial locations and activity status of the population of millipedes during the day. Elongate millipedes are active, coiled millipedes are inactive. Diagrams (from left to right) represent a burrow, a shrub canopy, and bunch grasses. (Modified from Wooten *et al.*, 1975 with permission by Academic Press Inc.)

Various kinds of beetles are among the more common arthropods encountered in deserts. Tenebrionid beetles exhibit diurnal patterns of activity and seasonal activity patterns (Smith and Whitford, 1976). Some species are nocturnal and others limit activity to early morning and evening hours (Cloudsley-Thompson, 1975).

Most desert arachnids live in burrows in the soil, e.g. wolf spiders, scorpions, whip scorpions, sun-spiders (solfugids), tarantulas, and most are nocturnal. Many desert arachnids also exhibit seasonal activity patterns. Perhaps the most extreme seasonality is exhibited by giant velvet mites, *Dinothrombium* spp. which emerge from their burrows in soil after heavy rains, feed voraciously, breed and return to their burrows (Tevis and Newell, 1962).

Among the most abundant arthropods in all hot deserts are the social insects, termites and ants. These insects avoid extreme conditions by living in below-ground nests which may have chambers extending 2–3 m below the soil surface. Termites and ants limit their surface activity to non-stressful times of the day (Whitford and Ettershank, 1975). Subterranean termites frequently limit surface activity to periods following rains (Johnson and Whitford, 1975). Species of liquid-feeding ants that continue foraging at mid-day remain above the soil surface in the vegetation when soil surface temperatures are above lethal limits (Shaffer and Whitford, 1981). Subterranean termites use thermal shadows in the soil as cues to locate large items of potential food during periods when hot, dry conditions curtail surface activity (Ettershank *et al.*, 1980). By feeding in large objects like dung pats and *Yucca* sp. logs, subterranean termites avoid the harsh conditions. Subterranean termites forage in litter accumulations at the soil surface in the early morning and evening when temperatures are below 30°C and the litter is moist. Based on field experiments, the threshold soil moisture for surface foraging by subterranean termites (*Gnathamitermes tubiformans*) is -5.4 ± 0.2 MPa (MacKay *et al.*, 1986).

6.2 PHYSIOLOGICAL AND MORPHOLOGICAL ADAPTATIONS

As with behavioral characteristics there are suites of morphological and physiological characteristics that contribute to the 'biological success' of species inhabiting deserts. Each successful species shares some adaptive characteristics with a number of other species. Successful desert-inhabiting species may or may not have some unique adaptive characteristics. Physiological adaptations that are widely shared traits of groups of desert organisms have been documented. Suites of adaptive physiological, morphological and behavioral characteristics of some desert-dwelling taxa have been extensively studied.

6.2.1 Perennial Plants

Unlike animals that can escape extreme temperatures and extremely low humidity by moving to moderate microclimate sites, perennial plants have no way to escape environmental extremes. Virtually the only avoidance mechanism available to perennials is to shed leaves during drought (drought deciduous species). However, producing new leaves is a water and energy intensive process. In addition drought deciduous plants are disadvantaged during the period immediately following a soil wetting rainfall because the developing leaf surfaces are very small and the plants cannot maximize photosynthesis when the required resources are most abundant. Most desert plants are not drought deciduous. However, there are a number of morphological and physiological characteristics of desert plants that allow them to survive extreme conditions and to respond rapidly during periods when water and nutrients are available.

In most deserts of the world, the vegetation is a mix of C_3, C_4 and CAM photosynthetic pathway plants. These photosynthetic pathways differ with respect to the intermediate molecule to which CO_2 is added. In the C_3 or Calvin Cycle photosynthetic pathway, carbon dioxide is added to a five-carbon compound (ribulose bisphosphate) to form two molecules of a three-carbon compound (phosphoglyceraldehyde). In the C_4 photosynthetic pathway, CO_2 is combined with phosphoenolpyruvate, a three-carbon molecule, to form malate or oxaloacetic acid (four-carbon molecules). In C_4 plants carbon dioxide is concentrated in bundle sheath cells which allow the plant to bind carbon dioxide at lower concentrations in the cell. This reduces water loss in hot environments. The CAM (crassulacean acid metabolism) photosynthetic pathway is a modification of the C_4 photosynthetic pathway in which the assimilation and the light reaction of the Calvin Cycle are segregated temporally. In CAM plants carbon dioxide uptake occurs at night and the carbon dioxide is assimilated into a four-carbon compound (either malate or oxalacetate) which is stored in the leaves in high concentrations. During the daylight hours, the carbon dioxide is released from the stored four-carbon compounds and enters the normal Calvin Cycle (Fig. 6.4).

The significance of different photosynthetic pathways in the adaptation of perennial plants to life in extreme desert environments is still debated In some deserts, such as the hot deserts of North and South America, one or more species of C_3 plants dominate the plant communities. According to the general physiological characteristics of such plants (Table 6.1) the C_3 plants should be the least adapted to hot desert environments.

Table 6.1
Comparison of the General Properties of C_3, C_4, and CAM Plants

	C_3	C_4	CAM
(1) Light saturation	1/4 full sunlight	Higher than full sunlight	Fixes C at night
(2) Optimum temperature	Around 25°C	Around 45°C	Around 35°C
(3) Water loss rate (g H_2O g^{-1} C fixed)	Around 600	Around 250	Around 50
(4) Photosynthetic rate (mg CO_2 dm^{-2} $leaf^{-1}$ h^{-1})	25	60	3

C$_3$ photosynthesis: CO$_2$ + RuBP → 2 PGA (3C)

Carbon dioxide plus a five-carbon molecule, ribulose biphosphate, yields two molecules of a three-carbon molecule, phosphoglyceraldehyde.

C$_4$ photosynthesis: CO$_2$ + PEP (3C) → OAA (4C) *or* M (4C)

Carbon dioxide plus a three-carbon molecule, phosphoenolpyruvate yields one of two four-carbon molecules, oxaloacetic acid or malate.

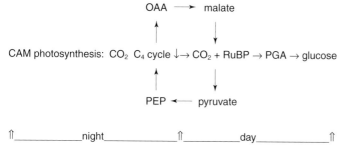

CAM photosynthesis: CO$_2$ fixed into oxaloacetic acid and malate during the night-time hours when the stomates are open and water loss by transpiration is minimal. During daylight hours, stomates are closed and carbon stored in oxaloacetic acid and malate is released where it reacts with ribulose biphosphate to form phosphoglycerate. Two phosphoglycerate molecules are biochemically converted into glucose and transported to other tissues.

Figure 6.4   Comparison of carbon fixation pathways found in desert plants.

As can be seen in Table 6.1, C$_3$ plants are less water efficient than C$_4$ plants, have a lower optimum temperature and lower rates of photosynthesis. C$_3$ plants light saturate at low light intensities. These properties of C$_3$ plants affect many of the growth and morphological characteristics of C$_3$ plants that are successful in hot temperate deserts.

Creosotebush, *Larrea tridentata*, a dominant C$_3$ shrub in North American Deserts, has been the subject of numerous studies focused on characteristics that contribute to the success of this shrub in a hot desert. In the Chihuahuan Desert, creosotebush produces both spring and late autumn cohorts of leaves. The numbers of live leaves produced varies with the availability of both water and nitrogen (Lajtha and Whitford, 1989). Leaves that are produced in the spring have slightly higher maximum longevity than leaves produced in the autumn. No leaves on *L. tridentata* survived for more than 16 months (Lajtha and Whitford, 1989). Leaf abscission is coincident with seasons or years of increased leaf growth and production. This pattern appears to be the result of redistribution of nutrients from old leaves to new leaves. Old nutrient-depleted leaves are then dropped.

Patterns of physiological carbon gain are different from patterns of leaf production and stem elongation. In creosotebush, *L. tridentata*, net photosynthetic assimilation of carbon peaked in July in shrubs exposed to natural rainfall and in shrubs exposed to summer drought imposed by 'rain-out' shelters. The pattern of physiological carbon gain is very different from the pattern of stem elongation and also very different from seasonal patterns of root growth. Stem elongation is greatest in the spring when ambient temperatures are above freezing and day time maxima < 35°C. Maximum root growth is in the autumn when ambient temperatures are moderate. Thus, carbon allocation patterns differ considerably coincident with the production of the spring and autumn cohorts of leaves. Although *L. tridentata* exhibits some growth during mid-summer, a large fraction of the terminal growth segments die (Fisher *et al.*, 1988).

Water-use efficiency defined as photosynthetic carbon gain per unit water transpired was positively related to leaf nitrogen with a correlation coefficient of 0.5 (Lajtha and Whitford, 1989). This is additional evidence for the importance of nitrogen as a factor that affects the physiology of this C_3 shrub. Both leaf nitrogen and leaf age affect net photosynthesis, water-use efficiency and nitrogen-use efficiency in creosotebushes. Leaf nitrogen and leaf age accounted for between 44% and 75% of the variance in these parameters. Perennial shrubs or trees growing in soils that are characterized by low concentrations of nutrients should be efficient at resorbing nutrients from leaves during senescence. *L. tridentata* resorbed between 48% and 72% of the leaf nitrogen from leaves prior to senescent abscission. In creosotebush, nitrogen resorption efficiency was greater in spring than in autumn.

The resilience of *L. tridentata* to summer drought is an important adaptation of this C_3 species. This resilience contributes, in large part, to its dominance in North American hot deserts that have different rainfall seasons. When creosotebushes were exposed to severe summer drought by 'rain-out' shelters, stem elongation was suppressed during the summer but the plants quickly recovered and added as much growth as nondroughted plants within a month or two after removal of the rain-out shelters. Reproductive effort as measured by flower production was affected by summer irrigation but not by summer drought (Whitford *et al.*, 1995). This response suggests that water from the preceding season was available to the droughted plants and supported the production of flowers and fruits.

Not only has the physiology of creosotebush adjusted to a variety of desert conditions, the morphology of creosotebushes contributes to its success in hot deserts. The stem and leaf orientation of *L. tridentata* maximizes light interception in the early morning hours when cool temperatures and relatively high humidity allow open stomata and efficient water use for photosynthesis. Leaf clusters on the stems of creosotebushes are inclined between 33° and 71° from the horizontal. The inclinations of leaf clusters are steeper in shrubs in the drier and hotter Mojave Desert than in the Chihuahuan Desert. Foliage clusters oriented toward the southeast exhibited less self-shading during spring mornings than foliage clusters that were oriented northeast. This effect was not observed at the summer solstice (Neufeld *et al.*, 1988).

Root distribution and canopy shape also confer advantages to established creosotebushes. Creosotebushes have deep roots (> 3 m) plus numerous fine shallow roots emanating from the root crown. This rooting pattern allows this shrub to use soil water from small rain events and to access soil nutrients via the shallow roots. Most of the soil nutrients are concentrated in the upper 10 cm of soil which are therefore available to the shallow roots. The deeper roots act as channels for stemflow water. Canopy morphology affects the fraction of intercepted rainfall that moves along stems to the root crown and then down root channels. Stem angle and stem length accounted for 41% of the variability in measured stem flow (Whitford *et al.*, 1997). Stemflow volumes were higher on stems with angles equal to or greater than 65° and little or no stemflow was collected from branches with stem angles equal to or less than 50°. Canopy morphology affects sub-canopy soil characterisitics. Shrubs with inverted cone mophologies do not accumulate litter layers under the canopy whereas those with a hemispherical shape develop litter layers and elevated nitrogen in subcanopy soils (DeSoyza *et al.*, 1997).

Examining the traits of *Larrea tridentata* that contribute to its success in hot deserts provides a suite of characteristics which may be available to other C_3 shrubs and small trees inhabiting hot arid environments. Although it is unlikely that any other species will share all of these traits, it is quite probable that some subset will apply. It is also likely that there are traits of other species not shared with creosotebush that confer some advantage to those species in hot, arid environments.

Perennial plants can be divided into functional groups based on shared morphological and physiological characteristics. Many arid adapted plants are succulents in which water is stored in stems or modified leaves (cladodes). Examples of succulents include the *Cactaceae* and many of the *Euphorbiaceae*. The primary advantage of succulence is that the stored water can be used during extended dry periods to keep the photosynthetic area alive. Many succulent plants utilize the CAM photosynthetic pathway. The CAM pathway reduces water loss per unit carbon gained.

In CAM photosynthesis, the stomates open at night. The extent of stomatal opening is governed by ambient temperature, plant water status, internal carbon dioxide concentration, duration of prior light exposure, ambient humidity, and age of the photosynthetic tissue. The carbon that is accumulated as malate during the night is stored in vacuoles and recycled into the photosynthetic process during daylight hours. The high water content of cell vacuoles dilutes the concentration of malate which allows sufficient malate accumulation to support carbon fixation into glucose during the day. Most of the stem succulents tend to have shallow root systems and are capable of exploiting water resulting from small rain events. Rapid rehydration allows optimum use of soil water. Many have spines and surface hairs that reduce tissue temperatures and also protect the plants from herbivores.

Although the shallow roots of succulents can allow rapid uptake of water and replenishment of tissue water storage even after light rain, they can also be pathways for water loss to drying soil. However, this avenue of water loss seems to be minimal in many succulents. For example, studies of two succulents from the Sonoran Desert (*Agave deserti* and *Ferocactus acanthodes*) lost very little water through the roots during drought (Jordan and Nobel, 1984). These succulents exhibited root elongation within 12 days after a large rainfall. The rapid root growth was balanced by root death during subsequent dry periods. Roots of these succulents decreased in volume during drought and exhibited little water loss to the surrounding soil (Jordan and Nobel, 1984).

Almost all plants with the CAM photosynthetic pathway are succulents but not all succulent plants use the CAM carbon fixation pathway. There are two major categories of succulent plants: leaf succulents and stem succulents. Leaf succulents occur in arid saline environments but are not restricted to saline soils. Many succulent *Chenopodiaceae* occupy saline soils. Examples of saline soil stem succulents include *Haloxylon, Hammada,* and *Anabasis* and saline soil leaf succulents of the genera *Suaeda, Seidlitzia,* and *Traganum* (Shmida, 1985). Ions are sequestered in cell vacuoles which allows processes within the cytoplasm to proceed without the interference of these ions. The ions stored in vacuoles of succulents also establish a higher water potential gradient between the soil and leaves. This may result in leaves becoming more succulent than they would under nonsaline conditions (Jennings, 1976).

Cactus-like plants are generally considered to be arid-adapted plants. These plants generally have succulent stems with large amounts of storage parenchyma and leaves are short-lived or absent. Genera that are subdominants or dominants in desert areas include *Opuntia* and *Carnegia* in North America and *Euphorbia* in southern Africa (Fig. 6.5). Because succulents cannot lower their tissue temperature by evapotranspiration, their tissues must tolerate higher temperatures than other plants in the same environment (Nobel, 1984). The high tissue water potentials of CAM plants limit their capacity to extract water from soil below –0.10 to –0.15 Mpa. Most stem succulents have roots concentrated near the soil surface where they can rapidly absorb water from small rain events. Because of the high water content of their tissues, few stem succulents can withstand long periods of below freezing temperatures. Stem succulents also die if desiccated. Recently established juvenile stem succulents are particularly susceptible to freezing and extended periods of drought. These characteristics explain the distribution of stem succulent CAM plants (Fig. 6.6)

Woody shrubs and small trees may use either C_3 or C_4 photosynthetic pathways. Many species are deep rooted; roots of some species have been reported at depths of >10 m. Deep-rooted shrubs frequently have some roots very near the surface of the soil. The separation of the functional roots into shallow and very deep roots is an essential feature of woody shrubs or trees that exhibit hydraulic lift. Hydraulic lift is a process in which water that is transported from deep in the soil to the root mass near the surface, diffuses into the soil at

Figure 6.5   A *Euphorbia* spp. succulent from northwestern South Africa.

Figure 6.6   The geographic distribution of CAM succulents.

night when the thermal characteristics of the soil favour diffusion out of the root system. During the daylight hours, that water is absorbed by the shallow roots and is used to support stomatal opening for a short period of time. This allows some photosynthesis to occur in plants when the upper 30 cm of the soil is too dry for plant roots to absorb moisture.

Deep-rooted shrubs tend to be phenologically predictable. For example mesquite (*Prosopis glandulosa*) produces new leaves during the late spring. Shortly after new foliage is added, the plants produce flowers and fruits. Although there is variation in the production of new stem leaders following wet and dry winter seasons, the timing of stem elongation and flowering remains relatively constant from year to year. Quantitative phenological measurements of mesquite in the Sonoran Desert of California showed that leaf production and stem elongation were rapid and of short duration (< 3 weeks). Mesquite is a C_3 shrub and like creosotebush produces two cohorts of leaves. The second cohort of leaves is only a small proportion of the first cohort and the survivorship of second cohort leaves is significantly shorter than of first cohort leaves in the California population (Nilsen *et al.*, 1987). When mesquite was subjected to drought stress, its leaf water potential responded to a large rain event as rapidly as unstressed plants (Ansley *et al.*, 1992). This rapid response in drought-stressed plants demonstrates the importance of shallow roots in addition to the deep roots of a desert shrub.

All of the Mojave Desert shrubs studied by Turner and Randall (1987) exhibited predictable phenology of leaf growth. Leaves were produced in all species of shrubs each year of the study despite large differences in the timing and amount of winter rainfall. Shrubs of the genera *Ambrosia*, *Grayia*, *Krameria*, *Larrea*, and *Lycium* initiated leaf production in February and March. However, flowering was considerably more temporally variable (March–May) and in some years shrubs that flowered did not produce fruits. Plasticity in flowering and fruiting is probably necessary in desert environments where seasonal rainfall may be too low to support sufficient carbon storage for flower and fruit production.

An interesting and important characteristic of desert shrubs is the capacity to allow branches to die and retain some branches as live stems. Studies of water relations of shrubs in the Tunisian Desert of North Africa showed that there were marked differences in water uptake patterns in two shrubs (Ourcival *et al.*, 1994). One shrub, *Artemisia herba-alba* is a shallow rooted shrub with the top fine roots at 2.5 cm depth. The other shrub, *Anthyllis henoniana* is a deep-rooted legume with the top fine roots at 8 cm depth. *Anthyllis* takes up newly available water more rapidly than *Artemisia* despite having fewer surficial roots. The main stem of *Artemisia* is divided into physiologically independent branches. Any one or several of the branches may die during an extended hot, dry period without resulting in the death of the rest of the plant. This variable stem death has been observed in other shrub species: *Artemisia campestris*, *Salsola vermiculata*, and *Rantherium sanveolens*. Shrubs that exhibit this physiological characteristic are considered to transport water in sectorial winding ascent (Waisel *et al.*, 1972) where water transported by the roots is absorbed by a branch that is not necessarily directly above that root. The result of sectorial winding ascent of water is that different sectors of the canopy have different leaf water potentials at the same soil water potential. This allows one sector of the canopy to die during drought thereby avoiding the death of the whole plant.

Desert perennial grasses exhibit very different responses to water availability and ambient temperatures. Although most desert grasses are C_4, some respond to soil moisture at temperatures that are too low for the growth of most desert grasses. A South African grass, *Eragrostis lehmanniana* (Lehmann's love grass) has spread rapidly throughout southeastern Arizona and southern New Mexico (Anable *et al.*, 1992). This species has been recorded producing flowering tillers and seeds in response to a large November rain event. None of the other grasses produced flowering tillers and only a few of the other grasses produced green leaves. At low ambient temperatures, most C_4 grasses are dormant but there are some species such as Lehmann lovegrass that are exceptions to that generalization.

The high temperature–high light intensity characteristics of C_4 plants would suggest that species with this photosynthetic pathway or the CAM pathway would dominate the vegetation in hot deserts. However, most of the studies of photosynthetic pathways of the flora of desert regions show that plants with the C_3 photosynthetic pathway dominate both numerically and in terms of cover and biomass (Syvertsen *et al.*, 1976; Zeigler *et al.*, 1981; Hattersley, 1983; Vogel *et al.*, 1986). Although the C_4 pathway is more common among grasses than in shrubs or herbaceous plants, many of the desert grasses are C_3. In winter rainfall areas, such as the Negev, Saudi Arabia, and the Namib, C_3 species have an advantage because they can grow at minimum temperatures far below those required for growth of C_4 species.

Hot desert perennial C_4 grasses exhibit differences in responses to extended drought. Some species apparently mobilize reserves from tillers in the center of the clump which results in the death of tillers from the center outward. After a drought, the surviving grasses are rings or partial rings of live tillers. This has been observed in dropseeds (*Sporobolus* spp.), three-awns (*Aristida* spp.), and fluff grass (*Dasyochloa* [*Erioneuron*] *pulchella*). In other species such as blackgrama, *Bouteloua eriopoda*, tiller death occurred at the periphery of the clump with the central tillers surviving the drought. In all of these species there is a positive relationship between size of grass clump and survivorship.

6.2.2 Reptiles

Maintaining water and electrolyte balance is a problem that herbivorous reptiles have solved by physiological traits. Several species of desert lizards, *Uromastix acanthinurus*, the chuckwalla, *Sauromalus obesus* and the desert iguana, *Dipsosaurus dorsalis*, share physiological traits that allow them to use plant parts as a food source (Grenot, 1968; Vernet *et al.*, 1988). Plant tissues have high concentrations of potassium that must be excreted in order for the animals to maintain electrolyte balance. Herbivorous lizards have salt-excreting glands that drain into the nares. These glands secrete specific salts against a concentration gradient thereby serving as extrarenal kidneys. The salts eliminated from the salt glands of herbivorous reptiles are mostly potassium bicarbonates. Insectivorous/carnivorous lizards excrete sodium chloride from the nasal salt glands (Vernet *et al.*, 1988).

Accumulating fat at the base of the tail is another adaptation of some herbivorous reptiles. Both *Uromastix* spp. from North Africa and *Sauromalus obesus* from North America accumulate large reserves of fat at the base of the tail during the brief period when flowers and fruits of ephemeral and perennial plants are available. These fat stores allow these species to survive extended drought periods when little or no suitable plant material is available.

The North American desert tortoise, *Gopherus agassizii*, is a herbivorous reptile that has no functional nasal salt gland. This species tolerates water and salt imbalances on a daily basis. When green vegetation is available, desert tortoises consume large quantities, which results in high concentrations of potassium salts in their plasma. The tortoises tolerate this osmotic imbalance and when ephemeral vegetation dries out, they shift to perennial grasses, which have low water content. When rainwater is available, the tortoises empty their bladders, drink copiously and store water in their bladders (Nagy, 1988). The capacity to tolerate large shifts in osmotic concentrations in the plasma and to store large amounts of water in the bladder allow this species to maintain an herbivorous life style in an extremly arid environment.

6.2.3 Invertebrates

Although there are few morpological or physiological specializations of invertebrates that are clearly different in desert dwelling species in comparison to related species inhabiting mesic environments, there are some that have been studied and which deserve mention. Tenebrionid or darkling beetles are numerous and conspicuous members of the invertebrate faunas of deserts throughout the world. For example, the Namib in southwest Africa is reported to support 90 genera (Koch, 1962). Desert tenebrionids are flightless with fused elytra enclosing an air space called the subelytral cavity. Most mesic species have functional wings and moveable elytra. Studies by Cloudsley-Thompson (1964) and Ahearn and Hadley (1969) demonstrated that the subelytral cavity was an important mophological adaptation that reduced water loss by evaporation from the integument. They hypothesized that evaporation was decreased because the abdominal spiracles open into the relatively humid subelytral cavity rather than to the outside air. By opening into the subelytral cavity, the spiracles are protected from the drying effects of wind. The subelytral cavity also reduces the vapor pressure gradient between the trachae and their external environment thereby reducing evaporative water loss via the spiracles. An alternative hypothesis was offered by Fiori (1977) and Slobodchikoff and Wisman (1981) who suggested that the subelytral cavity allows abdominal expansion which may be important for species that feed on detritus. The low nutritional value of dead plant material (detritus) requires the animals to ingest large volumes of material to meet energy needs. Thus an expandable volume digestive system would be adaptive. An alternative advantage of an expandable abdomen is the capability to store large quantities of water in the abdominal cavity. All of these characteristics of the subelytral cavity are adaptive for detritus feeding tenebrionid beetles that inhabit hot arid environments (Draney, 1993).

Tenebrionid beetles in the Namib generate a wax bloom (wax extruded from the pores in the elytra that forms a powdery or filamentous covering of the surface). Wax bloom is produced in response to high temperature and low humidity (McClain *et al.*, 1984). The wax bloom protects diurnal beetles against the high temperatures and radiant heat loads. Wax blooms may occur in tenebrionid beetles in other deserts but this has not been reported in the literature.

Perhaps the most interesting adaptation to desert environments is cryptobiosis. Animals that enter a cryptobiotic state exhibit marked reduction in body water content and extremely low metabolic rates. When water becomes available, these animals rehydrate quickly and return to normal activity. Cloudsley-Thompson (1991) reports one example of the larvae of an insect that exhibits cryptobiosis. The larvae of a chrinomid midge, *Polypedilum vanderplanki*, inhabit shallow rock pools in the semiarid regions of Nigeria. The rock pools fill and dry several times during the rainy season. When the pools fill, the desiccated larvae absorb water through their cuticle and begin feeding within an hour. Some Australian chrinomid larvae may also tolerate complete dehydration and rehydrate with no apparent physiological damage (Cloudsley-Thompson, 1991).

Cryptobiosis has been reported in soil mites (MacKay *et al.*, 1986). Cryptobiosis in Collembola and in Nematoda is referred to as anhydrobiosis (literally life without water). Soil nematodes lose water rapidly in drying soil and change into a coiled, dehydrated form (anhydrobiotic). When anhydrobiotic nematodes are placed in water, they rehydrate in less than one hour and resume all of their activities including reproduction (Demeure *et al.*, 1979). Collembola do not coil but do dehydrate and survive with immeasureable metabolism (Poinsot-Balaguer, 1984). Cryptobiosis or anhydrobiosis is an important adaptation for desert soil mesofauna because it allows these organisms to reproduce and expand their populations in response to available food sources rather than be constrained by the short pulses of suitable environmental conditions to gain sufficient energy to produce eggs and reproduce.

6.2.4 Large Mammals

Since large grazing animals like camels, oryx, sheep, etc., do not have the option of seeking moderate microclimates below ground, they are subjected to the full effect of the high radiative, thermally desiccating environment of the desert. These environmental effects can be summarized by examining the general energy budget equation for an animal:

$$M \pm K \pm R \pm C - E = S$$

where M is heat production by metabolism, K is energy gained from or lost to the environment by conduction, R is energy gained or lost by radiation, C is energy gained or lost by convection, E is energy loss by evaporation, and S is heat storage.

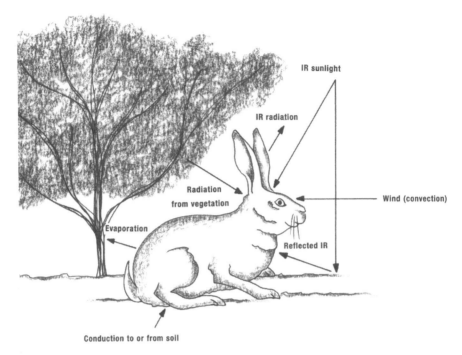

Figure 6.7   Diagram of the energy exchanges of a medium-size desert mammal.

When an organism is in thermal equilibrium, S, heat storage, will equal zero. These exchanges are summarized in Fig. 6.7.

Of the energy fluxes, radiative heating is the largest heat gain during the day and an avenue for heat loss at night. Many large desert animals store fat in localized areas of the body which enhances radiative and convectional heat fluxes (e.g. the camel's hump and fat-tailed sheep). Fat storage in the hump, or tail provides an energy reserve without the insulation that would result from subcutaneous fat storage. An insulative subcutaneous fat would reduce the rates of radiative heat flux from the body surface.

Some large desert mammals have labile body temperatures which allow the animals to store heat during the day without the necessity of expending water for evaporative cooling. These animals subsequently unload heat at night by radiation to the cool night sky and convectional cooling. In some species the body temperature is allowed to drop below 38°C, e.g. camels (Gauthier-Pilters and Dagg, 1981), eland and oryx (Fig. 6.8) (Taylor, 1969). In these animals the body temperature may fluctuate as much as 7–8°C, i.e. 33.9–42.1°C. Taylor (1969) reported that when dehydrated oryx were exposed to 45°C the animals did not sweat and the body temperature exceeded 45°C for 8 h. A body temperature of 42°C is lethal for most mammals. These large desert herbivores apparently survive such

Figure 6.8   An oryx, *Oryx gazella*, in the Kalahari Desert, South Africa.

high body temperatures by maintaining lower brain temperature by means of a countercurrent heat-exchange system at the base of the brain, which cools the blood going to the brain via the carotid artery. Taylor (1969) describes a carotid rete system that utilizes a countercurrent heat-exchange rete at the base of the brain to cool the blood going to the brain (Fig. 6.9). The venous return from the nasal sinuses is several degrees cooler than arterial blood from the heart. The countercurrent heat exchange within the rete cools the blood flowing to the brain allowing the brain to operate at sublethal temperatures while the main body mass is at a temperature that would kill brain tissue.

In addition to a labile body temperature, and modified fat storage, large desert herbivores may have one or more additional physiological characteristics that reduce the frequency of drinking or the dependence upon drinking for the replacement of water. Consider the avenues of water gain and loss in animals (Table 6.2). Water may be gained preformed in food, as an end product of metabolism or by drinking (free water). Metabolic water production varies directly as a function of metabolic rate and hence is a meaningful source of water only for small homeotherms which have a suite of water-conserving behavioral and physiological adaptations, e.g. North American kangaroo rats. Large grazing herbivores must obtain water by drinking or preformed in food. In an environment where sources of free water are sparse and unreliable, replacement of water lost in the urine, feces and by evaporation can be accomplished largely by preformed water in the food.

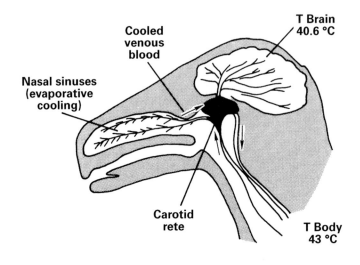

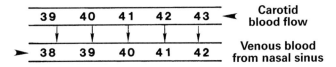

Figure 6.9   Diagram of the brain rete and the temperature changes in a countercurrent vascular system. (Modified from Taylor, 1969.)

Table 6.2
Water Balance in Mammals Listed in the Sequence of Relative Importance in
Most Mammals

Water gain	Water loss
Free water (drinking)	Evaporation from skin surface
Preformed water in food	Evaporation from respiratory surfaces
Metabolic water	Fecal water
	Excretory (urinary) water

However, preformed water in food depends on the availability of quality forage. Low quality forage (woody stems, stem cured grass) contains little water and has lower digestibility thereby increasing fecal water loss (Gauthier-Pilters and Dagg, 1981). Large herbivores must also be able to withstand marked water loss without explosive and lethal rise in body temperature. For example, camels can lose more than 25% of their body weight without serious physiological consequences and rapidly rehydrate when water becomes available. Camels also lose water slowly. The dense wool-like hair of the camel establishes a steep thermal gradient that reduces the skin surface temperature and also provides a steep humidity gradient. The steep temperature–moisture gradients across a dense fur (hair layer) reduces insensible evaporative water loss and reduces the rate of heat gain of an animal exposed to the intense thermal environment of a mid-day summer.

Kangaroos have a ruminant-like digestion and specialized dentition that allow them to utilize high-fiber, low-nutritive plants. Water loss via urine, feces and evaporation is reduced by a relatively efficient kidney, a capacity to recycle urea, and an ability to maintain plasma volume. Because of higher urea reabsorption from the glomerular filtrate, the euro is able to process lower quality food than the red kangaroo. The red kangaroo (*Macropus rufus*) moves over wide areas of the arid region, seeking fresh green shoots that result from scattered storms. The euros (*Macropus robustus*) are more sedentary since they can utilize the lower nitrogen content of dead plant materials.

The large Australian marsupials have evolved a reproductive strategy that allows them to cope with unpredictable climate and extended droughts. Joeys are born at about one month of age in kangaroos and attach to a teat in the marsupium. If the mother is unable to obtain sufficient nutrition, the young is aborted with little strain on the mother. Marsupials also undergo embryonic diapause in which a viable embryo is carried *in utero* in an arrested blastocyst stage. The embryo begins further development when stimulus from a suckling young disappears. In a drought, female marsupials may have a succession of young that die around 2 months of age and are replaced. Under severe conditions reproduction may cease completely (Denny, 1982).

In some large herbivores, urinary water loss may decrease as environmental stress increases. Camels, Bedouin goats and kangaroos conserve nitrogen by recycling urea back into the fermentation structures of the gut (Schmidt-Nielsen, 1959; Dawson, 1977) (Fig. 6.10). This allows these herbivores to use high C:N ratio (low quality, low protein, mostly cellulose) forage and maintain protein balance. Since the urea is returned to the gut rather than excreted, urinary water loss is reduced.

One important set of physiological characteristics of the black Bedouin goat is associated with the ability of that species to continue lactation on low water inputs and on low quality diets (Shkolnik *et al.*, 1972). The ability to continue lactation under water deprivation is in part due to the ability of this species to overhydrate in comparison to other species of goats and sheep.

The suites of adaptations of large herbivores may include physiological and morphological characteristics not described in the preceding discussion. It is important to re-emphasize that a large herbivore need not possess all of the characteristics described above to be 'adapted' to desert conditions. Although the suites of characteristics differ, the most successful large herbivores in arid ecosystems share a number of key features (Table 6.3).

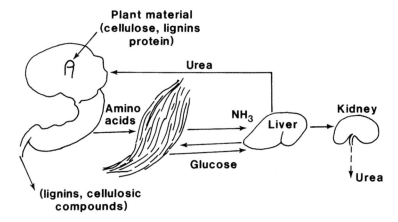

Figure 6.10   The resorption pathways of urea in herbivorous animals. (Modified from Schmidt-Nielsen, 1959.)

Table 6.3
Suites of Physiological and Morphological Characteristics of Three Large Desert Herbivores

Characteristic	Camel	Bedouin goat	Red kangaroo
Labile body temperature	Yes	?	Yes
Tolerate dehydration	> 20% of body mass	Yes	Yes
Thick wool-like hair	Yes	Yes	Yes
Resorption of urea into digestive fermentation structures	Yes	Yes	Yes
Specialized fat storage structure	Yes	No	No
Sweating under thermal stress	No	No	No
Rapid rehydration with infrequent drinking	Yes	Yes	Yes

6.2.5 Small Mammals

Small, 8–150 g, seed-eating rodents are found in all subtropical deserts except in extremely arid deserts. These seed-eating rodents share a number of morphological and physiological traits that coupled with the behavioral traits discussed in the previous section provide a suite of adaptations that allow granivorous rodents to be successful inhabitants of deserts. The adaptation most important for granivorous rodents is the ability to maintain water balance and survive on a diet of seeds during periods when green plants and free water are not available. In order for a mammal to maintain water balance inputs by drinking, preformed water in food, and metabolic water production must equal or exceed evaporative water loss (primarily via respiratory surfaces) fecal loss and urinary loss. For small granivorous rodents evaporative water is the most important loss route and metabolic water production the most important input (MacMillen and Hinds, 1983). Urinary water loss is reduced to minimal quantities in desert rodents because of the morphology and physiology of the kidney. Most but not all desert-adapted rodents have kidneys with elongate renal papillae. The ability to survive on a diet of seeds without access to preformed water is a trait shared by a large number of small desert rodents (Withers *et al.*, 1979; Hinds and MacMillen, 1985). Such rodents species frequently share the trait of an elongated rostrum which produces cooling of exhaled air via countercurrent vascularization of the nasal mucosa with the concomitant saving in respiratory evaporative water loss.

Not all desert rodents are granivorous, nor do the granivorous rodents restrict their diets to seeds. It has been suggested that reproductive success in North American desert rodents is tied to the availability of green vegetation implying that the preformed water in such vegetation fills a physiological requirement associated with reproduction that cannot be met by metabolic water alone. This added water need is probably that water required for lactation. Other authors have suggested that chemical compounds in green vegetation stimulate reproductive activity in kangaroo rats (Kerley and Whitford, 1994). The importance of green vegetation for lactation rather than as a 'stimulant to reproduction' is supported by data on kangaroo rats. The testes length of male kangaroo rats, *Dipodomys ordii*, varies with temperature and not rainfall and the presence of embryos in females varies with rainfall of the previous month (Kerley and Whitford, 1994). In a seasonal desert, onset of rainfall during the normal wet season should be a good 'prediction' of the availability of green vegetation during the period of lactation in rodents that have less than a 40-day gestation period.

Non-granivorous rodents require some characteristics in addition to those described for granivores. The chisel-toothed kangaroo rat, *Dipodomys microps* climbs into shrubs and harvests leaves of the halophyte *Atriplex confertifolia* (Kenagy, 1973). The epidermal tissues of these leaves are high in salts (electrolytes) but specialized photosynthetic parenchyma is low in electrolytes and high in starch. The lower incisors of *D. microps* are flattened and chisel shaped and are used to shave off the epidermal tissues in order to eat the inner tissue. The teeth of other kangaroo rats are rounded. The gerbilline rodent, *Psammomys obesus* of the Sahara and Middle East feeds on succulent parts of plants of the Chenopodiaceae, including some *Atriplex*. However, *P. obesus* consumes whole leaves but produces a more concentrated urine than *D. microps*. This example shows how differences in some characteristics of a suite of adaptations allows species to successfully use similar habitats and food. *D. microps* has morphological/behavioral traits and *P. obesus* physiological traits that all these species need to utilize leaf tissues with high electrolyte content.

There are sufficient physiological data for desert rodents to make comparisons of the metabolic rates of desert and nondesert rodents. Goyal *et al.* (1982) reviewed the available data and developed regression equations for metabolism of desert and nondesert rodents. For burrowing desert rodents the regression equation $M = 3.334W^{-0.295}$ with a correlation coefficient ($V = 0.82$) was derived from the data in comparison with $M = 8.853W^{-0.433}$ for nondesert rodents. The lower basal metabolic rate reduces pulmonary water loss, reduces gas exchange in burrow environments that are poor in oxygen and rich in carbon dioxide and reduces or eliminates heat storage in saturated atmospheres (Goyal et al., 1982).

In a study that examined metabolic relationships in nondesert and desert heteromyid (kangaroo rats, pocket mice) rodents, Hinds and MacMillen (1985) reported that metabolic rates of the nine xeric species was 26% below but essentially parallel to the metabolic rate, body mass relationship of other eutherian animals. Another difference in desert and nondesert mammals is their weight specific water loss. As with metabolism, desert mammals tend to have lower rates of water loss than nondesert animals.

6.2.6 Carnivores

Carnivorous mammals obtain preformed water in their food. Metabolic water contributes only a small percentage of the daily water requirements of intermediate and large size carnivores. For example Golightly and Ohmart (1984) reported that metabolic water contributed 18% of total daily water requirements of kit foxes (mean body weight = 2 kg) and 10% of the total daily water requirements of coyotes (mean body weight = 10 kg). They reported on energy to water requirement ratio, i.e. amount of prey required to meet energy needs divided by amount of prey to meet water requirements. For kit foxes this energy to water requirement ratio was 0.65 in winter and 0.57 in summer and for coyotes it was 0.30. Thus both of these species must either supplement their diet with succulent plant material as coyotes do or consume more food than is necessary for energy in order to meet their water requirements.

6.2.7 Birds

There are few unique physiological adaptations of desert birds despite the stress imposed on organisms that fly in hot environments (Dawson, 1984). Hyperthermia (elevated body temperature) results from the metabolic heat production when flying limits the duration of flights of desert birds (Hudson and Bernstein, 1981). White-necked ravens, *Corvus cryptoleucus*, either avoid flying during hot conditions or fly at higher altitudes where the air is cooler. Some birds may be able to withstand limited hyperthermia because of the anatomical arrangement of blood vessels into a countercurrent heat exchanger. This has been described as a rete mirable opthalmicum in which the venous blood is cooled by evaporation from the buccopharyngeal mucosa and from the cornea (Bernstein et al., 1979).

Both desert and nondesert birds pant to eliminate excess body heat. In some species, e.g. caprimulgids (nighthawks, poor wills, frog-mouths), panting is augmented by gular flutter (Dawson, 1984). Water turnover in desert birds appears lower than in non-desert birds of the same size but Dawson (1984) urges caution in accepting this generalization because of differences in the conditions of the studies and diets among the species studied.

Excretory water loss in birds is generally lower than in mammals because birds produce uric acid and uric acid salts that are eliminated as semisolids. Some desert birds possess nasal salt-secreting glands that allow excretion of salts with relatively little loss of water. Desert birds reported to have nasal salt glands include: ostrich, sand partridge, roadrunner and the Australian dotterel (Dawson, 1984). Salt-secreting glands are used by nestling roadrunners to compensate for the loss of water in evaporative cooling (Ohmart, 1972).

These examples, although not an exhaustive list, serve to emphasize the point that many desert organisms avoid desert conditions. They avoid desert conditions by being active, growing, and reproducing only during wet, relatively cool periods, living in burrows in the soil, having crepuscular or nocturnal activity patterns, and/or selecting moderate microclimates in plant canopies. These behaviors allow many species with no obvious physiological adaptations to the extremes of desert environments to be successful inhabitants of such environments. For other species, avoidance behaviors are coupled with suites of morphological and physiological characteristics that are adaptive in dry hot and/or dry cold environments. Large herbivores are able to survive in hot desert environments because they have the capacity to maintain brain temperatures when body temperatures exceed 40°C. This frees the large herbivores from the cost of thermoregulation by evaporation of water.

REFERENCES

Ahearn, G. A. and Hadley, N. F. (1969). The effects of temperature and humidity on water loss in two desert tenebrionid beetles, *Eleodes armata* and *Cryptoglossa verrucosa*. *Comp. Biochem. Physiol.* **30,** 739–750.

Anable, M. E., McClaran, M. P., and Ruyle, G. B. (1992). Spread of introduced Lehmann lovegrass (*Eragrostis lehmanniana* Nees.) in southern Arizona, USA. *Biol. Conserv.* **61,** 181–188.

Ansley, R. J., Jacoby, P. W., Meadors, C. H., and Lawrence, B. K. (1992). Soil and leaf water relations of differentially moisture-stressed honey mesquite (*Prosopis glandulosa* Torr). *J. Arid Environ.* 22, 142–159.

Bennett, A. F., Huey, R. B., Henry, J. A., Nagy, K. A. (1984). The parasol tail and thermoregulatory behavior of the Cape Ground Squirrel *Xerus inauris*. *Physiol. Zool.* **57,** 57–62.

Bernstein, M. H., Curtis, M. B., and Hudson, D. M. (1979). Independence of brain and body temperatures in flying American kestrels, *Falco sparverius*. *Am. J. Physiol.* **237,** R58–R62.

Burk, J. H. (1982). Phenology, germination and survival of desert ephemerals in Deep Canyon, Riverside County, California. *Madrono* **29,** 154–163.

Cloudsley-Thompson, J. L. (1964). On the function of the sub-elytral cavity in desert Tenebrionidae (Col.). *Entomol. Monthly Mag.* **60,** 3–7.

Cloudsley-Thompson, J. L. (1965). *Desert Life*. Pergamon Press, Oxford.

Cloudsley-Thompson, J. L. (1975). Adaptations of Arthropoda to arid environments. *Annu. Rev. Entomol.* **20,** 261–283.

Cloudsley-Thompson, J. L. (1991). *Ecophysiology of Desert Arthropods and Reptiles*. Springer Verlag, Berlin.

Costa, W. R., Nagy, K. A., and Shoemaker, V. H. (1976). Observations of the behavior of jackrabbits (*Lepus californicus*) in the Mojave Desert. *J. Mammal.* **57**, 399–402.

Cunningham, G. L. and Strain, B. R. (1969). An ecological significance of seasonal leave variability in a desert shrub. *Ecology* **50**, 400–408.

Davies, S. J. J. F. (1984). Nomadism as a response to desert conditions in Australia. *J. Arid Environ.* **7**, 183–195.

Dawson, T. J. (1977) Kangaroos. *Sci. Am.* **237**, 78–89.

Dawson, W. R. (1984). Physiological studies of desert birds: present and future considerations. *J. Arid Environ.* **7**, 133–155.

Demeure, Y., Freckman, D. W., and Van Gundy, S. D. (1979). Anhydrobiotic coiling of nematodes in soil. *J. Nematol.* **11**, 189–195.

Denny, M. J. S. (1982). Adaptations of the red kangaroo and euro (Macropodidae) to aridity. In W. R. Barker and P. J. M. Greenslade (eds), *Evolution of the Flora and Fauna of Arid Australia*, pp 179–184. Peacock Publications, Glen Osmond, South Australia

DeSoyza, A. G., Whitford, W. G., Martinez-Meza, E., and Van Zee, J. W. (1997). Variation in creosotebush (*Larrea tridentata*) canopy morphology in relation to habitat, soil fertility and associated annual plant communities. *Am. Midl. Nat.* **137**, 13–26.

Dimmitt, M. A. and Ruibal, R. (1980). Environmental correlates of emergence in spadefoot toads (*Scaphiopus*). *J. Herpetol.* **14**, 21–29.

Draney, M. L. (1993). The subelytral cavity of desert tenebrionids. *Florida Entomologist* **76**, 539–549.

Ettershank, G., Ettershank, J. A., and Whitford, W. G. (1980). Location of food sources by subterranean termites. *Environ. Entomol.* **9**, 645–648.

Fiori, G. (1977). La cavita sottoelitrale dei tenebrionidi apomorfi. *Redia* **60**, 1–112.

Fisher, F. M., Zak, J. C., Cunningham, G. L., and Whitford, W. G. (1988). Water and nitrogen effects on growth and allocation patterns of creosotebush in the northern Chihuahuan Desert. *J. Range Mgmt* **41**, 387–391.

Gauthier-Pilters, H. and Dagg, A. I. (1988). *The Camel: Its Evolution, Ecology, Behavior and Relationship to Man*. The University of Chicago Press, Chicago.

Golightly, R. T. Jr. and Ohmart, R. D. (1984). Water economy of two desert canids: coyote and kit fox. *J. Mammal.* **65**, 51–58.

Goyal, S. P., Ghosh, P. K., and Prakash, I. (1982). Energetic aspects of adaptation in the Indian desert gerbil *Meriones hurrianae* Jerdon. *J. Arid Environ.* **5**, 69–75.

Grenot, C. (1968). Sur l'excretion nasale de sels chez le lezard sharien: *Uromastix acanthinurus*. *C. R. Acad. Sci. Paris* **266**, 1871–1874.

Hattersley, P. W. (1983). The distribution of C3 and C4 grasses in Australia in relation to climate. *Oecologia* **57**, 113–128.

Hawke, S. D. and Farley, R. D. (1973). Ecology and behavior of the desert burrowing cockroach *Arenivaga* sp. (Dictyoptera, Poloyphagidae). *Oecologia* **11**, 263–279.

Hinds, D. S. and MacMillen, R. E. (1985). Scaling of energy metabolism and evaporative water loss in heteromyid rodents. *Physiol. Zool.* **58**, 282–298.

Hudson, D. M. and Bernstein, M. H. (1981). Temperature regulation and heat balance in flying white-necked ravens, *Corvus cryptoleucus*. *J. Exp. Biol.* **90**, 267–282

Huey, R. B, Pianka, E. R., and Hoffman, J. A. (1977). Seasonal patterns of thermoregulatory behaviour and body temperature of diurnal Kalahari lizards. *Ecology* **58**, 1066–1075.

Jennings, D. H. (1976). The effects of sodium chloride on higher plants. *Biol. Rev.* **51**, 453–486.

Johnson, K. A. and Whitford, W. G. (1975). Foraging ecology and relative importance of subterranean termites in Chihuahuan Desert ecosystems. *Environ. Entomol.* **4**, 66–70.

Jordan, P. W. and Nobel, P. S. (1984). Thermal and water relations of roots of desert succulents. *Ann. Bot.* **54**, 705–717.

Kay, F. R. and Whitford, W. G. (1978). Burrow environment of the banner-tailed kangaroo rat, *Dipodomys spectabilis*, in south-central New Mexico. *Am. Midl. Nat.* **99**, 270–279

Kemp, P. R. (1989). Seed banks and vegetation processes in deserts. In M. A. Leck, V. T. Parker and R. L. Simpson (eds), *Ecology of Soil Seed Banks*, pp. 257–281. Academic Press, San Diego.

Kenagy, G. J. (1973). Adaptations for leaf eating in the Great Basin Kangaroo Rat, *Dipodomys microps. Oecologia* **12**, 383–412.

Kerley, G. I. H. and Whitford, W. G. (1994). Desert-dwelling small mammals as granivores: intercontinental variations. *Aust. J. Zool.* **42**, 543–555.

Killingbeck KT (1992) Inefficient nitrogen resorption in a population of ocotillo (*Foquieria splendens*), a drought-deciduous shrub. *Southwest. Nat.* **37**, 35-42.

Killingbeck, K. T. (1993). Nutrient resorption in desert shrubs. *Rev. Chilena Hist. Nat.* **66**, 345–355.

Koch, C. (1962). The Tenebrionidae of southern Africa XXXI. Comprehensive notes on the tenebrionid fauna of the Namib Desert. *Ann. Transvaal Museum* **24**, 61–103.

Lajtha, K. and Whitford, W. G. (1989). The effect of water and nitrogen amendments on photosynthesis, leaf demography, and resource use efficiency in *Larrea tridentata,* a desert evergreen shrub. *Oecologia* **80**, 341–348.

Lee, A. K. and Mercer, E. H. (1967). Cocoon surrounding desert-dwelling frogs. *Science* **157**, 87–88.

Loveridge, J. P. and Craye, G. (1979). Cocoon formation in two species of Southern African frogs. *Sth Afr. J. Sci.* **75**, 18–20.

MacKay, W. P., Silva, S., Lightfoot, D., Pagani, M. S., and Whitford, W. G. (1986). Effect of increased soil moisture and reduced soil temperature on a desert soil arthropod community. *Am. Midl. Nat.* **116**, 45–56.

MacMillen, R. E. and Hinds, D. S. (1983). Water regulatory efficiency in heteromyid rodents: a model and its application. *Ecology* **64**, 152–164.

McClain, E., Praetorius, R. L., Hanarhan, S. A., and Seely, M. K. (1984). Dynamics of the wax bloom of a seasonal Namib Desert tenebrionid, *Cauricara phalangium* (Coleoptera:Adesmiini). *Oecologia* **63**, 314–319.

McClanahan, L. L., Shoemaker, V. H., and Ruibal, R., (1976). Structure and function of the cocoon of a ceratophryd frog. *Physiol. Zool.* **56**, 430–435.

Meltzer, K. H. and Whitford, W. G. (1976). Changes in O_2 consumption, body water and lipid in burrowed desert juvenile anurans. *Herpetologica* **32**, 23–25.

Moorhead, D. L., Fisher, F. M., and Whitford, W. G. (1988). Cover of spring annuals on nitrogen rich kangaroo rat mounds in a Chihuahauan Desert grassland. *Am. Midl. Nat.* **120**, 443–447.

Mulroy, T. W. and Rundel, P. W. (1977). Annual plants: adaptations to desert environments. *Bioscience* **27**, 109–114.

Nagy, K. A. (1988). Seasonal patterns of water and energy balance in desert vertebrates. *J. Arid Environ.* **14**, 201–210.

Neufeld, H. S., Meinzer, F. C., Wisdom, C. S., Sharifi, M. R., Rundel, P. W., Neufled, M. S., Goldring, Y., and Cunningham, G. L. (1988). Canopy architecture of *Larrea tridentata* (D. C.) Cov., a desert shrub: foliage orientation and direct beam radiation interception. *Oecologia* **75**, 54–60.

Nilsen, E. T., Sharifi, M. R., and Rundel, P. W. (1984). Comparative water relations of phreatophytes in the Sonoran Desert of California. *Ecology* **65**, 767–778.

Nilsen, E. T., Sharifi, M. R., Virginia, R. A., and Rundel, P. W. (1987). Phenology of warm desert phreatophytes: seasonal growth and herbivory in *Prosopis glandulosa* var. *torreyana* (honey mesquite). *J. Arid Environ.* **13**, 217–229.

Nobel, P. S. (1984). Extreme temperatures and thermal tolerances for seedlings of desert succulents. *Oecologia* **62**, 982–990.

Ohmart, R. D. (1972). Physiological and ecological observations concerning the salt-secreting glands of the roadrunner. *Comp. Biochem. Physiol.* **43A**, 311–316.

Ourcival, J. M., Floret, C., Le Floc'h, E., and Pontanier, R. (1994). Water relations between two perennial species in the steppes of southern Tunisia. *J. Arid Environ.* **28**, 333–350.

Phillippi, T. (1993). Bet-hedging germination of desert annuals: variation among populations and maternal effects in *Lepidium lasiocarpum*. *Am. Nat.* **142**, 488–507.

Poinsot-Balaguer, N. (1984). Comportment des microarthropodes du sol a climat mediterranean francais. *Bull. Soc. Bot. Fr.* **131**, 307–318.

Ruiball, R., Tevis, L., and Roig, V. (1969). The terrestrial ecology of the spadefoot *Scaphiopus hammondi*. *Copeia* 1969: 571–584.

Schmidt-Nielsen, K. (1959). The physiology of the camel. *Sci. Am.* **201**, 140–151.

Seymour, R. S. and Lee, A. K. (1974). Physiological adaptations of anuran amphibians to aridity: Australian prospects. *Aust. Zool.* **18**, 53–65.

Shaffer, D. T. and Whitford, W. G. (1981). Behavioral responses of a predator, the round-tailed lizard, *Phrynosoma modestum* and its prey, honey pot ants, *Myrmecocystus* spp. *Am. Midl. Nat.* **105**, 209–216.

Shkolnik, A., Borut, A., and Choshniak, I. (1972). Water economy of the Bedouin goat. *Symp. Zool. Soc. Lond.* **31**, 229–242.

Shmida, A. (1985). Biogeography of the desert floras of the world. In M. Evenari, I. Noy-Meir and D. W. Goodall (eds), *Ecosystems of the World. Volume 12. Hot Deserts and Arid Shrublands.* Elsevier, Amsterdam.

Shoemaker, V. H. (1988). Physiological ecology of amphibians in arid environments. *J. Arid Environ.* **14**, 145–153.

Slobodchikoff, C. N. and Wisman, K. (1981). A function of the subelytral chamber of tenebrionid beetles. *J. Exp. Biol.* **90**, 109–114.

Smith, W. E. and Whitford, W. G. (1976). Seasonal activity and water loss relationships in four species of *Eleodes* (Coleoptera: Tenebrionidae). *Southwest Entomol.* **1**, 161–163.

Syvertsen, J. P., Nickell, G. L., Spellenberg, R. W., and Cunningham G. L. (1976). Carbon reduction pathways and standing crop in three Chihuahuan Desert plant communities. *Southwest. Nat.* **21**, 311–320.

Taylor, C. R. (1969). The eland and the oryx. *Sci. Am.* **211**, 89–96.

Tevis, L. Jr (1958). A population of desert ephemerals germinated by less than one inch of rain. *Ecology* **36**, 668–695.

Tevis, L. Jr and Newell, I. M. (1962). Studies on the biology and seasonal cycle of the giant red velvet mite, *Dinothrombium pandorae* (Acari, Thrombidiidae). *Ecology* **43**, 497–505.

Turner, F. B. and Randall, D. C. (1987). The phenology of desert shrubs in southern Nevada. *J. Arid Environ.* **13**, 119–128.

Vernet, R., Lemire, M., Grenot, C. J., and Francaz, J-M (1988). Ecophysiological comparisons between two large Saharan lizards, *Uromastix acanthinurus* (Agamidae) and *Varanus griseus* (Varanidae). *J. Arid Environ.* **14**, 187–200.

Vogel, J. C., Fuls, A., and Danin, A. (1986). Geographical and environmental distribution of C3 and C4 grasses in the Sinai, Negev, and Judean Deserts. *Oecologia* **70**, 258–265.

Waisel, Y., Lipshitz, N., and Kuller, Z. (1972). Patterns of water movements in trees and shrubs. *Ecology* **53**, 520–523.

Westoby, M. (1981) How diversified seed germination behavior is selected. *Am. Nat.* **118**, 882–885.

Whitford, W. G. and Ettershank, G. (1975). Factors affecting foraging activity in Chihuahuan Desert harvester ants. *Environ. Entomol.* **4**, 689–696.

Whitford, W. G., Martinez-Turanzas, G., and Martinez-Meza, E. (1995). Persistence of desertified ecosystems: explanations and implications. *Environ. Monit. Assess.* **37**, 319–332.

Whitford, W. G., Anderson, J., and Rice, P. M. (1997). Stemflow contribution to the 'fertile island' effect in creosotebush, *Larrea tridentata. J. Arid Environ.* **35**, 451–457.

Withers, P. C., Lee, A. K., and Martin, R. W. (1979). Metabolism, respiration and evaporative water loss in the Australian hopping mouse *Notomys alexis* (Rodentia: Muridae). *Aust. J. Zool.* **27**, 195–204.

Wooten, R. C. Jr, Crawford, C. S., and Riddle, W. A. (1975). Behavioural thermoregulation of *Orthoporus ornatus* (Diplopoda: Spirostrepidae) in three desert habitats. *Zool. J. Linn. Soc.* **57**, 59–74.

Ziegler, H., Batanouny, K. H., Sankhla, N., Vyas, O. P., and Stichler, W. (1981). The photosynthetic pathway types of some desert plants from India, Saudi Arabia, Egypt, and Iraq. *Oecologia* **48,** 93–99.

Zwiefel, R. C. (1968). Reproductive biology of anurans of the arid southwest, with emphasis on adaptation of embryos to temperatures. *Bull. Am. Mus. Nat. Hist.* **140,** 1–64.

Chapter 7 | Primary Production

Primary production is the amount of solar energy converted into biomass via photosynthesis. That quantity of energy minus respiration and losses due to consumption by herbivores is *net primary production*. Net primary production is generally measured as above ground dry mass (biomass) at the end of the growing season. However, net primary production includes all leaf and stem production, root production, and increases in stem diameters of perennial dicots. Most of the literature values for net primary production are based on a subset of the actual total net primary production.

Plant production in an ecosystem determines the energy available to consumers and to decomposers. That part of the primary production that can be harvested directly or indirectly by herbivores is usually limited to the above-ground parts of a few species. In arid ecosystems, the biomass production estimates are generally made for those species of plants that provide forage for domestic livestock. Thus the above-ground net primary production that serves as forage for livestock is the production estimate that drives management decisions. Those variables that affect net primary production must be identified and their contribution to the variation in primary production understood to provide a basis for developing sustainable use of arid and semiarid rangelands.

If primary production varied according to the pulse-reserve paradigm, there would be a linear relationship between annual net above-ground production and growing season precipitation. Although there is no doubt that precipitation is the most important variable affecting primary production in deserts, it is not the only variable that affects production. The inability to predict annual above-ground net primary production (AAGNPP) and the lack of understanding of all of the variables that affect AAGNPP are major contributors to the failure of management systems to be sustainable. Understanding the complex interrelationships of environmental factors affecting net primary production and production of harvestable biomass is necessary if we are to avoid over-harvesting that leads to desertification.

7.1 MEASUREMENT OF NET PRIMARY PRODUCTION

Given the need for accurate measurement of AAGNPP and the considerable effort that has been expended to obtain such measurement, the absence of standard, accurate methods is surprising. The apparent simplicity of obtaining such data is deceptive. There are no universally accepted methods for measuring AAGNPP in any ecosystem. In deserts where the plant communities are composed of perennial dicots and moncots, and ephemeral herbs and grasses the problem of measuring AAGNPP is exacerbated. The variation in composition percentages of these different life forms affects the temporal patterns of production. Temporal and spatial variability of resource distribution across the landscape also affects patterns of production. Temporal and spatial variability change a superficially simple problem into a complex problem.

Ludwig (1986) reviewed many of the inherent problems in making accurate measurements of net primary production in deserts. The method least prone to large errors is the harvest method. Reasonable estimates of AAGNPP can be obtained by harvesting plants within a series of quadrats when the plants are at the peak biomass stage. In plant communities with species composition of various life forms, differences in growth rates and timing of maturation can lead to serious underestimates of biomass production. When harvesting perennial grasses, it is necessary to separate previous year's production from the current year's production. For perennial shrubs or trees, there may be different seasonal patterns of leaf senescence and new leaf production. These are important variables that must be considered when designing a harvest method for estimating AAGNPP in a desert ecosystem.

Biomass production estimates of perennial dicots are usually made by harvesting new leaders (terminal elongation of stems). However, harvesting new stems does not account for increased girth of stems. Conservative estimates of leaf production can be made by placing litter traps below the canopies of trees and shrubs and collecting the contents on a monthly basis. This method provides no estimates of stem production.

A plotless harvest technique has some advantages over fixed-plot techniques in that much larger areas can be sampled in short periods of time. In the plotless techniques, densities of plant species are estimated by measuring distances to the nearest individual plant from a point in each of four quarters around the point (Cottam and Curtis, 1956). By harvesting the plant in each quarter, the biomass produced can be estimated from the density data and the mass of the harvested plants.

Average mass of species a = number of individuals of species a m^{-2} × average mass of the harvested individual of species a

AGNPP = Σ mean biomass of species $a - n$

As with harvested plots, this method is most appropriate for grasses and herbaceous plants but is less accurate for trees and shrubs.

Accuracy of AAGNPP measurements can be improved by making a series of sequential harvests (Ludwig, 1986). Multiple harvests during a growing season can account for short pulses of growth and death that are missed by harvests at the end of the growing season. However, the additional costs of conducting sequential harvests frequently precludes this as an option.

Several groups of ecologists working in arid ecosystems have used nondestructive dimension analyses to estimate AAGNPP. Such measurements have the advantage of allowing sampling of the same quadrats and the same individual perennials over time. This is particularly important in experiments where relatively small plots are used to examine the long-term effects of variables such as rainfall and nutrients (irrigation and fertilization). For dimension analysis, a series of plants of varying sizes of each species are harvested after canopy size (usually 2 diameters) and height has been measured. The relationship between one or more of the dimension variables and dry mass is determined by regression analysis (Table 7.1). Repeated measurement of the dimensions of plants can be converted into biomass estimates per unit area from the regressions.

In order to estimate the growth increments for trees and shrubs, it is necessary to obtain estimates of length of new stems, leaf mass per unit new stem, leaf mass per unit length of old stem, and numbers of elongating stems per tree or shrub. Typically, nodes are marked and growth measured from that reference mark. However, there are problems of flaring errors (Ludwig, 1986) when dimension analysis is used for estimating AGNPP of perennial dicots. Flaring errors occur because of the number of steps involved in estimating the number of stems with terminal leaders, number of leaders per stem, number of leaves, flowers, or fruits per leader, and biomass–length relationships of the stem leaders, making the biomass estimates and the error inherent in the estimates made at each step. There are errors associated with the plant density estimates, with the canopy size and volume estimates and with the estimates of numbers of growth points per plant. The estimate of number of growth points per unit volume or canopy area of plant has a large variance. When the error terms are combined, the error of the estimate may be close to 100% (Ludwig, 1986).

Table 7.1
Examples of Best-fit Regression Equations Used to Estimate Biomass Production of Annual Plants. (Regressions from Ludwig *et al.* 1975 and Gutierrez and Whitford, 1987)

Species (life form)	Equation	r^2
Astragalus tephrodes (forb)	$B = 0.0062 + 0.0078\ V$	0.99
Bouteloua aristidoides (grass)	$B = 0.045 + 0.016\ V$	0.97
Euphorbia micromera (forb)	$B = 0.012 + 0.0033\ A$	0.87
Iva ambrosiaefolia (forb)	$B = 0.059 + 0.0022\ V$	0.91
Phacelia coerulia (forb)	$B = 0.12 + 0.006\ V$	0.98
Larrea tridentata (evergreen shrub)	$B = 1504\ A^2$	0.98
Flourensia cernua (deciduous shrub)	$B = 1425\ A$	0.95
Ephedra trifurca (evergreen shrub)	$B = 3249\ V - 481\ V^2$	0.99
Gutierrezia sarothrae (evergreen subshrub)	$B = 460\ V$	0.91
Zinnia acerosa (deciduous subshrub)	$B = 845\ A$	0.84

B, biomass in grams; A, canopy area in cm^2 for forbs and grasses; in m^2 for shrubs from the mean diameters. V, canopy volumes calculated from height and canopy area using volume of a spheroid or inverted cone depending on the morphology of the plant species.

7.2 COMPARISONS OF PRODUCTION ESTIMATES WITH MESIC ECOSYSTEMS

It is easy to pass off the productivity of desert ecosystems as insignificant when compared to the productivity of more mesic ecosystems. Based on the limited data used in global comparisons of primary production, it is evident that deserts are relatively unproductive places (Evenari *et al.*, 1976). The production figures quoted in most texts dealing with productivity are usually in the range of 0–250 g m^{-2} yr^{-1}. This range of values do indeed seem insignificant when compared to the 600–1500 g m^{-2} yr^{-1} for temperate forests and the 1000–3500 g m^{-2} yr^{-1} for tropical rain forests. In addition the above-ground standing crop biomass is generally low in comparison to mesic terrestrial ecosystems. Since deserts are characterized by short and frequently unreliable growing seasons because of variability in precipitation, cropping is not an option for obtaining food and fiber from the land. This largely reduces the harvestable food and fiber option to secondary production by domestic livestock. Harvestable production is therefore only a small fraction of the 0–250 g m^{-2} yr^{-1} and large land areas are required to sustain even low rates of offtake.

Although net primary production is low in areas where rainfall is less than 100 mm yr^{-1}, in many semiarid ecosystems, above-ground net production may approach that of temperate forests. For example, on a single watershed, Ludwig (1986) measured AAGNPP between 30 g m^{-2} and 592 g m^{-2} for a year in which the annual precipitation was 182 mm and the growing season rainfall a low 81 mm. On this desert watershed with its several vegetation communities distributed over the slope positions on the watershed, the AAGNPP varied from a low value within the range generally reported for scrub deserts to a high value that falls within the low end of the range for temperate forests. This example provides a cautionary note when generalizing about productivity of deserts. Although many landscape units in a desert may be relatively unproductive, there are often extremely productive units within the landscape. Productivity studies should include the entire spectrum of landscape units within the region of interest. Productivity estimates should be partitioned proportionately among the landscape units in order to obtain reliable estimates of AAGNPP.

The water-use efficiency (annual net primary production divided by annual evapotranspiration) of desert vegetation (0.1–0.3 g AAGNPP/1000 g water) is considerably lower than forests (0.9–1.8 g AAGNPP/1000 g water), and shortgrass prairie (0.2–0.7 g AAGNPP/1000 g water) (Webb *et al.*, 1978). For deserts, water use during the previous year was significantly correlated with the current year AAGNPP. The best correlation between water use and productivity was obtained by summing water use for both current and previous years (Webb *et al.*, 1978). The water-use efficiency relationship results from the higher proportion of ineffective rain events in arid regions, shrub species that utilize deep soil water, and rainfall redistribution processes.

In Australian mulga (*Acacia aneura*) banded landscapes, production of grasses and herbaceous plants in the groves averaged 180 g m^{-2} in response to 630 mm of rain two months prior to initiating measurements (Friedel, 1984). The long-term average annual precipitation in this area is 263 mm. Mulga litter production was estimated at 36 g m^{-2}. Summing litter production and herbage production provided an underestimate of AGNPP at these sites because it did not account for stem elongation and increases in girth of the mulga.

Comparison of the interannual variability in above-ground net primary production using data from eleven of the Long-Term Ecological Research sites across North America, showed that the greatest interannual variability in AGNPP (based on 6–23 years of measurement) occurred in grasslands. Forests were the least variable (Knapp and Smith, 2001). The interannual variability in AGNPP was not related to variability in precipitation. Maximum variability in AGNPP occurred in biomes where high potential growth rates of herbaceous vegetation were combined with moderate variability in precipitation. In this analysis, the desert site data were from a mixed shrub–grass site and the data for Chihuahuan Desert grassland was included in the analysis of variability in AGNPP as a grassland site. The interannual variability in AGNPP in Chihuahuan Desert shrublands should be lower than the variability in Chihuahuan Desert grasslands because the dominant C_3 shrubs and the C_3 ephemeral plants can utilize soil water from rainfalls during the months when low night-time temperatures preclude growth of the C_4 grasses. The mean AAGNPP reported for black-grama grassland (9 years of data) was 229.1 ± 21.3 g m^{-2} and for mixed desert grassland (10 years of data) was 184.5 ± 14.7 g m^{-2}. These values for AAGNPP are well within the range considered average for desert ecosystems.

7.3 RAIN USE EFFICIENCY

The effectiveness of precipitation varies considerably as was discussed in Chapter 2. For example, in the arid and semiarid zones of Tunisia, the proportion of rain events yielding less than 10 mm per month increased from 25% at the 500 mm average annual precipitation region to 50% at the 300 mm average annual precipitation region and to 60% in areas with average annual precipitation less than 200 mm. In extremely arid regions (< 100 mm average annual precipitation) that proportion is 80% (LeHouerou, 1984). However in the regions south of the Sahara, in the Sahel and the Sudan, this trend is less marked (LeHouerou, 1984).

Rain use efficiency has been proposed as a unifying concept in arid land ecology (LeHouerou, 1984). Rain use efficiency is calculated on the basis of biomass production per unit of rainfall. LeHouerou (1984) indicated that the assumption that AET (actual evapotranspiration) is equal to precipitation in arid and semiarid regions is valid only at the catchment basin or geographic area scale. The failure of rainfall at the plot or site level to be an adequate predictor of plant production or (AET) is the result of small-scale run-off, run-on patterns that modify soil water distribution spatially and that have an effect on patterns of nutrient availability. The exception to this generalization is sites located on deep, sandy soils or in extremely flat terrain. These landscape types are not widespread in arid and semiarid regions of the world.

LeHouerou (1984) suggested that in areas of sandy soils with average annual precipitation greater than 250–350 mm, there is a decrease in rain-use efficiency due to nutrient limitations. He reported that adding high nitrogen and medium phosphorus fertilizers might increase rainfall use efficiency two to five times. Occasionally potassium, sulfur, magnesium and calcium fertilization can produce an increase in net production. LeHouerou (1984) concluded that attempts at regional integration of production and rainfall have limited success because they fail to incorporate all of the variables that affect productivity, i.e. soil permeability, soil fertility, run-off amounts, vegetation structure, interception and distribution of rainfall in time and space. He also pointed out the need to consider the 'biological year' not just the calendar year for predicting productivity in seasonally arid areas. These variables are important considerations when interpreting results from productivity studies.

7.4 BELOW-GROUND PRODUCTIVITY

There are few studies of below-ground productivity in deserts. Measuring live root biomass is fraught with many difficulties and measuring root growth and root turnover is exceedingly difficult. In many desert communities the root/shoot ratios may be very large (Rodin and Bazilevich, 1967). Even data on root/shoot ratios are limited and subject to large errors of estimation. Bell *et al.* (1979) measured seasonal changes in root, shoot, and reproductive structure biomass in eight species of Mojave Desert winter annuals. Roots accounted for 12–22% of the total biomass until late in flowering when root biomass declined in all species. In these winter annuals, more biomass was allocated to reproduction (16–50%) than to roots. In comparison with annuals, root : shoot ratios reported for desert perennials exhibit considerable variation among species (Bell *et al.*, 1979). Root : shoot ratios for desert shrubs range from 0.15 to 8.0.

Using a technique based on the dilution of natural isotopes of carbon in structural carbon of root systems during the growing season, Caldwell and Camp (1974) were able to estimate turnover of the root systems in two North American Great Basin Desert shrub communities. In a *Ceratoides lanata* community they estimated root production of 186 g m^{-2} yr^{-1} and shoot production of 64 g m^{-2} yr^{-1}. In a sagebrush, *Atriplex confertifolia*, community they estimated root production of 443 g m^{-2} yr^{-1} and shoot production of 154 g m^{-2} yr^{-1}. Root turnover was three times greater than above-ground biomass production (Caldwell and Camp, 1974). Root turnover undoubtedly accounts for a large fraction of the carbon gain by desert perennial plants and is an important variable in carbon budgets. However, data are needed from many more desert areas before the contribution of root turnover in energy flow and nutrient cycling processes can be accurately assessed.

7.5 MODELS OF AGNPP

Models for predicting the responses of AAGNPP of desert ecosystems to abiotic variables were different from those that were applicable to forests and mesic ecosystems (Webb *et al.*, 1978). In mesic systems productivity varies with foliar standing crop and the most important abiotic variables are short-wave radiation and annual average temperature. For deserts and short grass prairie the important model variables were annual precipitation and annual potential evapotranspiration. The desert and short grass steppe model had the form:

$$AAGNPP = A \times FSC \, (\text{Sin}h \, [B_{max} \, (PPT - PET) \, ti])$$

where *FSC* is foliar standing crop in g m^{-2}, *ti* is the mean weekly temperature (°C), *PPT* is annual precipitation, *PET* is annual potential evapotranspiration. Sin*h* is a hyperbolic sine function. *A* and *B* are empirical parameters: *A* is a coefficient from the linear relationship of foliar standing crop and AGNPP; *B* is the shift in AGNPP that is dependent upon water availability and temperature.

Despite the inclusion of empirical parameters in the model, not all of the data from the International Biome Program measurements fit the model. There was reasonable correspondence between the model and actual estimated AAGNPP for a Sonoran Desert site, for two years of data from a Great Basin Desert site, and two of three years for a northern Chihuahuan Desert site. The AAGNPP of the Chihuahuan Desert site for the year that did not fit the model was approximately three times the predicted value. Obviously the failure of the model to predict AAGNPP for all years for which estimates were available suggests that variables in addition to those included in the model must be affecting primary production. Variables that may be important for predicting AAGNPP include: season of precipitation, precipitation event duration, intensity and storm depth; time lapse since last rainfall, and seasonal temperatures. The most important of these variables are distribution of storm depths and seasonal precipitation that together provide a measure of effective rainfall. Seasonal rainfall is especially important for predicting productivity of desert regions where growing season is determined by temperature as well as rainfall. For example in the northern Chihuahuan Desert, the growing season for the dominant C$_4$ grasses and herbaceous plants is very different from the growing season for the C$_3$ shrubs and ephemerals. Annual precipitation and potential evapotranspiration can completely mask important seasonal differences in distribution of storms of different depths. In the Chihuahuan Desert, the growing season for the C$_3$ component begins in November and continues through April. The summer growing season includes the remainder of the year. During the summer growing season, storm depth is important because most of the rainfall in the Chihuahuan Desert is from localized and intense convectional storms. If the summer rainfall has a preponderance of storms that produce less than 6 mm per event, there may be little if any soil moisture to support the growth of C$_4$ grasses and ephemerals. The model was much better at predicting AAGNPP for areas where most of the precipitation was from low intensity relatively long duration frontal storms. Low intensity rainfall has high potential for infiltration and storage therefore quantity of rainfall is a reasonable measure of water available for plant growth.

Considering the variables that affect AAGNPP in deserts, a model that is a slight modification of that described by Seely (1978) provides a more accurate prediction. This model is a modification of the Webb *et al.* (1983) model. It incorporates a run-off, run-on term, which is obviously necessary to account for the differences in productivity resulting from landscape position. That model has the general form:

$$AAGNPP = B \left([P \pm a] - PET \right)^{2Ti/10} (K \times FSC)$$

where B is a water use efficiency constant, P is precipitation in mm, a is water added or subtracted from P due to run-off or run-on, Ti is average ambient temperature, K is a constant that modifies FSC. FSC is the average foliar standing crop of perennials at the peak of the growing season.

The average foliar standing crop is a scalar reflecting the photosynthetic potential of the vegetation in an ecosystem. For example, dense shrubs lining a wash (ephemeral stream channel) have a higher photosynthetic potential per unit area than scattered shrubs on an upland surface or scattered clumps of bunch grasses on a basin grassland. Run-off, run-on decrease or increase effective moisture therefore are added or subtracted from precipitation depending upon the landscape position as a donor or recipient patch. A number of soil parameters affect the value of a, i.e. slope, soil bulk density, vegetal cover, landscape position and the intensity–duration characteristics of precipitation (P).

Another variable not accounted for in the model for AAGNPP is the time lag in nutrient availability in relation to rainfall events. Rainfall seasonality and recent history of rainfall events affects nutrient availability. Post-drought 'flush' of primary productivity is well known to desert pastoralists in North America and Australia. The large pulse of high quality plant growth following a drought breaking rain is an important part of the folk knowledge of desert pastoralists. This flush of growth is attributable to higher concentrations of available nitrogen and other nutrients in soils as a result of the low productivity during the drought.

Regression models provided good predictions of net above-ground production by winter ephemerals in winter-rainfall Mojave Desert (Turner and Randall, 1989). They found that net above-ground production of winter annuals was best predicted by September to March rainfall ($r^2 = 0.83$) but not with calendar year precipitation. The regression for calendar year precipitation accounted for only 12% of the variation in above ground production. They found that a nonlinear model provided the best fit for the net above-ground production of winter annuals using the September–March rainfall.

$$\log (AAGNP) = 1.98 \log (pptSM) - 26.2) - 2.75$$

where *AAGNP* is above-ground net production of annuals in g m^{-2} and *pptSM* is September to March rainfall in mm.

A similar model explained 86% of AGNP for perennial shrubs using the rainfall of October (year $n - 1$) to September. That model is:

$$AAGNP = 0.30 \, (pptOS) - 6.12$$

where *pptOS* is the rainfall from October the previous year to September of the year in which measurements are made.

Both models imply that some minimum amount of rainfall is required for new above-ground production during the spring growing season. For Mojave Desert annuals that minimum is 21 mm and for perennial shrubs, 37 mm. The concept of a minimum or threshold quantity of rainfall for the initiation of growth of desert vegetation is obviously important.

The data on production by Mojave Desert shrubs is particularly instructive. Rainfall amounts above the threshold but less than the long-term annual average rainfall, resulted in no significant differences in production of shrubs (Fig. 7.1). Four species of Mojave Desert shrubs exhibited large increases in biomass production at annual rainfalls of around 200 mm and four of the species exhibited little increase in production from that recorded with rainfall just exceeding the threshold for growth (Fig. 7.1). The absence of a linear response in shrub production with rainfall shows that the responses of individual species cannot

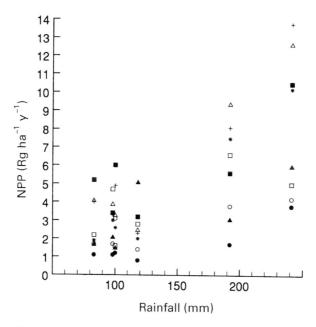

Figure 7.1   Net primary production (NPP) of eight species of Mojave Desert shrubs. ○, *Larrea tridentata*; Δ, *Ambrosia dumosa*; ●, *Atriplex confertifolia*; ▲, *Ephedra nevadensis*; □, *Grayia spinosa*; ■, *Krameria parvifolia*; +, *Lycium andersonii*; *, *Lycium pallidum*. (Data from Turner and Randall 1989.)

be generalized. This also suggests that primary production of desert vegetation does not respond as the pulse reserve model predicts.

Several short-term studies concluded that primary production varies directly with rainfall. An *Aristida pungens* (perennial grass)–*Retama retam* (shrub) community in the arid rangeland of Libya produced 200 g m^{-2} in a year with 200 mm rainfall and 95 g m^{-2} in a year with 171 mm of rainfall. Annual plants contributed 46 g m^{-2} and 30 g m^{-2}, respectively, during those years (Gintzburger, 1986). Net primary production can also vary with age of the community. In the Chaco region in Argentina, natural vegetation communities recolonize abandoned dryland farms. The net primary production of an 8 year old 'chaco' woodland (2.0–11.5 g m^{-2}) was higher than that of a 50 year old 'chaco' woodland dominated by *Prosopis flexuosa* (Braun-Wilke, 1982).

7.6 LANDSCAPE RELATIONSHIPS

The best example of how patch location and precipitation variables affect AAGNPP in a desert is provided by Ludwig (1986) in a summary of five years of data from a variety of landscape units on a watershed in the northern Chihuahuan Desert. The data from the alluvial fan unit were the data used by Webb *et al.* (1983) in their modeling of AAGNPP for desert ecosystems. Ludwig's summarization shows that some of the patterns were attributable to seasonal patterns of rainfall and storm characteristics. His data showed that productivity on the uplands was greatly affected by the characteristics of the rain events of the growing season of a particular year. In a year when a large fraction of the growing season rainfall was a small number of high intensity, short duration storms, productivity on the upland slopes was considerably lower than the productivity in another year of lower annual rainfall but with storms of lower intensity and longer duration (Fig. 7.2). When the growing season rainfall was primarily high intensity, short duration storms, run-off from the upland units emptied into the drainage channels and initiated flow in the large drainage channels of the ephemeral streams (arroyos). The transmission loss water filled the sediments and provided a water source for high productivity of the channel margin vegetation (Fig. 7.2). Similarly the variability in AAGNPP on the basin slopes was not directly the result of amount of rainfall but largely due to the run-on water from the upslope landscape units. This clearly demonstrates the necessity of including a run-off, run-on term in any model of productivity for desert ecosystems. An interesting comparison with the AAGNPP data shown in Fig. 7.2 can be made by examining data for a desert grassland located 10 km from the watershed studied by Ludwig (1986). The desert grassland is topographically flat with sandy soils. The soils have high infiltration characteristics and little if any potential for run-off or run-on. On this type of landscape unit, the timing and quantity of rainfall are the important determinants of productivity. The temporal pattern of rainfall determines which group of species responds.

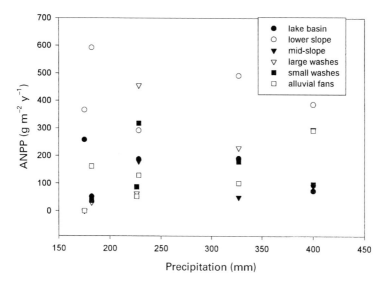

Figure 7.2   The relationships between annual precipitation and annual above-ground net primary production (AAGNPP) on a variety of landscape units on a watershed in the northern Chihuahuan Desert. (Data from Ludwig 1986.)

Cool season rainfall results in no productivity by the C_4 grasses that are the dominant perennials in this system. Cool season rainfall does stimulate production by winter–spring ephemerals (Table 7.2) Despite the absence of run-off or run-on, there was not a good correlation between annual rainfall or growing season rainfall and AAGNPP.

The importance of run-off and run-on as a variable affecting primary production is not limited to North American Deserts. An analysis of AAGNPP for watersheds in the Negev Desert, Israel showed that production in *Hammada* communities at the base of the watershed was more a function of water gain by run-off floods than of actual precipitation (Evanari *et al.*, 1976). It had been shown that in dry years the soils of the *Hammada* communities received only 21% of the rainfall as run-on but in wet years this increased to 31% of the total precipitation.

Here the importance of saturated soil increasing run-off is the explanation for the differences. In addition, much of the productivity in the Negev Desert communities in wet years was production by ephemeral plants. In relatively dry years there was little or no production of ephemeral plants. In the *Hammada* community, ephemeral plants accounted for nearly 64% of the biomass production in the wettest year of the study. In comparison, the production of ephemerals contributed an even larger percentage of the total biomass production in the *Zygophyllum dumosum* community on low slopes at near average rainfall for that area (Table 7.3).

Table 7.2

Annual and Seasonal Above-ground Net Primary Production (g m^{-2} yr^{-1}) in a Desert Grassland for Three Consecutive Years. Annual rainfall and summer (growing season) rainfall are provided for water input comparisons. Data for shrubs primarily the subshrub *Gutierrezia sarothrae*. Winter–spring annuals and summer annuals are ephemeral herbaceous plants. (Data from Pieper *et al.*, 1983.)

Year	Annual rainfall (mm)	Summer June–Sept rainfall (mm)	AAGNPP (g m^{-2} yr^{-1})			
			Total	Shrubs	Winter–Spring annuals	Summer annuals
1970	109	66.1	140	19	4	45.8
1971	206	104	315	1	1	106.2
1972	324	200	264	7	20	45.8

Table 7.3

Annual Above-ground Net Primary Production in Relationship to Rainfall in Three Plant Communities in the Negev Desert, Israel. *Hammada scoparia* communities are on the floodplains, *Zygophyllum dumosum* communities are on low sloping hills, and *Artemesia herba-alba* communities are on steep hillslopes. (Data from Evenari *et al.*, 1976)

Hammada scoparia		*Artemesia herba-alba*		*Zygophyllum dumosum*	
Ann. rain[a] (mm)	Biomass (g m^{-2})	Ann. rain (mm)	Biomass (g m^{-2})	Ann. rain (mm)	Biomass (g m^{-2})
94	192 (8)	38	38 (0)	52	38 (38)
163	420 (64)	164	229 (51)	93	211 (71)
54	39 (0)				
134	225 (51)				

Ann. rain, annual rainfall in the year of the measurements; biomass, production. The numbers in parentheses are the percentage of the biomass production contributed by ephemeral plants.

Differences in AGNPP in different locations receiving the same annual rainfall were reported by Evenari *et al.* (1976). They concluded that the relationship between annual precipitation and primary production disappears in landscapes where topography and geomorphology affect infiltration, soil water storage, and run-off. The topographic position and geomorphological history of a landscape unit is an important determinant of the abundance and species composition of the subdominant plants in the community. They reported AGNPP of 250 kg ha^{-1} and 350 kg ha^{-1} for two Negev Desert communities with the shrub, *Zygophyllum dumosum* as a dominant and annual rainfall of 93 mm. The AGNPP of an *Artemesia herba-alba* community in the same region produced 440 kg ha^{-1} and a *Hammada scoparia* community in that region produced 1430 kg ha^{-1} in response to annual rainfall of 93 mm. A mixed *A. herba-alba–Z. dumosum* community had an AGNPP of 470 kg ha⁻¹ with 93 mm of rainfall. The *Z. dumosum* and *A. herba-alba* communities are on hillslopes and the *H. scoparia* community occupies a floodplain.

An *A. herba-alba* community in Algeria produced an AGNPP of 177 kg ha⁻¹ with 300 mm of annual rainfall and an *A. herba-alba* community in the northern Negev in Israel produced 360 kg ha⁻¹ with 164 mm of rainfall (Evenari *et al.*, 1976). These studies clearly demonstrate the importance of landscape position and geomorphology–soil characteristics as variables affecting AGNPP when rainfall is equal at all of the sites.

A long-term data set that provided information on productivity of severely degraded ecosystems in contrast with ecosystems that were in relatively good condition is that reported by Floret *et al.* (1983). Their studies were in a winter rainfall, scrub desert region of Tunisia. During the six years of their study, rainfall varied between 85 mm yr⁻¹ and 431 mm yr⁻¹. Their data showed a strong seasonality component to productivity, especially that of ephemeral plants (Fig. 7.3). They attributed much of the variation in AAGNPP to the seasonal distribution of rainfall. They also suggested that during wet years, nutrient availability, *not* water was the principal factor limiting productivity. On the most degraded site, all of the AAGNPP was contributed by ephemeral plants (Fig. 7.3). The fraction of AAGNPP contributed by ephemerals varied considerably among years and among sites (Fig. 7.4). Ephemeral plants accounted for all of the AAGNPP on the degraded erosion slopes.

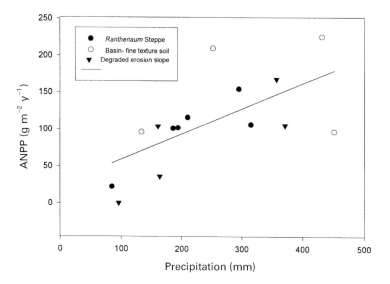

Figure 7.3   The relationship between rainfall and annual above-ground net primary productivity on three scrub desert sites in Tunisia. (Data from Floret *et al.*, 1983.)

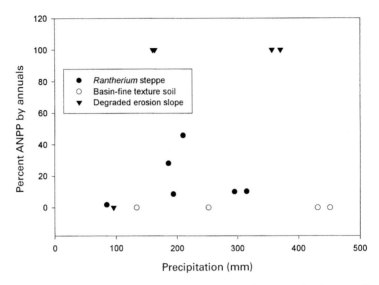

Figure 7.4   Percentage of annual above-ground net primary production contributed by ephemeral plants on three scrub sites in Tunisia. (Data from Floret *et al.*, 1983.)

Another example of the importance of topographic position affecting productivity was demonstrated by the measurements of production on two grassland types in western Spain (Corona *et al.*, 1995). Above-ground production was 126 g m^{-2} in the grassland on the upper slope and 304 g m^{-2} on the lower slope. Not only did biomass production differ with landscape position but the fraction of that AAGNPP that was grass, forb, and legumes also varied with position. Grasses dominated the lower slope position while forbs (herbaceous plants) dominated the upper slope positions.

All of the available empirical data show that position of the landscape unit in reference to the surrounding watershed or landscape is an important variable that must be incorporated into models of productivity in desert ecosystems. These studies also clearly demonstrate that water inputs are not the only variable that must be considered. Nutrient limitations can occur depending on the recent past history of rainfall and distribution of productivity among perennials and ephemerals. Temporal lags in nutrient availability occur when rainfall stimulates microbial activity, which can either mobilize nutrients or immobilize nutrients. The details of nutrient mobilization and immobilization are provided in Chapter 9.

7.7 PRODUCTIVITY IN EXTREME DESERTS

At the extreme of the aridity gradient, productivity responds to rainfall as predicted by the pulse-reserve paradigm. In a study of productivity along a rainfall gradient in the Namib Desert (one of the most arid regions of the world), Seely (1978) reported that productivity was highly correlated with rainfall and that the relationship between rainfall and productivity could be expressed by a simple equation:

$$Y = b(P - a)$$

where Y is productivity (g m^{-2}) P is precipitation (mm), a is an 'ineffective precipitation' term that includes water losses by evaporation and run-off or the minimal rainfall necessary to stimulate germination and b represents photosynthetic water-use efficiency (mg dry matter per g of water transpired).

In Seely's study, the rainfall gradient was from 12.5 to 95 mm yr^{-1} and productivity ranged from 0.75 to 50 g m^{-2}. In comparison to productivity estimates for a rainfall gradient in mesic West Africa, the slope of the line fitted to the data from the Namib was considerably steeper. Seely attributed this to the differences between the ephemeral plants of the dry Namib and the perennial grasses of the more mesic areas of West Africa. Seely (1978) argued that the perennial grasses may use rainfall more effectively because they have a developed root system that allows immediate response to soil moisture and also that production may simply be higher because of the longer growing season in the more mesic areas.

7.8 PRODUCTIVITY LINKED TO RAINFALL

The considerable literature that reports AAGNPP varying directly with rainfall is based on incomplete measurements of primary production. Most production studies have been conducted using harvest plot methods where grasses and herbaceous plants are harvested from unit area quadrats. These studies ignore productivity by the shrub and tree component if that component is part of the community being studied. LeHouerou et al. (1988) reviewed and analyzed data from a large number of studies of productivity in semiarid to arid ecosystems in a variety of locations around the world. They factored in a rain use efficiency factor ([REU], the quotient of annual primary production divided by rainfall). They also incorporated a production to rainfall variability ratio, which is the coefficient of variation in annual production divided by the coefficient of variation in annual rainfall (CVP/CVR). These figures were then used in an empirical formula to predict production.

$$P = (1 + CVP/CVR \times V - r/V) \, r \times REU$$

where P is annual production, REU is rainfall use efficiency (defined above), V is annual precipitation, r is the long-term mean annual rainfall, CVP = coefficient of variation of production, and CVR is the coefficient of variation of rainfall.

LeHouerou et al. (1988) used this equation to predict production for arid rangelands in Azizia, Libya. They used the production estimates from this equation to predict the carrying capacity for livestock for a variety of rainfall distributions. This equation predicts forage production *not* annual above-ground net primary production. A review of the published studies cited by LeHouerou *et al.* (1988) showed that in every study, the data were for harvested plots where 'forage species were harvested'. In none of these studies were any measurements of the productivity of woody perennials included. In some of the studies the plant material harvested was even more limited, being confined to perennial grass species only. This analysis provides a model that may be quite useful in estimating forage production for livestock that do not use woody perennials. However, in many desert regions, most of the livestock forage is derived from woody perennials.

The review of LeHouerou *et al.* (1988) shows a closer linkage of annual net primary production and precipitation than found in studies of where perennial shrubs were a part of the ecosystem. The linkage between annual precipitation and AAGNPP was also much lower in studies where the dominant vegetation was shrubs or small trees. Reviews that purport to demonstrate a close linkage between AAGNPP and rainfall in deserts must be considered with caution and with a critical examination of the sources of data to determine if the data are for the whole ecosystem or for some subset of the vegetation. The studies that have examined productivity of whole desert ecosystems have shown that annual rainfall accounts for less than half of the variation in AAGNPP of the ecosystem. A simple relationship between rainfall and above ground net primary production seems to be superficially realistic. However, when all the variables that affect primary production are taken into account, it is obvious that productivity in desert ecosystems is far more complex than rainfall equals an incremental increase in biomass.

Grass production in tussock grass communities dominated by *Festuca pallescens* in Patagonia, Argentina, exhibited considerable seasonal variation between years (Defosse *et al.*, 1990). The average annual precipitation in the region is 374 mm with 67% of that precipitation in winter as snow or rain. Seasonal production ranged from 32.7–35.0 g m^{-2} and varied with temperature and precipitation. A study in which AGNPP of shrubs and grasses were measured separately, showed that the annual shrub production was associated with the annual cumulative precipitation (Jobbagy and Sala, 2000). Annual grass production was not associated with annual precipitation. Grass AGNPP peaked in spring and there was no grass production in late summer and early autumn. The mean AGNPP for the 10 years of the study was 56 g m^{-2} yr^{-1} with a coefficient of variation of 26%. In this Patagonian steppe the AGNPP of grass was associated with seasonal temperatures as well as seasonal rainfall. This study emphasizes the importance of separating functional types of vegetation in productivity studies. It also demonstrates the necessity of using both seasonal and annual scales of analysis when attempting to relate AGNPP in deserts to climatic variables.

7.9 PRODUCTIVITY OF SPECIES AND FUNCTIONAL GROUPS

The problem of estimating productivity of woody perennials has been described in detail (problem of multiplicative errors). However, there are a sufficient number of studies of some species of shrubs, like creosotebush (*Larrea tridentata*) to allow examination of some of the variables that affect productivity of desert shrubs and trees. The importance of geomorphic position was clearly shown in a study of productivity of *L. tridentata.* Burk and Dick-Peddie (1973) reported that three of their study sites had well-developed, indurated (cemented) calcium carbonate layers at depths between 15 cm and 35 cm. In two years with above-average rainfall, the shrubs on shallow soils had the highest productivity recorded in their study. However, in the second successive wet year, productivity was essentially the same as in a year with average rainfall. The pattern of decrease in productivity in the second successive wet year was not the same at all locations. The variations in productivity were not as great among years on sites where there was no indurated calcium carbonate layer. This study demonstrates that depth of soil can affect productivity patterns of perennial shrubs because of the differences in water storage capacity of the soil available to shrub roots. Their work also showed that productivity of a perennial shrub was not directly linked to rainfall. Their data are consistent with patterns of productivity in creosotebush linked to nitrogen availability in the soil.

Detailed studies of the productivity of *L. tridentata* in southeastern Arizona showed that growth measured as addition of new nodes, occurred only at soil temperatures above 17°C (Chew and Chew, 1965). They also reported that stem elongation was largely independent of rainfall. Growth independent of rainfall is characteristic of other desert shrubs and small trees such as mesquite (*Prosopis* spp.) and palo verde (*Cercidium* spp.). The study by Chew and Chew (1965) also demonstrated that productivity in a desert ecosystem can be nearly as high as the productivity of a forest. In a year in which rainfall in late summer was consider-ably above the long-term average, they estimated AAGNPP of 1400 kg ha^{-1}. They also reported an estimate of 90 kg ha^{-1} for a year with an annual rainfall of 435 mm (nearly double the long-term average annual precipitation for the area). The Chew's study documents the relative independence of productivity and annual rainfall in this species.

Studies of productivity of *L. tridentata* have not been limited to harvest or growth increment measurements. This evergreen shrub has been the subject of intensive physiological studies of photosynthesis and carbon allocation patterns (Reynolds *et al.*, 1980): 'In the model, 14 compartments are defined, two for each of seven plant organs or developmental stages of interest: (1) leaves, (2) stems, (3) roots, (4) early reproductive buds, (5) mature reproductive buds, (6) flowers, and (7) fruits.' The two compartments per organ represent the separation of carbon into a nonlabile and labile pool. Labile carbon assimilates are translocated between organs by two-way flows from each organ pool through a common labile pool (the leaf pool). The leaf pool receives the currently produced photosynthate. Control of photosynthesis is modeled as a function of leaf water potentials (stomatal resist-ance) radiation (photon flux) and leaf age. Allocation to organ structures is made by an allocation submodel that incorporates the known physiological parameters. The model behavior was examined on a one-year simulation using environmental conditions for southern New Mexico. The growth patterns and reproduction patterns were similar to those reported for *L. tridentata* in this region by Ludwig and Whitford (1981). Although the growth patterns are realistic, prediction of biomass production requires scaling up from physiology at the leaf level to the foliage of the whole plant and to the assemblage of plants within a landscape unit. Physiological based models of production serve as rigorous synthesis frameworks that identify gaps in our understanding of the responses of desert plants to weather patterns, and variability in available soil resources.

Studies of productivity of *L. tridentata* in the Mojave Desert, where rainfall is predominantly from winter frontal storms, showed considerable between-year variation even between years with essentially equal rainfall. There were no clear patterns of productivity related to seasonality of rainfall in the Mojave Desert (Turner and Randall, 1989). When data for all the shrubs in the creosotebush community were combined there was a positive correlation between annual net primary production and growing season net primary production. When the data were examined by species, for several of the species there were very small differences in biomass production at rainfall varying between 80 mm and 192 mm. The only consistent pattern for all of the shrubs in this ecosystem was significantly higher production in a year in which the annual rainfall was double the long-term average. Here also, there was no close correlation between annual rainfall and productivity of shrubs.

A comparison of AAGNPP of several Mojave Desert shrubs in response to a wet growing season (246.7 mm) and a dry growing season (58.1 mm) demonstrated the importance of phenological responses of shrubs with different life histories (Table 7.4). The growing season in the Mojave Desert is October through May. Adequate moisture conditions during the wet year allowed the drought-deciduous *Ambrosia* and *Lycium* species to retain their leaves into the summer

Table 7.4
Estimated Net Above-ground Biomass Production of Five Species of Perennial Shrubs in the Mojave Desert, Nevada, USA (data from Bamberg et al., 1976)

Species		Life history	Biomass (Δ) (g m^{-2})
Ambrosia dumosa	(dry)	Drought deciduous	2.8
Ambrosia dumosa	(wet)		10.1
Krameria parvifolia	(dry)	Hot season shrub	5.2
Krameria parvifolia	(wet)		10.5
Larrea tridentata	(dry)	Evergreen shrub (C$_3$)	2.1
Larrea tridentata	(wet)		5.4
Lycium andersonii	(dry)	Drought deciduous	2.7
Lycium andersonii	(wet)		12.0
Lycium pallidum	(dry)	Drought deciduous	0.8
Lycium pallidum	(wet)		5.5

(mid-July) and to greatly increase AAGNPP in comparison with the dry year. The evergreen shrub, *L. tridentata*, replaced old leaves with new foliage during the wet year. The leaf biomass during the wet year was lower than in the dry year in this species and there was very little difference in AAGNPP in the two years. There was also very little difference in AAGNPP in *Krameria parvifolia* in the wet and dry year (Table 7.4). These small changes in AAGNPP were attributed to the adaptations of these species to grow under hot, dry conditions. Differences in the physiology of the shrub species are reflected in the phenological patterns of growth exhibited in years with very different rainfall inputs (Bamberg *et al.*, 1976).

A study of water sources used for growth of desert plants in southern Utah found that all species used winter–spring recharge precipitation for spring growth but utilization of summer rains was dependent on life form (Ehleringer *et al.*, 1991). Succulent perennials and annuals were dependent on summer precipitation and winter–spring rainfall stored water. Several species of woody perennials did not respond to summer rainfall. Herbaceous and other woody perennials utilized both summer rainfall and winter–spring rainfall. This study shows that productivity of different life forms is dependent on the seasonal distribution of rainfall as well as the amount of rainfall.

An irrigation study in the Mojave Desert provided evidence of the importance of previous year rainfall as a determinant of productivity (Johnson *et al.*, 1978). Irrigation stimulated growth and reproduction in perennial shrubs and succulents on rocky slopes and a sandy piedmont (bajada). No annuals grew in response to irrigation on the piedmont slopes but several species of annuals grew on the rocky slope. The following year there was an inverse relationship between the previous year's water inputs and abundance of annual plants. Summer irrigation stimulated higher leaf production and fruiting in creosotebush (*L. tridentata*) (Johnson *et al.*, 1978).

Experimental studies using irrigation to simulate rainfall and 'rain-out' shelters (Fig. 7.5) to simulate drought provide some insights into the mechanisms that account for the absence of close linkage between AAGNPP of desert shrubs and precipitation. In studies where *L. tridentata* were irrigated and fertilized during the growing season, the irrigated shrubs had higher growth rates than the controls but only half the rate of growth of shrubs that were irrigated and fertilized with nitrogen (Fisher *et al.*, 1988). Irrigation with 6 mm of water per week was more effective than irrigation with 25 mm once each month. The combination of small weekly irrigation with nitrogen fertilization produced the largest growth increments. In a study using 'rain-out' shelters to produce summer season drought, *L. tridentata* exhibited rapid recovery from drought with the first rainfall following the removal of the rain-out shelter covers (Whitford *et al.*, 1995). The annual net production of plants exposed to complete summer drought was equal to those that were irrigated and to the unmanipulated controls by early November. Leaf litter fall during summer months was significantly reduced in shrubs that were exposed to summer drought. However, there were no differences in leaf litter fall among

Figure 7.5 'Rain-out' shelters constructed over-trenched plots in a creosotebush, *Larrea tridentata*, shrubland, Chihuahuan Desert, New Mexico.

treatments during the remainder of the year. One additional interesting result of this experiment was that after five consecutive years of summer drought, perennial grasses and perennial herbaceous plants disappeared on the drought plots. Perennial grasses increased on the irrigated plots but there were no differences in perennial herbaceous plants between the irrigated and control plots.

These experimental studies show that nitrogen availability is a critical factor in the growth of creosotebush. The imposition of summer drought resulted in higher total soil N and higher nitrate levels in the soils of the drought plots in comparison to the control and irrigated plots (Whitford *et al.*, 1995). This nitrogen was available to support rapid growth of the shrubs with the first rains following the drought. This study also provides an insight into the traditional pastoralist knowledge concerning a growth flush following drought. Imposed drought resulted in accumulation of N in the soil. When rainfall occurred, there was no nitrogen limitation. The experimental studies also showed remarkable drought tolerance of this desert shrub. After five consecutive years of growing season drought, productivity of the creosotebushes was the same as the controls and irrigated plants.

Not all desert vegetation responds to irrigation and nitrogen fertilization. In a coastal desert area of Chile, biomass production of alien species responded to small rain event (5 mm month^{-1}) supplementation but native species did not (Gutierrez, 1992). None of the species responded to nitrogen fertilization. It was suggested that the lack of response was due to higher thresholds of water inputs required for germination and growth of the annuals in this community. The absence of a fertilizer response was attributed to the low water inputs. This study supports the threshold rainfall event concept and also provides support for the idea that nutrient limitations are evident only when there is sufficient water for growth of the species.

The AAGNPP of annual plants on sand dunes was correlated with soil moisture and organic matter through the effects of these parameters on species diversity (Kutiel and Danin, 1987). Species diversity was related to organic matter content and soil moisture. There was a relationship between the number of species and AAGNPP described as an arched curve ranging from 150 g m^{-2} in plots with five or less species to 350 g m^{-2} in plots with 10–15 species. Peak AAGNPP of approximately 400 g m^{-2} occurred in plots with eight to ten species of annuals. At the low end of the productivity scale, production was due to a small number of species that are adapted to greater environmental stress including sand movement. At the high end of the scale most of the phytomass was accounted for by one or two species that appear to be better competitors for the available water and nutrient resources. The spatial variability in AAGNPP of desert annuals is related to soil nutrients, soil moisture resulting from organic matter accumulation under perennials and the modification of the microclimate under shrub canopies. The morphologies of canopies of desert shrubs can affect the productivity of annuals under the shrub canopies (Halvorson and Patten, 1975; DeSoyza et al., 1997). In the Sonoran Desert production of annuals was 60 g m^{-2} under shrub canopies and 31 g m^{-2} in open areas (Halvorson and Patten, 1975). In the Chihuahuan Desert, biomass production of annual plants was 24 ± 10 g m^{-2} under creosotebush (L. tridentata) canopies and 4 ± 1g m^{-2} in open spaces between shrubs (Parker et al., 1982). In the Mojave Desert, the production of annuals under shrubs was 1.5–5 times higher than in the intershrub spaces (DeSoyza et al., 1997). Soils under shrub canopies have higher concentrations of available nutrients, increased water infiltration and water content, and less temperature fluctuation than soils in intershrub spaces (Parker et al., 1982; DeSoyza et al., 1997).

The importance of growth form differences affecting productivity was documented in a study of mesquite production in the Sonoran Desert of California. The AAGNPP of tree form mesquite (P. glandulosa) was 5300 g m^{-2} and for shrub form mesquite was 2200 g m^{-2}. The productivity for the whole community was 365 g m^{-2} (Sharifi et al., 1982). This level of production is extremely high for an area that receives an annual precipitation of approximately 70 mm. This high level of productivity is the result of the deep-rooted dominant shrub reaching the water table. The roots of both morphological types reach the water table at this location. Possibly because water is not limiting in this environment, most of the photosynthate was allocated to woody tissues in trunk and branches (51.5%) and only 33.6% was allocated to leaves (Sharifi et al., 1982).

REFERENCES

Bamberg, S. A., Vollmer, A. T., Kleinkopf, G. E., and Ackerman, T. L. (1976). A comparison of seasonal primary production of Mojave Desert shrubs during wet and dry years. *Am. Midl. Nat.* **95**, 398–405.

Bell, K. L., Hiatt, H. D., and Niles, W. E. (1979). Seasonal changes in biomass allocation in eight winter annuals of the Mojave Desert. *J. Ecol.* **67**, 781–787.

Braun Wilke, R. H. (1982). Net primary productivity and nitrogen and carbon distribution in two xerophytic communities of central-west Argentina. *Plant Soil* **67**, 315–323.

Burk, J. and Dick-Peddie, W. A. (1973). Comparative production of *Larrea divaricata* Cav. on three geomorphic surfaces in southern New Mexico. *Ecology* **54**, 1094–1102.

Caldwell, M. M. and Camp, L. B. (1974). Belowground productivity of two cool desert communities. *Oecologia* **17**, 123–130.

Chew, R. M. and Chew, A. E. (1965). The primary productivity of a desert shrub (*Larrea tridentata*) community. *Ecol. Monogr.* **35**, 335–375.

Corona, M. E. P., Cuidad, A. G., Criado, B. G., and de Aldana, R. V. (1995). Patterns of aboveground herbage production and nutritional quality structure on semiarid grasslands. *Commun. Soil Sci. Plant Anal.* **26**, 1323–1341.

Cottam, G. and Curtis, J. T. (1956). The use of distance measures in phytosociological sampling. *Ecology* **37**, 451–460.

Defosse, G. E., Bertiller, M. B., and Ares, J. O. (1990). Above-ground phytomass dynamics in a grassland steppe of Patagonia, Argentina. *J. Range Mgmt* **43**, 157–160.

DeSoyza, A. G., Whitford, W. G., Martinez-Meza, E., and Van Zee, J. W. (1997). Variation in creosotebush (*Larrea tridentata*) morphology in relation to habitat, soil fertility, and associated annual plant communities. *Am. Midl. Nat.* **137**, 13–26.

Ehleringer, J. R., Phillips, S. L., Schuster, W. S. F., and Sandquist, D. R. (1991). Differential utilization of summer rains by desert plants. *Oecologia* **88**, 430–434.

Evenari, M., Schulze, E. D., Lange, O. L., Kappen, L., and Buschbom, U. (1976). Plant production in arid and semi-arid areas. In O. L. Lange, L. Kappen and E.-D. Schulze (eds), *Ecological Studies, Analysis and Synthesis*, Vol. 19. *Water and Plant Life*, pp. 439–451. Springer-Verlag, Berlin.

Fisher, F. M., Zak, J. C., Cunningham, G. L., and Whitford, W. G. (1988). Water and nitrogen effects on growth and allocation patterns of creosotebush in the northern Chihuahuan Desert. *J. Range Mgmt* **41**, 387–390.

Floret, C., LeFloc'h, E., and Pontanier, R. (1983). Phytomasse et production vegetale en Tunisie preshaarienne. *Acta Ecol./Ecol. Plant.* **4**, 133–152.

Friedel, M. H. (1984). Biomass and nutrient changes in the herbaceous layer of two central Australian mulga shrublands after unusually high rainfall. *Aust. J. Ecol.* **9**, 27–38.

Gintzburger, G. (1986). Seasonal variation in above-ground annual and perennial phytomass of an arid rangeland in Libya. *J. Range Mgmt* **39**, 348–353.

Gutierrez, J. R. (1992). Effects of low water supplementation and nutrient addition on the aboveground biomass production of annual plants in a Chilean coastal desert site. *Oecologia* **90**, 556–559.

Gutierrez, J. R. and Whitford, W. G. (1987). Chihuahuan Desert annuals: importance of water and nitrogen. *Ecology* **68**, 2032–2045.

Halvorson, W. L. and Patten, D. T. (1975). Productivity and flowering of winter ephemerals in relation to Sonoran Desert shrubs. *Am. Midl. Nat.* **93**, 311–319.

Jobbagy, E. G. and Sala, O. E. (2000). Controls of grass and shrub aboveground production in the Patagonian steppe. *Ecol. Appl.* **10**, 541–549.

Johnson, H. B., Vasek, F. C., and Yonkers, T. (1978). Residual effects of summer irrigation on Mojave Desert annuals. *Bull. Calif. Acad. Sci.* **77**, 95–108.

Knapp, A. K. and Smith, M. D. (2001). Variation among biomes in temporal dynamics of aboveground primary productivity. *Science* **291**, 481–484.

Kutiel, P. and Danin, A. (1987). Annual-species diversity and aboveground phytomass in relation to some soil properties in the sand dunes of the northern Sharon Plains, Israel. *Vegetatio* **70,** 45–49.

LeHouerou, H. N. (1984) Rain use efficiency: a unifying concept in arid land ecology. *J. Arid Environ.* **7,** 213–247.

LeHouerou, H. N., Bingham, R. L., and Skerbek, W. (1988). Relationship between the variability of primary production and the variability of annual precipitation in world arid lands. *J. Arid Environ.* **15,** 1–18.

Ludwig, J. A. (1986). Primary production variability in desert ecosystems. In W. G. Whitford (ed.), *Pattern and Process in Desert Ecosystems,* pp. 5–17. University of New Mexico Press, Albuquerque.

Ludwig, J. A. (1987). Primary productivity in arid lands: myths and realities. *J. Arid Environ.* **13,** 1–7.

Ludwig, J. A. and Whitford, W. G. (1981). Short-term water and energy flow in arid ecosystems. In D. W. Goodall and R. A. Perry (eds), *Arid Ecosystem Dynamics,* vol. 2, pp. 271–299. Cambridge University Press, Cambridge.

Ludwig, J. A., Reynolds, J. F., and Whitson, P. D. (1975). Size-biomass relationships of several Chihuahuan Desert shrubs. *Am. Midl. Nat.* **94,** 451–461.

Parker, L. W., Fowler, H. G., Ettershank, G., and Whitford, W. G. (1982). The effects of subterranean termite removal on soil nitrogen and ephemeral flora. *J. Arid Environ.* **5,** 53–59.

Pieper, R. D., Anway, J. C., Ellstrom, M. A., Herbel, C. H., Packard, R. L., Pimm, S. L., Raitt, R. J., Staffeldt, E. E., and Watts, J. G. (1983). *Structure and Function of North American Desert Grassland Ecosystems.* New Mexico State University Agricultural Experiment Station, Las Cruces, NM Special Report 39.

Reynolds, J. F., Strain, B. R., Cunningham, G. L., and Knoerr, K. R. (1980). Predicting primary production for forest and desert ecosystem models. In J. D. Hesketh and J. W. Jones (eds), *Predicting Photosynthesis for Ecosystem Models,* Vol. II, pp. 160–207. CRC Press, Boca Raton, Florida.

Rodin, L. E. and Bazilevich, N. I. (1967). *Production and Mineral Cycling in Terrestrial Vegetation.* Oliver and Boyd. London.

Seely, M. K. (1978). Grassland productivity: the desert end of the curve. *South Afr. J. Sci.* **74,** 295–297.

Sharifi, M. R., Nilsen, R. T., and Rundel, P. W. (1982). Biomass and net primary production of *Prosopis glandulosa* (Fabaceae) in the Sonoran Desert of California. *Am. J. Bot.* **69,** 760–767.

Turner, F. B. and Randall, D. C. (1989). Net production by shrubs and winter annuals in Southern Nevada. *J. Arid Environ.* **17,** 23–26.

Webb, W., Szarek, S., Lauenroth, W., Kinerson, R., and Smith, M. (1978). Primary productivity and water use in native forest, grassland, and desert ecosystems. *Ecology* **59,** 1239–1247.

Whitford, W. G., Martinez-Turanzas, G., and Martinez-Meza, E. (1995). Persistence of desertified ecosystems: explanations and implications. *Environ. Monit. Assess.* **37,** 319–332.

Chapter 8 | Consumers, Consumption, and Secondary Production

Much of ecology, and desert ecology is no exception, has focused on feeding relationships within assemblages of organisms and/or the relationships of plants to the animals that feed on them. Major differences in the structure of ecosystems are a function of the proportion of net primary production that is consumed as live tissue and that which is processed as dead material. Energy flow through the consumer populations sets the constraints on the number of trophic levels in each ecosystem. Energy is analogous to the currency of ecosystems. In order to understand ecosystem dynamics, considerable effort has been expended on understanding trophic relationships and the flow of energy through ecosystems.

A large amount of the recent work on competition, predation, symbiosis, coevolution and even sociobiology focuses on characteristics of the sources of energy required by the subject organisms, the ways in which these organisms obtain that energy, and the organisms with whom they interact either directly or indirectly in that pursuit. These are all fascinating topics and relate either directly or indirectly to important processes in desert ecosystems.

8.1 SECONDARY PRODUCTION

In their review of production by desert animals, Turner and Chew (1981), because of limited data, confined their analysis to several species of mammals, lizards, birds and one species of grasshopper. In order to estimate secondary production it is necessary to have detailed information on birth and death rates of the consumers, changes in mass of all of the animals alive during the time period chosen for the estimation, and the energy equivalent of the animal tissues. Turner and Chew (1981) had to make a number of assumptions concerning survival of neonate animals, etc. in order to estimate secondary production for a small number of small mammals. Similar problems were encountered in estimation of secondary production of a single lizard species (*Uta stansburiana*) and bird communities in southeastern Arizona. In a study of secondary production by a creosotebush specialist desert grasshopper *(Bootettix punctatus)* in southern Nevada, Mispagel (1978) was limited to samples from 4–12 shrubs. He acknowledged that approximately 15% of the early instars and up to 30% of the adults avoided capture by the vacuum sampling method employed. The errors associated with sampling animal populations and the assumptions that have to be made in order to estimate secondary production demonstrate that an accurate estimate of secondary production for a desert ecosystem is virtually impossible. In order to make some estimate of secondary production in deserts, Turner and Chew (1981) used production efficiency estimates for desert animals (efficiency of converting biomass consumed into animal tissue). They estimated that less than 10% of the primary production is consumed as live tissue and estimated that secondary production was only 1% of the above-ground net primary production. That estimate is probably reasonable for most desert ecosystems but is probably low for desert grasslands where much more than 10% of the above-ground net primary production may be consumed by herbivores.

One of the most comprehensive and detailed studies of consumption of net primary production and energetics of consumers was a study of a small mammal community in the Chihuahuan Desert of southeastern Arizona. Small mammals were estimated to use 2% of the net primary production. Since not all of the net primary production is available to consumers, it was estimated that small mammals used 5.5% of the available net primary production. Because of the large proportion of seed consumers in the Chihuahuan Desert small mammal community, the estimate of seed biomass consumed by small mammals was 86% (Chew and Chew, 1970).

An attempt to relate biomass production of canopy arthropods to growing season rainfall yielded very different results for the arthropod faunas on creosotebush and on mesquite (Fig. 8.1). The biomass of canopy arthropods was very low during a year with below-average growing season precipitation. The biomass of creosotebush canopy arthropods peaked in May during an above-average growing season rainfall year. The biomass of mesquite canopy arthropods peaked in September of that year (Fig. 8.1). The next year the biomass of canopy arthropods on creosotebush declined rapidly between May and July and stabilized at a level equal to that of the insects on mesquite. The biomass production of canopy arthropods was less than 3% of the estimated net above-ground primary production of those shrubs.

One important feature of secondary production in desert ecosystems that is rarely addressed is the role of spatial heterogeneity in providing refugia for episodic species. Episodic species are characterized by high reproductive rates and high rates of population increase during periods of high primary production. These are 'immigrant' species that can account for more than half the biomass of a consumer group at times when productivity is high (Whitford, 1976) (Fig. 8.2). In the Chihuahuan Desert, species of cricetid rodents that were rare or absent from the small mammal community during dry years, migrated into marginal habitats during wet years and accounted for more than half the small mammal biomass in the second successive 'wet' year. The species that migrated into marginal habitats exhibited explosive population increase in those habitats. These episodic species survive as small 'seed populations' in more mesic habitat patches in the landscape. In landscapes where refuge patches are absent or widely scattered, there is

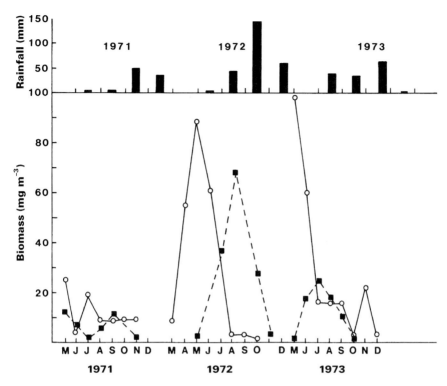

Figure 8.1   Biomass of arthropods per unit volume of shrub on creosotebush, *Larrea tridentata* (○), and mesquite, *Prosopis glandulosa* (■). (Data from Whitford 1975.)

no source of episodic species to colonize suitable habitat during favorable years. In these landscapes, secondary production of small mammal and lizard populations lag behind the production peak and a larger fraction of that production is shunted to the decomposer organisms in the food web. Landscape heterogeneity in terms of 'source' and 'sink' units probably contributes to the temporal tracking of primary productivity by other vertebrate and invertebrate consumer groups.

In a review of consumers in ecosystems, Chew (1974) concluded that in most ecosystems consumers process only a small fraction of the net primary production as live material. Even in grass-dominated rangeland, domestic livestock were estimated to remove only 30–45% of the available forage and ungulates on African grasslands were estimated to consume 20–60% of the net primary production. In deserts, the best estimates are approximately 10% of the net primary production consumed as live material.

Chew (1974) argued that consumers play important roles in ecosystems as regulators of ecosystem processes rather than as movers of energy. He suggested that in order to predict ecosystem behavior, it is essential to know how consumers cause

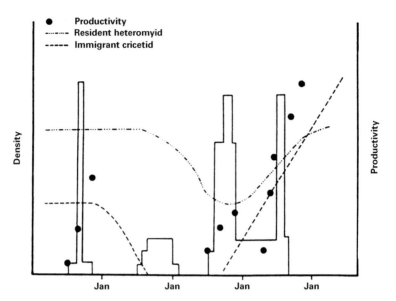

Figure 8.2   The relationship between rainfall (bars) and changes in population densities of resident heteromyid rodents and immigrant cricetid rodents in northern Chihuahuan Desert ecosystems on a small watershed. (From Whitford, 1976.) With permission from Allen Press, Inc.

departures from the linear transfers of energy depicted in most conceptual models of energy flow in ecosystems. Although he did not use the term ecosystem engineer, Chew (1974) called for studies of consumer effects on ecosystem properties and processes. The ways in which consumer species affect ecosystem properties and processes has recently been called 'ecosystem engineering' in a call for focusing studies of consumers on their roles in ecosystems (Lawton, 1994; Jones *et al.*, 1994). There are a number of indirect ways in which consumers affect ecosystem processes such as energy flow and nutrient cycling. Animals that build nests in the soil, burrow, dig caches, or dig for food affect the patch structure and productivity of ecosystems. Galleries, tunnels and chambers provide for bulk flow and increased infiltration of water into soil. Pits and burrows function as litter and seed traps. These traps become nutrient-rich hot spots and 'safe' germination sites for some plant species (Whitford and Kay, 1999; Whitford, 2000). It has been suggested that 'all ecosystems are probably modulated and modified to a significant extent by at least one species of engineer' (Lawton, 1994).

The ways in which consumers can affect ecosystem processes include many direct and indirect mechanisms. Consumers modify growth forms (pruning or grazing), serve as carbon sinks (sucking herbivorous insects), shift plant carbon allocation patterns to reproductive structures or to roots, consume seeds thereby affecting seed banks, transport and disperse seeds and fruits, and directly or indirectly affect plant species diversity. Consumers also affect productivity and plant species composition by their effects on patterns of soil nutrients and/or water infiltration. The many ways in which consumers can potentially affect ecosystem structure and function calls for examination of consumers and consumption as rate regulators and process modifiers rather than as transformers of energy.

Another approach to energy flow and consumer effects on ecosystems that has received considerable attention is theoretical and empirical studies of food webs (Polis, 1994). Food web studies have described trophic relationships and the many facets of food web interactions that affect biological community structure and ecosystem function. These include keystone predation and herbivory, intermediate disturbance hypothesis, intraguild predation, apparent competition, and the importance of indirect effects. Theoretical (modeling studies) food web research makes predictions that 'omnivory should be relatively rare in those webs that have persisted in nature' (Polis, 1994). In desert ecosytems, omnivory appears to be the rule rather than the exception. Many if not most vertebrates switch from plant to animal foods seasonally or during exceptionally dry or wet periods. Many desert carnivores include significant quantities of seeds or fruits in their diets (example, coyotes and foxes). Many small and intermediate size desert mammals feed on arthropods and plant materials. Most desert seed-eating birds are opportunistic and consume large quantities of arthropods when they are available. Many desert ant species are omnivorous, feeding on arthropods, arthropod excretions and plant exudates. Polis (1994) contends that real webs exhibit 2–4 orders of magnitude more species and trophic links than considered by theoretical food webs. Simplified (modeled) food webs greatly underestimate complexity, connectivity, looping trophic interactions, and omnivory. The relative importance of these components of food webs can only be evaluated by incorporating the abiotic regulation of productivity and productivity patterns with understanding and documentation of the natural history of consumers and the long-term variability in feeding relationships of consumer species.

8.2 FOLIAGE CHEWERS AND BROWSERS: VERTEBRATE AND INVERTEBRATE

Desert herbivores require three basic resources from the foliage: water, energy, and nutrients. The availability of these resources varies temporally and spatially and frequently asynchronously, that is peak water and energy content may not occur at the same time as peak nutrient content, etc. A further complicating factor is the temporal and spatial variability in the concentration of a variety of compounds that serve as feeding deterrents. The foliage of most desert woody perennials contains chemicals that reduce the palatability or digestibility of these tissues. In contrast leaves and stems of grasses contain few if any such compounds. However, some species of grasses have alkaloids in the seedling leaves that are not present in leaves of mature grasses (Crawley, 1983). Perennial herbaceous plants probably contain toxins or compounds that reduce digestibility and some annual herbs contain toxins, etc., whereas others do not. The compounds range from antioxidants, such as dihydroguaiaretic acid in creosotebush, *Larrea tridentata,* to alkaloids, such as those in *Astragalus* spp. which poison herbivores, and feeding deterrents such as tannins that reduce the digestibility of the tissues. Toxins and feeding deterrents tend to be concentrated in the epidermal layer and not in the mesophyll or fluids in conductive tissues. Rhoades (1977) reported on the antiherbivore chemistry of *Larrea* spp. from Arizona and Argentina. His data showed that the resinous coating on the leaves of *L. tridentata* and *L. cuneifolia* repelled leaf-chewing insects. The resins were mainly phenolic agylcones which have tannin-like protein-complexing and starch-complexing properties and which reduce the digestibility of protein. The resin concentrations were higher in the younger leaves than in mature leaves.

The foliage chewing-browser food webs usually contain only one or two large or intermediate size mammals, i.e. jack rabbit and prong-horned antelope in the Chihuahuan Desert; cape hare and oryx in the Namib (Seely and Louw, 1980). Water contents of perennial plant tissues vary both seasonally and diurnally and in temperate deserts, many species are winter deciduous and/or enter a temperature-induced dormancy. The feeding behavior of herbivores may reflect these changes in degree of hydration of plant tissues and the species composition of the landscape unit. For example, the most abundant nondomestic mammalian herbivore in North American deserts, black-tail jack rabbits, *Lepus californicus*, feeds on herbaceous plants and grasses when these are available and succulent. During the dry winter in the Chihuahuan Desert, jackrabbits consume the new stems of *L. tridentata*, selecting those plants with the highest water content in the stems (Steinberger and Whitford, 1983). Black-tail jackrabbits appear to be the only mammal that consumes creosotebush. They consume parts of the stem and drop the remaining stem, leaves and twigs to the ground. During the dry winter months, jackrabbits and cottontails clip stems of creosotebush, tarbush, snakeweed (*Gutierrezia* spp.), mormon tea (*Ephedra trifurca*) and Torrey ephedra (*Ephedra torreyana*). Rabbits and woodrats (*Neotoma* spp.) peel the bark and cambium from mesquite stems during the winter probably to obtain water. Rabbits consume parts of the stem of perennial shrubs and drop the remaining stem and leaves to the ground. Stem water contents of shrubs that were pruned by rabbits were higher than the stem water contents of neighboring unbrowsed shrubs. All the creosotebushes that received irrigation water during the late summer were completely pruned by rabbits during the following winter. Shrubs not receiving spill-over irrigation water were not pruned by rabbits (Steinberger and Whitford, 1983). In the Mojave Desert, during the summer dry season, *L. tridentata* shrubs on irrigated plots were heavily pruned by jackrabbits (Johnson *et al.*, 1978). By consuming the small stems the jack rabbits avoid the resinous leaves which are dropped on the ground as wastage. During the winter in the Chihuahuan Desert and summer in the Mojave Desert, young stems of woody plants are one of the few sources of water for herbivores that are dependent on preformed water in food to maintain water balance (Fig. 8.3a and b).

Figure 8.3a and **b**   Creosotebushes, *Larrea tridentata*, that have been pruned by black-tailed jackrabbits, *Lepus californicus*, in the Chihuahuan Desert, New Mexico.

Creosotebushes that have a history of being browsed by jackrabbits have a high probability of being browsed again (Ernest, 1994). In her study, Ernest (1994) found that where creosotebushes were pruned by investigators, jackrabbits continued to prune only those with a history of pruning by jackrabbits. The jackrabbits did not feed on the shrubs pruned by the investigator. These data were interpreted as enhanced resistance to pruning by jackrabbits by an induced response to herbivory called constitutive resistance (Ernest, 1994). However, differences in water status of plants selected by rabbits and those pruned by investigators were not measured. The difference in stem water content among individual shrubs is a more logical explanation of the pruning patterns by rabbits. Considering the observation that pruning of creosotebushes is highest during the driest periods in different deserts and that pruning rates are higher in dry years than in wet years, it is likely that the water status of the plant is the basis of selection by jackrabbits rather than any differences in chemical profiles as constitutive resistance.

Additional evidence that rabbits select plant tissues for water content during dry seasons was supplied by a study of rabbit herbivory on cacti (*Opuntia violacea*). Both jackrabbits and desert cottontails (*Sylvilagus auduboni*) consumed cladodes of *O. violacea* during dry seasons when there was little annual or perennial herbaceous biomass available (Hoffman *et al.*, 1993). Rabbits did not feed on cacti with three or fewer cladodes and individual plants with more than 20 cladodes were browsed more frequently than were small individuals. Plant size and not spinescence was correlated with rabbit browsing. Rabbits avoided browsing on new cladodes and more than 80% of the cladodes produced in May survived for six months. About 51% of the *O. violacea* plants were browsed during the study. It was concluded that although heavy browsing had a negative effect on cladode, flower, and fruit production of some individuals, rabbit herbivory had no long-term effect on growth, reproduction, and recruitment in populations of this cactus (Hoffman *et al.*, 1993).

Jackrabbits prune branches from the same shrubs within one or two years. The creosotebushes that are pruned by jackrabbits exhibit compensatory growth. The production of new stem and leaf biomass on pruned shrubs is equal to or greater than the biomass production of unpruned shrubs. Compensatory growth was reported in Great Basin (cold North American desert) shrubs that were completely eaten by black-tailed jackrabbits. Most of the above-ground plant of winterfat (*Ceratoides lanata*) and green rabbitbrush (*Chrysothamnus viscidiflorus*) were consumed by jackrabbits during the winter. At the end of the following growing season there were no differences in biomass of *C. viscidiflorus* in protected and browsed plots and the new growth of browsed *C. lanata* was significantly greater than the new growth of protected plants (Anderson and Schumar, 1986). Not all Great Basin shrubs exhibit compensatory growth in response to stem clipping. Sagebrush (*Artemisia tridentata*) died after several bouts of clipping but five other species exactly compensated for lost tissues. Loss of up to 90% of the buds of the previous years growth did not limit production of new growth (Wandera *et al.*, 1992).

Defoliation and stem consumption of woody shrubs in arid regions does not always result in compensatory growth. Defoliation of rapid-growing deciduous shrubs resulted in increased resistance to insect attack the following year. Leaf nitrogen and phosphorus concentrations decreased and leaf phenol and condensed tannin concentrations increased in the defoliated shrubs the following year (Bryant et al., 1991). Defoliation of slow-growing shrubs resulted in lowered resistance to insect attack during the following year. In slow-growing shrubs, concentrations of leaf nitrogen and phosphorus increased and concentrations of phenols and tannins decreased in the year following defoliation. These results were interpreted as supporting the hypothesis that plants maintain a balance of carbon and other nutrients rather than the hypothesis that defoliation results in long-term induction of chemical deterrents in the affected plants (Bryant et al., 1991). South African Karoo shrubs exhibited a similar variable response to browsing removal of meristematic tissues. A palatable, rapid regrowth, deciduous species, *Osteospermum sinuatum*, exhibited no change in polyphenols and tannins. Shrubs that exhibit slower regrowth responses developed higher levels of tannins and polyphenols the year after simulated browsing (Stock et al., 1993). These results were interpreted as passive alterations in plant chemistry rather than an active defense response to herbivory.

Milton (1991) has argued that the origin of spinesence in the Karoo Desert of Southern Africa is based on the selectivity by mammalian herbivores of plants with higher water status. The selectivity of plants with higher water or nutrient status by mammalian herbivores may override the importance of toxins as feeding deterrents. These studies clearly demonstrate the importance of not placing too much emphasis on one set of physiological or morphological characteristics of plants when attempting to interpret variation in the behavior of herbivores.

Spinescence is not a complete herbivore deterrent in plant species that produce moist, nutrient-rich foliage. Several African *Acacia* spp. and other spinescent genera of plants were readily eaten by large and intermediate-size ungulates. Kudu (*Tragelaphus strepsiceros*), a large (180 kg) antelope, bit off short ends of branches despite the thorns. Kudu reduced their eating rates to avoid damage by the thorns. The medium-size impala (*Aepyceros melampus*) restricted its bite size to single leaves or leaf clusters. The spinescent plant species had higher crude protein in their foliage than the unprotected species. The acceptability of spinescent species to the antelopes was essentially the same as the palatable species without thorns or spines (Cooper and Owen-Smith, 1986). Spines provided little protection to mesquite (*Prosopis glandulosa*) in the Sonoran Desert where jackrabbits were the largest herbivores. Shoots were entirely clipped off by rabbits. Rodents (mostly wood rats, *Neotoma* spp.) girdled main shoots causing death of the stem complex if the stem was completely girdled. Mesquite spines offer no protection from insect herbivory and leaves and terminal meristems were consumed by several species of insects (Nilsen et al., 1987).

In the winter-rainfall Negev Desert Dorcas gazelles (*Gazella dorcas*) concentrate their feeding on the succulent growing types of shrubs in summer (the dry period in the Negev) to get water. In winter during the rainy season, Dorcas gazelles use leaves and pods of *Acacia* spp. and other shrubs and the time spent foraging in a patch is correlated with shrub density (Baharov and Rosenzweig, 1985). Water content of potential forage also influences forage choice in arid zone kangaroos (Denny, 1982). Availability of green succulent foliage may be a critical factor in the reproduction of some animals, e.g. desert granivorous rodents. Thus for foliage-feeding herbivores availability of water may determine the plant species or plant tissues consumed.

The quality of perennial grasses varies considerably among species and seasonally. Some grasses at the end of the growing season resorb virtually all of the labile nutrients, storing these materials in the roots. In such species, the standing dead material is of little value to most herbivores usually because of the low nitrogen content. Such species are suitable forage for most herbivores only when green and growing. Other species of grasses have silica contents, which reduce palatability to herbivores. The presence of large quantities of senescent material in grass clumps can also reduce the availability of new green material to grazing herbivores. Since the phenology of perennial grasses varies with rainfall and temperature, climatic variability imposes unpredictability in the temporal and spatial availability of quality forage for desert herbivores. Most desert herbivores utilize both grasses and shrubs in order to be successful in desert environments.

In most arid and semiarid regions, grasshoppers make up the most abundant and largest biomass of herbivorous insects that chew foliage. There are other groups of insects that feed on foliage and on occasion these insects can be locally abundant. One example of such insects are the crysomelid beetles (*Zygogramma tortusa*) which occasionally reach outbreak population size and defoliate a large percentage of their preferred host plant, tarbush, *Flourensia cernua* (Estell *et al.*, 1996). However, crysomelid beetles and other leaf chewers generally make up only a small proportion of the insect herbivore biomass.

In a comparison of grasshoppers of the Sonoran Desert in North America and the Monte Desert in Argentina, ground-dwelling species had broader diets than vegetation-inhabiting species (Otte and Joern, 1977). One species, *Trimerotropis pallidipennis,* which is common in both deserts, had a broad diet, which was attributed to the fact that this grasshopper lives in a wide range of desert habitats. Otte and Joern (1977) reported that grasshoppers that feed primarily on annual plants, are forced, during times of absence or low densities of annual plants, to feed on perennials like *Larrea* and *Franseria*, genera with high concentrations of feeding deterrents and toxins. The last plants to be affected by drought are shrubs and trees. Switching from feeding on grasses and herbaceous plants to shrubs may be necessary for grasshoppers to remain in water balance during droughts. Shrub- and grass-inhabiting desert grasshoppers tend to be food specialists and to occur at low densities. Such species are generally cryptic and infrequently taken by predators. The more generalized soil-loving species like *Trimerotropis pallidipennis,* also tend to be cryptic by matching the roughness and color patterns of the soil (Whitford *et al.*, 1995).

Shrubs may also be pruned by insects that utilize dead stems as oviposition sites. Pruning can result in compensatory growth by a shrub. Compensatory growth was demonstrated in a study of the effects of a twig-girdling beetle (family Cerambycidae) on mesquite. Cerambycid beetles chew a complete girdle through the bark, cambium and phloem of terminal branches of mesquite. The girdle cuts off the supply of water and nutrients and the mobilization of sap to the portion of the stem above the girdle (Figs 8.4 and 8.5). The beetles oviposit in the girdled stem thereby avoiding the defense mechanisms of the plant that would encase the developing larva in a resinous tomb. The larvae feed on the dead wood of the girdled stem. In a study of the production of new stems on mesquite shrubs that had been subjected to natural girdling or simulated girdling of 40–80% of the stems, there were no differences in shoot and leaf growth of girdled plants compared to controls (Whitford *et al.*, 1978). Lateral nodes below the girdle assumed apical dominance and produced new stem leaders. Frequently more than a single terminal stem developed below a girdle. A similar response was reported for the effects of livestock herbivory on twig dynamics of a Sonoran Desert shrub, *Simmondsia chinensis* (Roundy and Ruyle, 1989). This shrub also regrew from lower lateral or apical buds after the terminal twig was browsed.

Figure 8.4   Dead terminal stems on a mesquite, *Prosopis glandulosa,* shrub resulting from girdling of stems by a beetle (Cerambycidae), Chihuahuan Desert, New Mexico.

Figure 8.5   Close-up of a girdled mesquite, *Prosopis glandulosa*, stem.

8.3 CANOPY INSECT COMMUNITIES

Although there are few published reports on the taxonomic composition and relative abundance of insects on desert plants, there are limited data for several desert regions. There are many similarities in the insect faunas on desert plant canopies. The assemblages of insect populations on desert plants vary according to plant type as suggested by the 'appearance' hypothesis (Rhoades and Cates, 1976; Feeny, 1976). Perennial plants are 'apparent' meaning that they are always present and a seemingly reliable source of food for herbivorous insects. Because of this 'apparency' perennial plants must protect their leaves and to a lesser extent their roots and stems from being consumed by herbivores. Foliage may be protected by toxins, such as alkaloids, feeding inhibitors, such as tannins, or simply have a C:N ratio that is too large for most foliovores to survive on (i.e. quality is too low).

Despite the chemical protection of the foliage of desert shrubs, such shrubs support a diverse insect fauna. The Chihuahuan Desert drought-deciduous shrub, *Flourensia cernua*, is a well-documented example of food quality and chemical defenses against herbivory. In *F. cernua*, the fiber fraction (fiber and lignin) and tannins increase in concentration during growth and maturation of foliage. Total phenolic content does not vary with growth stage but did vary among years. Phenolic concentration varied from 0.06 g g^{-1} leaf tissue to 0.96 g g^{-1} leaf tissue during the three years that plant chemistry was measured (Estell *et al.*, 1996).

Despite the feeding deterrent chemistry of the foliage of this species, *F. cernua* supports a diverse arthropod fauna dominated by sucking herbivores. Chewing herbivores, mostly chrysomelid beetles and Lepidoptera, accounted for 12–22% of the arthropod biomass. Sucking herbivores mostly Homoptera: coccids, *Lygus* sp. and thrips were 47–63% of the biomass, omnivorous ants contributed 1–20% of the biomass, and predators (wasps and spiders) varied between 2% and 14% of the arthropod biomass (Schowalter, 1996). The insect fauna of two species of snakeweed (*Gutierrezia* sp.), a small shrub that is toxic to livestock and has a high concentration of phenolics, is dominated by sucking herbivores (Homoptera: Acanaloniidae, Aphidae, Cicadellidae, Eriococciddae, Membracidae, Pseudococcidae, and Psyllidae) and chewing herbivores (Coleoptera: Chrysomelidae, 12 species and Curculionidae, 12 species) (Foster *et al.*, 1981).

An example of the diversity of the insects that utilize desert shrubs as a food source or as a hunting substrate (predatory arthropods) was provided by a qualitative study of the faunas associated with North American and South American species of *Larrea* (Schultz *et al.*, 1977). There were notable similarities in the faunas on *Larrea* in both areas. For example both *L tridentata* and *L cuneifolia* supported one or more species of cryptic grasshoppers (branch mimics and/or light-dappled foliage coloration). Each species of *Larrea* supported several species of insects that prefer to feed on flowers, several kinds of beetles that feed on foliage or flowers (examples: tenebrionids, scarabs, and buprestids). Flowers of both species may be taken by fungus-growing ants. Several species of ants tend homopterans on both *Larrea* species. Some differences were also recorded. *L. tridentata* supported a walking stick (Phasmidae), a katydid (Tettigoniidae) and several lepidopterans that were not represented in the fauna on *L. cuneifolia*.

Desert plants are no exception to these general hypotheses with respect to secondary compounds and leaf palatability. Herbivores that can avoid the chemical deterrents in the epidermal cells of leaves are those with piercing and sucking mouth parts that feed on phloem sap. These are primarily Homopteran insects but include some Hemiptera. In general the most abundant herbivorous insects on desert perennials are species in these two orders of insects. Sucking herbivores generally account for more than half of the numbers and biomass of insects on shrubs and surprisingly, on perennial grasses The most numerous taxa in all reports are species of insects that feed on cell sap or phloem sap (sucking herbivores). Species of Hemiptera and Homoptera dominate the canopy insect faunas (Lightfoot and Whitford, 1991; Milton, 1993; Boldt and Robbins, 1994; Schowalter, 1996; Schowalter *et al.*, 1999). Sucking herbivorous insects are generally the most abundant taxa on desert plants but occasionally large numbers of chewing insects dominate the fauna (Table 8.1). The predominance of sucking herbivores feeding in the foliage of desert shrubs is usually attributed to the concentrations of feeding deterrents (terpenes) or toxins in the epidermal cells of the leaves. However, data from current studies and from the US/IBP Desert Biome studies (Tables 8.1 and 8.2) clearly show that the foliage-feeding insect fauna of grasses has the same general composition as the fauna found on shrubs.

Table 8.1

Sucking Herbivores, Percentage of Total Biomass of Arthropods Collected by D-Vac Suction Collector, on Shrubs and Grasses During a Wet Year in the Northern Chihuahuan Desert. (Data from Whitford, 1975)

	% of total biomass						
Species	Mar	Apr	May	June	July	Aug	Sept
Ephedra trifurca (shrub, evergreen)	–	–	23.4	–	29.0	31.6	15.2
Prosopis glandulosa (shrub, deciduous)	–	–	57.7	81.5	87.6	84.7	56.8
Flourensia cernua (shrub, deciduous)	84.5	–	83.7	47.0	53.8	14.3	61.0
Fallugia paradoxa (shrub, deciduous)	35.7	–	49.0	2.4[a]	6.5[a]	0[a]	14.5
Larrea tridentata (shrub, evergreen)	–	70.9	99.4	–	9.0[b]	57.6	75.0
Hilaria mutica (perennial grass)	–	90.3	–	–	22.8	–	24.4
Panicum obtusum (perennial grass)	–	95.7	–	–	25.6	94.6	10.8

[a] Samples on these dates contained large numbers of Chrysomelid beetles.
[b] Sample biomass of chewers skewed by two large grasshoppers.

Table 8.2

Biomass of Sucking Herbivores Collected by D-Vac Suction Collector, on Shrubs and Grasses During a Wet Year in the Northern Chihuahuan Desert

	Biomass (g 100 m^{-3} shrub or 100 m^{-2} grass)						
Species	Mar	Apr	May	June	July	Aug	Sept
Ephedra trifurca	–	0	0.1	–	0.2	0.2	0.2
Larrea tridentata	–	4.9	0.5	–	4.3	1.1	0
Prosopis glandulosa	–	0	–	3.5	6.2	2.5	0.3
Fallugia paradoxa	1.1	–	3.3	0.3	0.3	0	2.2
Flourensia cernua	16.3	–	5.3	4.5	0.4	0.1	0.6
Gutierrezia sarothrae	–	–	6.9	–	2.3	0	0.7
Hilaria mutica	–	0.71	0[a]	0[a]	0.23	0	0.32
Panicum obtusum	–	2.1	0[a]	0[a]	1.6	1.0	0.1

[a] No green foliage present on this sampling date.

The presence of terpenes and other feeding deterrent compounds in the leaves of shrubs is an important factor selecting for specialized, monophagous species of insects that feed on a shrub. In a study where the insect fauna was classified by feeding niche breadth where possible, 25–28% of the insects on desert *Baccharis* spp. were polyphagous, 13–16% were oligophagous, and 2–13% were monophagous (Boldt and Robbins, 1994). Insects were categorized as polyphagous if they were recorded from plants of several families; oligophagous if they were recorded only from plants in the family Asteraceae, and monophagous if they were recorded only from plants in the genus *Baccharis*. Of the insects collected from desert *Baccharis* spp. 66% fed on leaves and stems, 20% fed on external or internal parts of flowers, 11% were stem or root borers or gall formers, and 3% were leaf miners. This proportional breakdown of feeding strategies of canopy insects is probably fairly general (Milton, 1993; Schowalter, 1996). The comparison of insect faunas on a drought-deciduous palatable shrub (palatable to hares and sheep) (*Osteospermum sinuatum*) and an evergreen unpalatable shrub (*Pteronia pallens*) showed that the most common insect on the palatable shrub was a leaf beetle (*Cassida* spp., Chrysomelidae) which fed on new growth at any season (Milton, 1993). The most common insects on the unpalatable shrub were scale insects and plant hoppers that were tended by three species of ant. Liquid feeding, homopteran tending, ants are numerous in samples of insects collected from desert plants. The most abundant species of ants in most desert areas are species that tend sucking insects (Whitford *et al.*, 1999). These species compete for rich food resources such as concentrations of aphids on *Yucca* flowers (Van Zee *et al.*, 1997).

Populations of phytophagous insects are thought to be influenced by the predictability of plant resources in space and time, the diversity of plant species within a community and the spatial and temporal relationships of individual plants within patches (Cates, 1981). Long-lived perennials are examples of predictable resources and ephemeral or annual plants are unpredictable resources. The more predictable plant species differ in terms of their allocation of energy to chemical defenses and in the types of products that serve as defensive chemicals (Cates and Rhodes, 1977a). Even polyphagous herbivorous insects exhibit preferences. In his review, Cates (1981) found that the six polyphagous species of insects, for which data were available, all exhibited preferences for the least abundant annual and perennial herbaceous species of plant in the community. In a comparison of herbivorous insects on mesquite, *Prosopis* spp., the larvae of some Lepidoptera preferred mature leaves of a species with low alkaloid contents whereas other species of larvae exhibited no correlation between alkaloid content and feeding preference. In general, polyphagous herbivores prefer mature leaves despite the known or postulated lower nutritional value, lower water content, and high concentration of digestibility-reducing substances (like tannins) than younger leaves (Cates and Rhodes, 1977b). This pattern was without exception for polyphagous herbivores feeding on woody perennials. However, there were exceptions to this generalization in polyphagous insects feeding on annuals

(Cates, 1981). Although predictability in space and time is an important variable acting as a determinant of feeding preferences in insects, those preferences are shaped by a suite of other factors including type and concentration of feeding deterrent compounds, and water and nutrient content of leaves.

Time lags in peak populations of herbivorous insects in relation to productivity of desert plants have been documented in studies of a mirid bug (*Capsodes infuscatus*) and the geophyte (*Asphodelus ramosus*) in the Negev Desert. This insect feeds on inflorescence meristems, flowers, and fruits and in some years destroyed more than 95% of the potential fruit production. The abundance of the mirid bug was related to fruit production but with a one year time lag. Years in which the geophyte produced a large number of inflorescences, the mirid bugs exhibited high per capita reproduction. The following year, the high population of mirids destroyed most of the fruit production and had low per-capita reproduction (Ayal, 1994). This kind of time lag is thought to be common among desert insect herbivores that specialize in using ephemeral resources.

It is generally observed that defoliation of desert plants is a very rare occurrence (Watts *et al.*, 1982). Defoliation outbreaks typically last one season. This is true of the most famous (or infamous) insect, the migratory locust (*Schistocerca gregaria*) in the Sahel of Africa and probably true of most other 'outbreak' species. One species of desert shrub, *Flourensia cernua*, experiences loss of up to 30% of its foliage to lepidopteran larvae and to adult chrysomelid beetles (*Zygogramma tortuosa*) (Schowalter, 1996). The frequent defoliation of this species has been attributed to the higher N content of the foliage than is found in other desert shrubs (Estell *et al.*, 1996). In more than 30 years of field research in the northern Chihuahuan Desert, I have observed only one episode of massive defoliation. That defoliation episode was by a species of Meloid beetle feeding on mesquite (*P. glandulosa*) foliage. Active defoliation by these beetles lasted less than one week and covered an area of approximately 10 km². Much of the vegetation in deserts produces foliage even in drought years (drought-resistant woody perennials). However, because drought affects the physiology of the plants, foliage insect populations may suffer dramatic reductions during stress periods.

The complex factors affecting feeding preferences and host plants for insects has led to controversy concerning the regulation of insect abundance (and diversity) on plants: top-down or bottom-up) (Hunter and Price, 1992; Matson and Hunter, 1992). Bottom-up regulation of insect abundance is via the quality of the plant tissues used as food by the insects. Quality varies inversely with concentrations of toxins and feeding deterrents and directly with carbohydrate and protein content. Top-down regulation is by a myriad of arthropod and vertebrate predators. The abundances of herbivorous insects on creosotebush (*L. tridentata*) in the Chihuahuan Desert, were reduced by the presence of arthropod and avian predators (Floyd, 1996). However, avian predators depressed herbivorous insect populations in only one of the two years of the study. Despite this inconsistency, this manipulative study, using exclusion cages and hand removing predatory arthropods, provides direct evidence that some part of the regulation of insect abundance is 'top-down'.

There are two conflicting hypotheses for the foliage quality mechanisms affecting abundance of insects on plants: (1) the plant stress hypothesis: when plants are stressed by changes in weather patterns, they become a better source of food for invertebrate herbivores, and (2) the plant vigor hypothesis: vigorous plants are more suitable hosts for phytophagous insects. The plant stress argument has been expanded to suggest that stressed plant tissues break down and mobilize nitrogen thus making nitrogen more readily available to arthropods than if feeding on unstressed plants (White, 1984). Studies of insect populations on creosotebush (*L. tridentata*) shrubs that were irrigated and fertilized with nitrogen provided evidence that high quality shrubs supported larger more diverse insect populations than untreated controls (Lightfoot and Whitford, 1987). Simply irrigating the shrubs had no effect on the abundance of foliage arthropods. The leaf-chewing insects exhibited little response to changes in plant quality, but the sap-sucking insects accounted for the differences in abundance among treatments. Creosotebushes growing in high nutrient status sites were larger, had denser foliage, greater foliage production, higher concentrations of foliar nitrogen and water and lower concentrations of foliar resin than low nutrient shrubs. The abundance of both herbivores and predators were higher on the high nutrient status shrubs (Lightfoot and Whitford, 1989). This same pattern was reported comparing creosotebush canopy arthropods on roadside plants and distant plants (Lightfoot and Whitford, 1991). They concluded that the positive relationship between creosotebush productivity and foliage arthropod abundance and biomass is contrary to the hypothesis that physiologically stressed plants provide better quality foliage to insect herbivores.

A more definitive test of the plant stress and plant vigor hypotheses was provided by a study of insect populations on creosotebush shrubs exposed to complete elimination of summer rainfall by 'rain-out' shelters and creosotebushes that were irrigated at double the annual rainfall and untreated controls (Schowalter *et al.*, 1999). Of the 44 arthropod groups evaluated, only two taxa (including a lepidopteran folivore) exhibited significant negative response to water availability. These two taxa were the only ones that responded as predicted by the 'plant stress hypothesis'. Ten taxa (including eight phytophages) responded positively to water availability, thereby supporting the plant vigor hypothesis. One phytophagous insect species exhibited a nonlinear response (variable response). That species supported neither hypothesis. These studies indicate that the numerical responses of insects to changes in plant quality is complex and cannot be generalized. The abundance and diversity of insects on any given desert plant species will vary with both 'top-down' and 'bottom-up' characteristics of the plant assemblage and the predator–parasitoid populations at that time. Thus the numerical responses of plant foliage insects is a function of patch characteristics and patch dynamics in response to a variable environment.

In the Arabian Desert of Abu Dhabi, the structure of desert insect communities was studied by light-traps placed in several different plant communities (Tigar and Osborne, 1999). The Arabian Desert insect communities were dominated by moths and butterflies (Lepidoptera) which accounted for 37.4% of the total arthropod biomass and beetles (Coleoptera) which accounted for 42.6% of the total arthropod biomass. Scarab beetles (Scarabaeoidae) accounted for 42.6% of the total insect biomass. One species of dung beetle, *Scarabeus christatus*, represented 33.7% of the total insect biomass (Tigar and Osborne, 1999). Scarab beetles use ungulate dung as oviposition sites. The abundance of one species of scarab beetle reflects the specialization of that species on the dung that is most abundant in these habitats. Because light traps sample only those mobile species that fly at night, the data on the arthropod communities are skewed to those species. Sucking herbivores (Homopterans) fly infrequently and hence are undersampled by this technique.

8.4 SURFACE ACTIVE ARTHROPODS

There are interesting similarities in the surface-active arthropod faunas of arid and semiarid regions (Table 8.3). The surface-active arthropod fauna is numerically dominated by ants in most desert areas (Crawford, 1988; Tigar and Osborne, 1997). Ants accounted for more than 75% of the surface-active insects captured in the extremely arid desert of Abu Dhabi (United Arab Emirates) (Tigar and Osborne, 1997). Even in areas where vegetation is sparse or virtually absent there are likely to be a few colonies of ants. In addition to ants, the arthropod fauna is a diverse. Ground-dwelling beetles are generally the next most abundant group of arthropods in desert regions. The most abundant species of ground-dwelling beetles are detritivores in the family Tenebrionidae. In the northern Chihuahuan Desert, the species richness and abundance of tenebrionid beetles was higher in sand dune habitats than in areas with sandy loam soils and vegetation cover of creosotebush (*L. tridentata*), mesquite (*P. glandulosa*) or mixed grasses (Crawford, 1988). The higher diversity and abundance of tenebrionids and several other taxa in the sand dune system was attributed to soil texture and the microenvironments characteristic of sand dunes. Crawford (1988) concluded that sand is a poor conductor of heat, which means that even shallow depths provide survivable temperatures on hot or cold days. In addition, the high infiltration rate associated with deep sands allows water to accumulate below the layer of intense evaporation. The sand burrowers included camel crickets, scorpions, and tenebrionid beetles (Crawford, 1988). Other studies have indicated that sandy soils are important for sand-swimming cockroaches (Edney *et al.*, 1974) and tenebrionid beetles (Seely and Mitchell, 1987). Two other factors affect the abundance of detritivores in sand dune environments: trapping and burial of wind blown detritus and the availability of thermal gradients that allow detritivores to select microsites where temperatures are optimum for the microbial gut symbionts to break down cellulose (Crawford, 1988). In shrub, shrub–grass mosaic, and desert grassland habitats, tenebrionid beetles feed on the refuse piles around the nests of seed-harvesting ants or on fine litter trapped behind litter dams. In these habitats, there is sufficient vegetation to provide moderate microclimates for the beetles during the hottest parts of the day.

The taxonomic composition and relative abundance of surface-active arthropod fauna in desert grasslands of New Mexico was very similar to that reported by Crawford (1988) for sand dune habitats in the northern Chihuahuan Desert (Table 8.3). The taxonomic composition and relative abundance were also similar to those reported for pit-fall trap grids in a creosotebush (*L. tridentata*) shrubland on a coarse sandy-loam bajada in the northern Chihuahuan Desert (Whitford, 1974). The taxonomic composition and relative abundance of the surface-active arthropod fauna of Abu Dhabi was similar to that reported for the Chihuahuan Desert (Tigar and Osborne, 1997). Data from pit-fall trap sampling of surface-active arthropods in arid western Australia showed that, excluding ants, beetles (Coleoptera) were the second most abundant arthropods. Orthoptera (crickets and cockroaches) were the next most abundant detritivores. Spiders were the most abundant predators (Lobry de Bruyn, 1993). These limited data suggest that desert surface-active

Table 8.3
Average Numbers of Surface-active Arthropods (Excluding Ants) per Hectare in Desert Grassland Habitats in the Northern Chihuahuan Desert for Three Consecutive Years with Annual Rainfall Amounts
(Data from the Jornada Long-Term Ecological Research Project)

Mean annual rainfall (mm)	343.4	297.7	250.3
Taxa	No. ha^{-1}	No. ha^{-1}	No. ha^{-1}
Arachnida			
Araneae (predators)	24.0	18.9	18.1
Scorpiones (predators)	16.5	19.6	16.3
Solfugae (predators)	15.3	3.3	2.7
Coleoptera			
Carabidae (predators)	16.5	7.8	5.7
Scarabidae (detritivores)	10.8	2.4	8.7
Tenebrionidae (detritivores)	584.4[a]	98.7	38.3
Orthoptera			
Gryllacridae (detritivores)	41.3	13.3	11.5
Polyphagidae (detritivores)	12.3	5.4	7.7
Gryllidae (detritivores)	4.5	12.0	11.0
Hymenoptera			
Mutilidae (adults – pollen, larvae – parasitoids)	93.1	15.3	25.2

[a] Most of the tenebrionids captured during this year were of one small body-size species (*Araeoschizus decipiens*).

arthropod faunas are dominated by coleopteran and orthopteran detritivores and by three groups of arachnid predators (spiders, scorpions, and solfugids).

Arid ecosystems may support large populations of collembolans. Epigeic collembolans feed on grasses and herbaceous plants and on plant litter. Temporal and spatial patterns of epigeic collembolans studied by pit-fall trapping on a watershed in the northern Chihuahuan Desert revealed no significant correlations between abundance and species diversity and either weekly or long-term precipitation patterns (Loring *et al.*, 1988a). During the cold-dry season (November through March) *Brachystomella arida* (Hypogastruridae) and *Cryptopygus ambus* (Isotomidae) were the most abundant collembolans on the watershed. During the warm-wet season (July through September) and the warm dry season (May–June), the epigeic collembolan assemblages were dominated by *Seira bipunctata* and *Entomobrya* sp. (Entomobryidae). The sminthurid, *Sminthurides pumilis*, was most abundant collembolan on the ephemeral lake bottom and in the mesquite shrubland at the lake periphery. The hypogastrurid, *B. arida* was most abundant in the black-grama grassland at the base of the desert mountain. The entomobryid, *S. bipunctata,* was the most abundant species on the creosotebush (*L. tridentata*) bajada and *Entomobrya* sp. was the most abundant species on the lower slopes of the watershed (Loring *et al.*, 1988a). The abundance of collembolans on this watershed suggests that the importance of these primitive insects may be missed in most studies of consumers in deserts.

Several groups of 'sit and wait' arthropod predators are undersampled or missed by pit-fall trapping, which is the most common method of studying ground-dwelling arthropods. Burrowing spiders and scorpions are globally distributed in arid and semiarid areas. The burrows of spiders and scorpions ensure a protective microclimate in a permanent nest. Burrowing spiders and scorpions are nocturnal predators and generally live their entire lives in a permanent burrow. Australian mygalomorph burrowing spiders are opportunistic feeders. Many of the Australian burrowing spiders feed on ants and in some habitats feed on termites. Large species take curculionid beetles and even small vertebrates (Main, 1982). Termite-feeding burrowing spiders locate burrows under tree canopies where litter accumulations attract harvester termites. In semiarid *Acacia aneura* banded woodland, large burrowing spiders concentrate their burrows in the interception zone of the groves and in the groves. Large burrowing spiders are frequent prey of the varanid lizard, *Varanus gouldii* which excavate the burrows to capture the spiders (Whitford, 1998). There is no evidence of lizards excavating spider burrows in the Chihuahuan Desert of North America. In the Chihuahuan Desert, the burrowing spider, *Geolycosa rafaelana* (Lycosidae) is contagiously distributed in the grass and grass–shrub mosaic habitats on a watershed. The abundance of *G. rafaelana*, was experimentally shown not to be related to food availability. The populations of this burrowing spider were regulated by parasitic wasps that entered the burrows, paralyzed the occupant, and laid eggs on the paralyzed spider (Conley, 1985).

The foraging strategy of desert scorpions is similar to that of burrowing spiders, i.e. they forage primarily from the entrance of their burrows. In the Negev

Desert, the scorpion, *Scorpio marus palmatus,* feeds mostly on isopods, ants and beetles (Shachak and Brand, 1983). These scorpions position burrows in favorable hunting locations near ant colonies or in patches of isopod burrows. Food availability and avoidance of predators are two factors that affect the spatial distribution of the desert scorpion, *Plaruroctonus mesaenis* (Polis *et al.*, 1986). In this species, the 0–1-year-old scorpions and intermediate age scorpions located burrows in areas with creosotebush (*L. tridentata*) cover whereas the adults were randomly distributed with respect to the shrubs. This distribution pattern was not attributable to prey availability because there was no difference in prey abundance in high and low vegetation cover areas. The high ground cover (leaf litter etc.) reduces the probability that the small and younger scorpions will be eaten by the older conspecifics (Polis *et al.*, 1986). The spatial distribution and abundance of scorpions and burrowing spiders appears to be a function of predator or parasitoid avoidance more than the availability of prey.

Desert ant communities share several features: numerical dominance by small generalists (omnivores), moderate densities of one or more species of seed-harvesting ants, and varying numbers of species that are primarily liquid feeders or fungus feeders (Chew, 1976; Schumacher and Whitford, 1976; Whitford, 1978; Breise and McCauley, 1977; Chew and DeVita, 1980; Marsh, 1985; Heatwole and Muir, 1991; Nash *et al.*, 1998; Whitford *et al.*, 1999; Rojas and Fragoso, 2000). In the Chihuahuan Desert of central Mexico, there were 14 species of omnivorous ants, nine species of granivorous ants, four species of predators, three species of honeydew feeders, and two species of fungus growers. Species richness varied between 11 and 26 species of shrubland and grassland vegetation types but neither abundance nor biomass were related to vegetation type (Rojas and Fragoso, 2000). The total species richness at the Mapimi site in Mexico (32 species) was considerably lower than reported for a watershed (50 species) in the northern Chihuahuan Desert, New Mexico (MacKay, 1991). Differences in species richness may be related to the heterogeneity of the landscape and to the sampling methods used.

In most deserts, the seasonal activity of ant species varies with the phenological patterns of host plants of honey-dew producing insects, flowering and seed production patterns, and the availability of moderate thermal environments for predatory species and omnivores. Peak diversity and abundance of ants foraging on plants occurs when the plants are flowering. Flowers attract insect prey and provide nectar for ants. In the hyperarid Namib, there was a close relationship between mean annual rainfall (15 mm yr^{-1} to 95 mm yr^{-1}) and ant species richness. In the Namib, most of the ants are seed feeders or honeydew-nectar feeders (Marsh, 1985). In the hyperarid Atacama Desert, ants are limited to ephemeral washes and are associated with the shrubs that occur in this habitat (Heatwole, 1996). In the North American Chihuahuan Desert abundance of active nests of seed-harvesting ants was correlated with seed drop in ephemeral plants. Abundance and species richness of active nests of omnivores and liquid-feeders were correlated with rainfall with several days lag following the event (Schumacher and Whitford, 1976; Whitford, 1978). Although, ant species richness is higher in the arid zone of Australia than in North American deserts, trophic group composition is similar with the exception that predatory ponerines are absent in North American deserts (Briese and McCauley, 1980).

8.5 LARGE HERBIVORES: GRAZERS

The principal use of desert rangelands is production of domestic livestock. There is a voluminous literature on grazing strategies, stocking rates and range management that is beyond the scope of a study of desert rangeland ecosystems. However, there are a number of attributes of grazing by large herbivores that parallel the attributes of native small grazers and invertebrate herbivores that are considered here.

Consumption of grasses and herbaceous plants by grazers can have a number of effects on vegetation and soils. Effects on vegetation include changes in growth and reproduction of individual species of plants, changes in species composition at the community level, and modification of patch relationships at the landscape scale.

The effects of grazers on plant production varies depending on the physiological characteristics of the species being utilized. For some species of grasses, there appears to be a beneficial effect of grazing on net primary production. Several field and laboratory studies have shown that primary productivity may be stimulated by grazing under some conditions (Dyer and Bokhari, 1976; McNaughton, 1979). Stimulation of primary productivity by herbivore grazing led to the development of the 'grazing optimization conceptual model' (Hilbert *et al.*, 1981) (Fig. 8.6). Compensatory growth of grasses and herbaceous plants in response to defoliation by herbivores has been hypothesized to vary with the level of stress experienced by plants when defoliated and by the length of time for recovery.

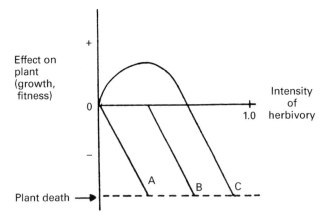

Figure 8.6 The grazing optimization conceptual model. The curve shows the change in production due to grazing. (McNaughton, 1979.) With permission by the University of Chicago Press.

Experimental defoliation of a grass species showed partial compensation when the plants were growing at high growth rates at the time of defoliation to over-compensation when the plants were stressed and growing at low rates. Time available for recovery of above-ground biomass recovery had a major impact on the outcome of defoliation. With short recovery time, regrowth did not compensate for loss to consumption. Longer time for recovery was suggested to result in a more positive response to grazing (Oesterheld and McNaughton, 1991).

A number of mechanisms have been proposed to explain increased net primary productivity in grazed plants. These mechanisms include: (1) increased photosynthetic rates in the remaining plant tissue or in the new tissue produced after grazing; (2) increased tillering or production from lateral buds following the removal of apically dominant plant parts; (3) opening of the canopy and increased light penetration; (4) shifts in photosynthate allocation to production of new leaves or photosynthetic tissue; (5) conservation of soil moisture because of the removal of transpiring plant tissue (Hilbert *et al.*, 1981). The most controversial and frequently cited effect of grazing is the stimulation of plant growth by saliva or components of the saliva of herbivores on plant tissues (Dyer and Bokhari, 1976) These studies suggest that bite wounds on plant tissues absorb some growth stimulating compounds in the saliva of grazing animals. Although this is an interesting phenomenon, the documentation has been limited to too few species of plants and herbivores to warrant considering this as a widespread mechanism for compensatory growth of grazed plants. It is also misleading to extrapolate from the results of these few studies to the responses of plant communities or landscapes to grazing by large herbivores. The effects of large herbivores on ecosystems should not be confused with the effects of large herbivores on individual plants or species of plants. Large herbivores graze at the patch scale, are managed at the paddock or pasture scale, and have the greatest impact at the ecosystem or landscape scale. Grazing by livestock and species of wildlife must be examined at the appropriate spatial scale in order to understand grazing as an ecosystem process.

Species of herbivores exhibit clear preferences for certain species of plants or growth forms of plants. Even under light grazing pressure from small populations of a grazing herbivore, some species of plants may be heavily defoliated whereas other species will not be eaten at all. Grazing by herbivores affects plants in complex ways that are reflected in the growth and allocation physiology of the grazed plants. These changes in physiology of grazed plants affect the competitive relationships with nongrazed plants, and their responses to changes in soil nutrients and physical properties of soils resulting from the activities of the grazers. The preference of a grazer for a plant species may change seasonally. Plant species that are palatable to a grazer in one season may be unpalatable to that species in another season. For example, some grasses are taken by grazers only when green whereas other species are taken either as green forage or as senescent material (Coppock *et al.*, 1983a,b) Species preferences are also affected by the soil characteristics of the patch or mosaic. The preference of a grazing herbivore for a particular species frequently varies with nutrient status. At the landscape scale there are several feedbacks between the large herbivores and the vegetation. The degree of dietary overlap among herbivores is dependent on the species composition of the vegetation and the preferences of the herbivores. Therefore the impact of grazing herbivores on an ecosystem varies both temporally and spatially and with the composition of the herbivore community.

Selection of preferred species by large herbivores changes the competitive relationships of plants that make up the patches or mosaics of a rangeland ecosystem. Livestock forage in a landscape mosaic by selecting habitat at the macroscale (De Vries and Schippers, 1994). Cattle spent more time foraging in landscape units where the vegetation patches provided more nutrients such as phosphorus. The migratory patterns of large native herbivores in Africa has been related to productivity and to concentrations of sodium, phosphorus and magnesium in the plants of selected landscape units (McNaughton, 1988). These studies suggest that the largest impact of large herbivores should be on the mosaics in landscape units characterized by higher available soil nutrients. Grazing may benefit some species by fertilizing them with urine or dung or by removing or weakening competitors. However, most plants that are heavily grazed are negatively impacted by the grazing process (Belsky, 1986, 1987). Increased growth rates, and increased seed production by some plant species that are browsed or grazed may be due to the species position on a spectrum of normal plant regrowth patterns. Overcompensation and herbivore-stimulated growth may simply be the response of plants at one end of this spectrum to damage of meristematic tissue. Physical-climatic variables that can damage growing plant tissues include fire, wind, freezing, and heat. Rapid regrowth probably evolved in response to all types of damage not just grazing and browsing by herbivores (Belsky *et al.*, 1993).

The impact of livestock on vegetation is not limited to the effects of massive defoliation. Selective browsing can have important impacts on the reproduction and structure of individual plants in some species (Kerley *et al.*, 1993). *Yucca elata* is a common perennial in the Chihuahuan Desert. Cattle browse the leaves, which may make up as much as 20% of the diet in winter and spring and intensively utilize the young inflorescence stalks and flowers. Cattle consumed 98% of the inflorescence in paddocks that were grazed during the flowering period. Since this plant is exclusively pollinated by a single moth species (the yucca moth *Tegeticula yuccasella*), loss of most of the inflorescences can result in the local extinction of the pollinator (Kerley *et al.*, 1993). This has implications for the long-term survival of *Y. elata* populations. Consumption of the leaves by cattle also induces production of new ramets from the caudex or from the below-ground tuber. This results in clones of multiheaded *Y. elata*. The impacts of grazers on the plant community can really only be assessed by careful consideration of the life history characteristics of the plants and the characteristics of the landscape units in which the plants are distributed.

Large herbivores impact soil properties. The extent and characteristics of those impacts on soils depend on the spatial and temporal concentrations of herbivores. This is especially important in the development of piospheres around water points. A piosphere is a concentric pattern of decreasing impact of livestock on the vegetation and soils on a gradient extending from a water point (Andrew and Lange, 1986). Some investigators have found that the effects of livestock on soil physical properties is independent of stocking rate (Greenwood *et al.*, 1997). Greenwood *et al.* (1997) reported that soil porosity was lower in all grazed paddocks compared with the soils of an ungrazed paddock. The consequences of activities of large herbivores are frequently the most dramatic effects of animals that are recorded in desert rangelands (Table 8.4). Thus the direct effects of grazers on vegetation may be less important than the indirect effects on vegetation through changes in soil properties.

Table 8.4
Potential Effects of Herbivores on Soil Properties and Processes

Impact	Effects	Consequences
< plant and litter cover	> area of bare soil	> insolation > raindrop splash erosion > soil temperature < infiltration > run-off > potential for soil erosion
	< input of litter < root growth	< soil organic matter < soil cohesion > potential leaching to groundwater
Input of dung and urine	Change spatial and temporal distribution of nutrient inputs Change quality of nutrient inputs	> heterogeneity of soil nutrients > organic matter turnover
Compaction (> bulk density)	< macropores < total pore space < rain water infiltration > run-off	< soil water-holding capacity < plant available water Less favorable soil microclimate for soil biota < environment quality for seedling establishment and root growth > soil erosion potential
Trampling (break-up soil aggregate structure)	< size of soil aggregates Formation of soil surface crusts	> wind and water erosion potential > loss of surface soil < infiltration > run-off Loss of part of soil seed bank < quality of environment for germination and seedling establishment

The seed pods of many trees and shrubs that are numerically dominant in arid landscapes are consumed by native and domestic herbivorous vertebrates. The fruit pods of *Prosopis flexuosa* (Fabaceae) are consumed by domestic, exotic animals (horses and cattle), medium-size, herbivorous rodents (*Lagostomus maximus* Chinchillidae) and (*Dolichotis patagonum* (Caviidae)), and gray foxes (*Pseudalopex griseus*). Large numbers of *P. flexuosa* seeds were found in cattle and horse dung. Seeds that passed through cattle had lower viability but higher germination rates than seeds that passed through horses. Seeds passing through the herbivorous rodents had lower viability but higher germination rates than seeds collected from mature pods on the trees. Seeds passing through gray foxes did not differ in viability and germination from seeds collected from mature pods on the trees (Campos and Ojeda, 1997). Seeds of ten species of legumes and one species of grass survived passage through the digestive tract of cattle. Seeds of tussock grasses did not survive passage through cattle (Gardener *et al.*, 1993). Dispersal of problem shrubs such as mesquite (*P. glandulosa*, Fabaceae) into grassland habitats is one result of the introduction of domestic livestock into the desert rangelands of North America. Mesquite seedlings established from feces of domestic cattle, white-tailed deer (*Odocoileus virginianus*) and coyotes (*Canis latrans*) in semiarid grasslands and these consumers of mesquite fruits are considered vectors for the spread of mesquite into arid and semiarid grasslands (Kramp *et al.*, 1998)

Other examples of animal dispersal of seeds of desert plants reviewed by Campos and Ojeda (1997) include dispersal of seeds of prickly pear cacti (*Opuntia* sp.) by jackrabbits (*Lepus* spp.) and by collared peccaries (*Pecari tajacu*). Some carnivores that exhibit seasonal consumption of fruits and seeds are effective dispersal agents. These carnivores include: badgers (*Taxidea taxus*) and coyotes (*Canis latrans*). *Acacia* fruits are consumed by a variety of animals that are probably effective dispersal agents for the seeds.

In some desert plants, such as *Opuntia* sp. cacti, vertebrate consumption of fruits (frugivory) is the only mechanism of primary dispersal of seeds since all of the fruits are removed within a one-month period after ripening. In dense *Opuntia* dominated scrublands in Mexico, 100% of the fruits were consumed by birds and large mammals. The intensity of fruit removal was faster in areas with sparser cactus populations and in the driest year of the study (Montiel and Montana, 2000).

Mammals are not the only seed dispersers. Large birds that consume fruits can be effective dispersal agents. Australian emus (*Dromarius novaehollandiae*) are primary dispersal agents of nitre bush fruits. The salt-rich pericarps are removed as the fruits pass through the birds' digestive system. In addition the fruit pit is scarified which enhances germination (Noble, 1975).

8.6 GRANIVORY

With the exception of the feeding behavior of domestic livestock, no other feeding behavior has attracted more research effort in arid and semiarid ecosystems than seed feeding. Questions about the relative importance of seed-eating rodents, ants and birds have generated a large number of studies in North America and in other arid regions of the world (Kerley and Whitford, 1994). Interest in small mammals as granivores has been stimulated by the extensive research on North American heteromyid rodents, the best-studied desert small mammals. The diversity and abundance of heteromyid rodents, which are largely granivorous, has led to granivory being considered one of the key adaptations of small mammals to desert environments. A recent study of community structure in small mammals across four continents suggests that granivores are most abundant in some Northern Hemisphere temperate deserts (Kelt *et al.*, 1996). The patterns of species composition included: low α diversity (two to four species per site), high β diversity (high species turnover between sites) and local coexistence of only 20–30% of the species in the regional pool. Deserts in the northern hemisphere had more granivores except for the Turkestan Desert where there were more foliage feeders. Omnivorous rodents dominated the small mammal communities in South America, Australia, and the Thar Desert of the Indian subcontinent (Kelt *et al.*, 1996). However, the demonstration that the process of granivory has major impacts on desert plant communities has focused attention on this process as critical in the functioning of desert ecosystems (Brown and Heske, 1990; Heske *et al.*, 1993; Guo *et al.*, 1995). Seed-harvesting ants, seed-feeding insects such as bruchid beetles, and birds also impact both the quantity and species composition of the seed reserves in the soil.

The evaluation of the relative importance of various groups of seed feeders has been done by the use of seed bait experiments in which removal of seeds by different groups of seed feeders are measured (Brown *et al.*, 1975; Mares and Rosenzweig, 1978). The interpretation of the results of these experiments must be made with caution because the technique has some serious limitations (Parmenter *et al.*, 1984). The removal of seeds by ants and rodents does not reflect the time-specific seed requirements of these organisms because both taxa cache seeds for later consumption and caching is a common response to abundant seeds (Parmenter *et al.*, 1984). Birds require much longer time to locate dense patches of seeds because they rely on predictable, high quality food patches and spend little time locating new resources (Mares and Rosenzweig, 1978). Thus seed tray techniques measure the ability to locate and harvest seeds *not* granivory.

The simplicity of the seed tray technique has allowed for a number of studies in arid regions. In North America and in Israel seed removal by rodents was considerably greater than seed removal by ants or birds (Fig. 8.7). However in a recent study, it was found that seed removal by ants was significantly higher than seed removal by rodents in Chihuahuan Desert grasslands but not in shrub-dominated communities (Kerley and Whitford, 2000). In other desert regions seed removal by rodents was considerably lower than seed removal by harvester ants (Kerley and Whitford, 1994).

The seed tray studies in North America were conducted in Chihuahuan Desert shrublands that had been desert grassland in the late 19th century. In a comparison of granivory in mesquite (*Prosopis glandulosa*) dunes, cresotebush (*Larrea*

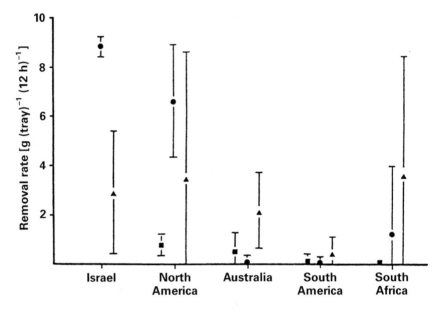

Figure 8.7   Comparison of seed removal from seed trays by rodents (●), ants (▲), and birds (■) in desert areas for which there are published data. (From Kerley and Whitford, 1994.) Reprinted with permission of CSIRO Publishing.

tridentata) shrublands and desert grasslands using seed trays with large and small (< 1 mm diameter), grassland ants removed 14.2 g of seeds in 12 h. This rate of seed removal is the highest for any desert area in the world (Kerley and Whitford, 2000). Small native fire ants, *Solenopsis xyloni*, recruited rapidly to the seed trays and dominated the assemblage of ants visiting the trays. *S. xyloni* and *Pheidole* spp. removed large quantities of the small seeds and the larger seed harvester, *Pogonomyrmex desertorum*, removed larger seeds.

Seed removal by ants was higher than by rodents in both grazed and ungrazed desert grassland sites. Seed removal by ants was lower at the shrubland sites than at the grassland sites. Rodents removed more seeds than ants in the creosotebush shrubland sites and slightly more than ants at the mesquite sites. The creosotebush sites had higher numbers of granivorous rodents than the mesquite and grassland sites. These patterns of seed removal by ants and rodents are a feature of vegetation changes resulting from desertification in the past 150 years (Kerley and Whitford, 2000). These results suggest that some of the differences in patterns of granivory in North American deserts in comparison with other desert areas of the world may be an artifact of the population responses to granivorous rodents to vegetation changes.

In a study encompassing four years of monthly sampling of desert rodents, Whitford (1976) reported six species of seed-feeding heteromyid rodents and eight species of omnivorous rodents (Fig. 8.8). One species, *Onychomys torridus*, is primarily carnivorous–insectivorous and two additional insectivorous species, the desert *Notiosorex crawfordii* and grasshopper mouse, *Onychomys leucogaster,* occur in the region but were not taken on the study areas. In a list of 16 species of rodents from the Australian arid zone, Morton and Baynes (1985), list five species that would be described as granivores but point out that in these species insects and plants make up part of the diet. This is also true of North American heteromyid rodents which consume variable quantities of insects and plant parts depending upon availability (Reichman, 1975). In that study, stomach contents were compared with cheek pouch contents. Cheek pouch contents were

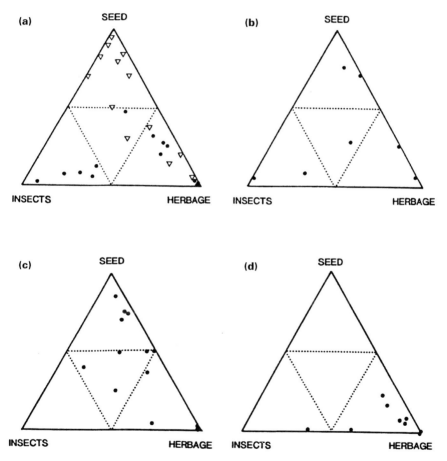

Figure 8.8 Diets of small mammals for (a) North American desert (∇, heteromyid rodents; ●, nonheteromyids), (b) South American Atacama Desert, (c) Australian deserts, (d) South African Karoo. (From Kerley and Whitford 1994.) Reprinted with permission of CSIRO Publishing.

generally used to obtain dietary information for heteromyid rodents. The seed-eating rodent species cited most frequently as a granivore (*Dipodomys merriami*) included insects as 15.5% of its diet and green vegetation made up 6.1% of its diet (Reichman, 1975). This is only one example of the variability in the diet of 'granivorous' rodents.

Apparent 'plague' numbers of omnivorous rodents occur during particularly favorable conditions whereas the population densities of the more 'k' adapted heteromyids remain relatively constant. A conceptual model for this pattern of rodent abundance includes migration by omnivorous rodents into formerly unfavorable habitats and high, aseasonal reproduction by these species as factors accounting for the plague numbers of cricetid (murid) rodents following two wet years in succession (Whitford, 1976). Thus, although there are proportionately more species of 'granivorous' rodents in North American deserts, in the Chihuahuan Desert which has a climate most similar to the Alice Springs area, e.g. total rainfall, summer wet season and summer temperatures, the rodent assemblages of Australian deserts and the Chihuahuan Desert share many similarities.

Comparative studies of granivory have been done using uncovered and screen mesh covered seed trays to measure relative rates of seed removal as an index of which group of granivores is most important in a given area (Brown *et al.*, 1975; Mares and Rosenzweig, 1978; Morton, 1985; Kerley, 1991). In the Monte Desert of Argentina, seed removal rates were much higher from microsites under shrub canopies than in open, exposed microhabitats (De Casenave *et al.*, 1998). De Casenave *et al.* (1998) state that comparisons of seed removal rates among geographically disparate sites must be made with caution because the sampling effort among microhabitats is frequently not reported. However, their study confirmed the conclusions of Mares and Rosenzweig (1978) that ants are the most important seed consumers in the Monte Desert and the total impact on seed reserves seems to be lower in South American and Australian deserts than in other desert areas of the world. In the Chilean matorral (semiarid shrubland) of South America, seed removal in sparse habitat was primarily by ants with little seed removal during the part of the year when ants were not active (Vasquez *et al.*, 1995). In dense habitat seed removal by ants was equaled by seed removal by birds and mammals removed very little seed. Seed removal rates were correlated with abundance of ants in these habitats (Vasquez *et al.*, 1995). Despite the differences in granivory in South American and Australian deserts, the seed reserves in these deserts are essentially equal to the seed reserves in North American deserts (Marone and Horno, 1997). Understanding granivory with respect to seed reserves will require rigorous studies on the impact of poorly studied granivores such as birds and on the effects of bacteria and fungi on seed reserves. The assessments of impacts of granivores on seed reserves will also require estimation of germination losses.

Seed removal by mammals is consistently low in southern hemisphere sites. This pattern of seed consumption may be related to the paucity of granivorous rodents in southern continents (Kerley and Whitford, 1994) and/or to the more omnivorous food habits of South American small mammals (Kelt *et al.*, 1996). In

South American and Australian deserts, ants were the dominant granivores. Marsh (1985) reported that most of the ants in the Namib were seed-harvesters and that the structure of the ant community was similar to that in Australia. Marsh's data suggest that ants are the most important seed consumers in the Namib. Kerley (1991) showed that in the Karoo (South Africa) ants removed considerably more seeds from trays than either birds or rodents. However, it is possible that the relative importance of ants and rodents as granivores has been misunderstood, especially in regions where rodent populations experience periodic explosive increases. Using a conservative energetics approach, it was estimated that seed consumption by rodents would be 30 times higher at a population episodic peak than was estimated by seed dish studies during periods of low rodent population densities. During peak populations of rodents in the Australian deserts, rodent consumption of seeds was estimated to be higher than the estimates for North American rodents and much higher than seed consumption by desert ants (Predavec, 1997).

What is it about North American desert rodents that are different from rodents in other deserts? It is certainly not physiological because there are numerous reports of a wide variety of desert rodents having kidneys with long loops of Henle that allow maximal concentration of dissolved solids in the urine. Most if not all live in burrows where the microclimate is relatively humid and temperatures are moderate and most are nocturnal. One feature of heteromyid rodents that is different from most of the other groups of desert rodents are cheek pouches. Cheek pouch capacity is directly proportional to the body mass of the rodent. It has been estimated that a heteromyid rodent can transport a mass of seeds with the equivalent energy requirements of an individual for one day (Morton *et al.*, 1980; Vander Wall *et al.*, 1998). The cheek pouches allow heteromyid rodents to remain away from the burrow collecting seeds and fruits and storing them in the cheek pouches for later transport to the home burrow or to cache. The larger heteromyids collect seeds with all of the investing structures whereas the smaller species may husk the seeds and place only the edible parts in the cheek pouch (Leman, 1978). The overwhelming advantage of cheek pouches is the ability to collect and store quantities of seeds efficiently during periods of seed abundance. This allows the 'resident' individuals to survive long periods when there is little or no suitable seed production. Reichman (1975) suggested that heteromyids eat vegetation and insects as encountered but collect seeds for transport in cheek pouches at the same time. Thus cheek pouch contents do not necessarily reflect diet. External cheek pouches appear to be a feature of North American heteromyid rodents that is not shared by other rodent families. Seed-eating rodents that have no external cheek pouches must make numerous trips between a seed source and home burrow in order to build up a store of seeds or numerous trips to a cache site if seeds are to be cached.

One study in Israel showed that the removal of seeds by rodents was nearly four times greater than seed removal by ants (Abramsky, 1983) (Fig. 8.7). The rodent community studied by Abramsky is dominated by gerbelline rodents (species with no external cheek pouches). These data suggest that external cheek pouches are not essential for rodents to specialize on seeds as a primary food source. Body size has an important effect on the diet composition of 'granivorous' rodents. Gerbilline rodents use green vegetation during the winter rainy season when seed reserves are lowest and green vegetation is available. Small gerbillines continue to consume seeds as a large fraction of their diet during winter whereas the large body size gerbilline (*Meriones crassus*) consumed green vegetation as the largest fraction of its diet (Degen *et al.*, 1997). Seeds are the largest component of the diet of small gerbils and green vegetation is the largest component of the diet of a large gerbil. These differences have been attributed to the need for high quality, digestible food for small animals and the ability of larger animals to get by with lower quality food. They state that there is an inverse relationship between body mass and the ratio of seed:green vegetation intake in other Old World and New World granivorous rodents.

In order to supply their metabolic needs, small desert rodents that have no cheek pouches must spend time foraging that is inversely related to the abundance of seeds or other food resources. When seeds, vegetation, and insects are sparse, small mammals are exposed to higher probability of being killed by predators. Predation has been suggested as one of the more important factors involved in structuring desert rodent communities (Price and Brown, 1983; Kotler, 1984; Bowers, 1990). Owls and snakes and several kinds of mammals are known to prey on desert rodents (Bouskila, 1995; Meserve *et al.*, 1996; Gutierrez *et al.*, 1997). Predator exclusion studies have concluded that predation is the primary factor affecting the abundance of rodents and indirectly their long-term effects on vegetation (Meserve *et al.*, 1993, 1996; Bock and Bock, 1994). Granivory during periods of depleted seed banks therefore exposes desert rodents that do not hoard seeds to greater risk of predation.

Most desert 'seed-eating' rodents are climate 'responders' not climate 'integrators'. Desert small mammal populations exhibit large changes in abundance in relation to rainfall and food availability. The exception to this generalization appears to be the dominant North American heteromyid rodents (kangaroo rats and pocket-mice) (Whitford, 1976). Seed storage by heteromyids allows them to become climate 'integrators', i.e. species that maintain relatively stable populations despite environmental fluctuations. Seed-harvesting ants have the advantage of sociality that allows them to be successful utilizers of seeds in desert regions and to be 'climate integrators'. Although individual workers transport single seeds, the energy cost:energy benefit ratio is high when compared with the few seeds that a rodent could carry and the energetic costs of travel for that rodent. The lower metabolic costs for seed transport in seed-harvesting ants, allows them to accumulate seeds in their nests as a hedge against times with little or no seed production. Thus ants are the most successful environmental 'integrators' among the desert granivores.

8.7 IMPACTS OF RODENTS AND ANTS ON ECOSYSTEM STRUCTURE AND PROCESSES

A long-term study initially established to examine competition between seed-eating ants and rodents, has produced a rich data set that provides important insights into the impacts of seed consumption on the structure of a desert ecosystem (Brown and Davidson, 1977; Brown and Heske, 1990; Heske *et al.*, 1993). The study was designed with plots from which ants and rodents were excluded, plots from which rodents were excluded, plots from which ants were excluded, plots from which large-body size rodents were excluded, and plots that excluded neither group of granivores. The results from the first few years of the experiment showed that kangaroo rats preferentially removed large seeds resulting in a shift in the abundance of plants that produce small seeds. Ants removed more small seeds than large seeds producing a complementary effect on the vegetation in the plots (Brown and Davidson, 1977). A summary of the data after the experiment had been in place for 10 years showed that there was no evidence for either competition or facilitation between the granivorous rodents and granivorous ants based on the densities of these taxa on the experimental plots (Samson *et al.*, 1992). Examination of the plots after 15 years showed that the interactions between ants and rodents were indirect and mediated through the vegetation rather than by direct competition. The most notable result was the positive correlation between abundance of one seed-eating ant species (*Pheidole rugulosa*) and the percentage grass cover on the experimental plots (Valone *et al.*, 1994).

An analysis of the long-term trends in vegetation on these experimental plots led to the conclusion that kangaroo rats (genus *Dipodomys*: Heteromyidae) were functioning as a keystone guild in North American desert grasslands (Brown and Heske, 1990). The kangaroo rats were shown to affect the density and species composition of the grasses in the experimental plots, especially the tall tussock-forming grasses. These data were interpreted as one explanation for changes in vegetation resulting from desertification processes. The mechanisms invoked for the effects of kangaroo rats on the grass component of the vegetation was granivory and/or soil disturbance (Heske *et al.*, 1993). However, these mechanisms were shown not to be feasible for the vegetation changes that were recorded over the long term of the study (Kerley *et al.*, 1997). The kangaroo rat behavior of cutting the flowering tillers from tussock grasses appears to be a more logical mechanism for the vegetation trends recorded by Brown and Heske (1990). Characteristically cut grass tillers were absent from all rodent and medium-size kangaroo rat exclosures. Tillers cut by rodents were abundant in large-size kangaroo rat and rabbit exclosures (Fig. 8.9). This indicated that medium-size kangaroo rats (*Dipodomys ordii, D. merriami*) were responsible for cutting the flowering tillers. Kangaroo rats exhibit preferences for tillers of certain species of grasses. *Dipodomys* spp. cut very few tillers from Lehmann lovegrass (*Eragrostis lehmanniana*), the species that exhibited increased cover on plots with kangaroo rats. Because kangaroo rats preferentially cut flowering tillers of potential competitors of Lehmann lovegrass,

Figure 8.9 Flowering tillers of a desert grass, *Sporobolus* spp., cut by rodents in the Chihuahuan Desert, New Mexico.

the increase in *E. lehmanniana* was attributed to the competitive advantage of this species over other tall grasses as it was subjected to less tiller pruning (Kerley *et al.*, 1997). Because tillers are pruned from the plants before the seeds are mature, tiller pruning must have a dramatic effect on the soil seed bank of preferred plant species.

Grass tiller and seed harvesting are not the only rodent activities that have implications for vegetation structure and productivity. Heteromyid rodent cache pits are traps for seeds and organic matter and are safe germination sites for some plant species. Burrows, runways and fecal deposition areas affect heterogeneity in soil seed distribution, soil nutrient concentrations, and organic matter turnover. Soil disturbance by the degu (*Octodon degus*) was concluded to be the primary factor opening areas under shrubs to colonization by competing plant species and by exotic herbs and facilitating the establishment of some annuals that are restricted to nutrient rich areas under shrubs (Gutierrez *et al.*, 1997).

There are several important lessons to be learned from the extensive and intensive studies of granivory in deserts. Obviously long-term experiments are required to provide an accurate assessment of impact of animal consumption on ecosystem structure and/or ecosystem processes. The long-term experiment clearly showed that neither the hypothesized indirect competition nor facilitation occurred between ants and rodents. The most important effect of ants and rodents on vegetation may be by establishment of soil fertility gradients around ant colonies or rodent burrow systems, or by the production of safe germination sites. Granivory is not the only direct consumption effect on vegetation as shown by the studies of gramnivory by desert seed-eating rodent species. Although the rodents consume very little of the grass stem (eating only a small portion of the base of the cut tiller), they contribute a large quantity of high nutrient content litter to the plant detritus pool. These studies document the complexity of what superficially appear to be simple effects of consumers on ecosystems.

8.8 VERTEBRATE PREDATORS

The most abundant vertebrate predators in most desert ecosystems are lizards. Not all lizards are predators; some like the North African *Uromastix* spp. and the desert iguana, *Dipsosaurus dorsalis,* are herbivores. There are important differences in the relative abundance and species richness of lizards in different deserts of the world. Algerian Sahara lizard communities vary from five to seven species in three families depending on habitat characteristics (Grenot and Vernet, 1972). There are 18 species representing four families of lizards reported for the entire Saharan region (Lambert, 1984). The Saharan lizard communities are depauperate in comparison with lizard faunas of other deserts. Lizard communities in North and South American deserts have fewer species than deserts in southern Africa and Australia. It has been suggested that these differences reflect the paucity of termites in North American deserts in comparison to the rich termite fauna of southern Africa and Australia (Pianka, 1985). An intercontinental comparison shows that each of these areas supports five families of lizards but only one family is found in all three areas. North American deserts and the Kalahari support 12 genera and 13 genera, respectively, whereas Australian deserts support 24 genera. At the species level these differences are magnified. North American hot desert sites have six to eleven species of lizards depending on the complexity of the vegetation and landscape position, the Kalahari between 11 and 17 species depending on the site and in the Great Victoria Desert of Western Australia lizard species richness varied from 15 to 42 species (Pianka, 1985).

Desert lizard communities have been divided into two general classes: widely foraging or sit and wait predators (Huey and Pianka, 1981). Sit and wait lizards tend to feed on active prey species that are moving through the environment. The prey taken by sit and wait predators are mostly lepidopterans, coleopterans, and orthorpterans (Huey and Pianka, 1981). Sit and wait predators divide space and time, foraging at different times of day and choosing different microhabitats to sit and wait. The energetics of predation by a widely foraging lacertid lizard, *Eremias lugubris*, and a sit and wait species, *Eremias lineoocellata,* in the Kalahari Desert were found to differ considerably in terms of energy expenditure and foraging efficiency (Nagy *et al.*, 1984). Energy expenditure during the active period was 12 times resting in *E. lineoocellata* and three times resting in *E. lugubris*. Foraging efficiency defined as metabolizable energy gained while foraging divided by total energy spent while foraging was higher in the wide forager (2.0) than the sit and wait predator (1.6). *E. lineoocellata* was foraging for much shorter periods (2.75 h day^{-1} vs. 10.24 h day^{-1}) than *E. lugubris*. Although the sit and wait predator made a greater profit on its energy expenditure (46% vs. 36%), the wide-foraging *E. lugubris* grew nearly twice as fast as *E. lineoocellata* (Nagy *et al.*, 1984). These apparent contradictions based on energetics demonstrates that variation in food availability and/or differences in predation rate introduce variables not accounted for in the energy-based model.

Energetic studies comparing sit and wait and widely foraging lizards provide some interesting insights into secondary production by predators (Anderson and Karasov, 1988). The wide-foraging North American lizard, *Cnemidophorus tigris* (Teiidae) and the ambush forager, *Uta stansburiana* (Iguanidae) did not differ in the proportion of the energy budget expended on reproduction but the total reproductive effort (which includes metabolism associated with reproduction) was lower in *C. tigris*. *C. tigris* produces smaller clutches of larger eggs than *U. stansburiana* but the number of eggs laid per unit time was equal for these species. It was concluded that the higher rates of production allow *C. tigris* to reach larger size than sit and wait iguanid lizards at the same age (Anderson and Karasov, 1988). Daily maintenance energy expenditures of widely foraging lizards appear to be about 1.3–1.5 times greater than those of sit and wait foragers. However, food gains of widely foraging lizards have been estimated at 1.3–2.1 times greater than sit and wait foragers (Huey and Pianka, 1981).

Prey selection by lizards is largely a function of foraging mode. Widely foraging lizards tend to forage in patches that support large numbers of relatively sedentary prey. The prey taken by widely foraging lizards in North America, the Kalahari, and Western Australia is predominantly termites. The availability of termites at the surface is unpredictable in time and space. This requires a predator to move rapidly through the environment, testing potential patches until a rich patch is found. In North American deserts, teiid lizards (*Cnemidophorus* spp.) are the widely foraging lizards. Termites are reported to make up nearly 30% of the stomach contents of *Cnemidophorus tigris*, and 17% by volume of the stomach contents of *Uta stansburiana*, a moderately wide foraging species (Huey and Pianka, 1981). Termites accounted for 4% or less by volume of the stomach contents of sit and wait predators.

In spinifex grassland habitats of central Australia, the diverse lizard fauna is thought to have developed with the diversity of termites in these grasslands. It is argued that the species-rich termite fauna utilize the spinifex grassland in a number of ways. This diversity of foraging behaviors in the termites allows for coexistence of a number of lizard species using termites as their primary prey (Morton and James, 1988). In a study of five species of the skink genus *Ctenotus* in central Australia, the high dietary overlap in these species was accounted for by the high proportion of termites in the diet (James, 1991). However, each species of *Ctenotus* ate different genera or foraging guilds of termites. During a relative dry year, termites accounted for a larger portion of the prey items than during a relatively wet year. Other prey were Araneida (spiders), Blattodea (cockroaches), Chilopoda (centipedes), Coleoptera (beetles), Formicidae (ants), Hemiptera (bugs), Orthoptera (grasshoppers and crickets) and Thysanura (silver fish). Based on this two-year study, James (1991) concluded that *Ctenotus* spp. lizards were somewhat opportunistic in prey selection and that the abundance of prey has a large effect on lizard diets. A relationship between rainfall (with a one year lag) and population size of lizards has been documented and related to prey abundance (Whitford and Creusere, 1977).

A relatively long-term study of desert lizard communities related changes in biomass, secondary production and diversity to annual precipitation and growing season precipitation (Whitford and Creusere, 1977). There were differences in the responses of lizards in two different landscape units on a watershed. The species richness of the lizard communities ranged from 6 to 11 species in a grassland–shrub mosaic at the base of the watershed and from six to eight species in a creosotebush (*L. tridentata*) shrubland on the upper bajada (piedmont). Both communities were dominated by active foraging species (*Cnemidophorus* spp.). Recruitment of juveniles was virtually absent during the late summer and autumn of years with below-average summer rainfall. The recruitment of juveniles was extremely high during an exceptionally wet growing season. As a result of two successive 'above-average precipitation' growing seasons, several species of lizards immigrated into areas that they had not occupied during average or below-average rainfall years. During a drier-than-average growing season following the wet years, the immigrant species disappeared in the creosotebush community but remained present in the grass–shrub mosaic habitat at the base of the water-shed. These differences in the responses of lizards were attributed to productivity differences in the landscape resulting from soil water storage and retention. Lizard species that immigrated into marginal habitat originated from more mesic moun-tain slope areas and toe-slope grassland. The total linear distance from these source habitats to the base of the watershed was approximately 3 km. Immigration is less likely to be a factor affecting the species richness and productivity of lizard communities in habitats that are distant from source areas (Whitford and Creusere, 1977). These responses of desert lizard communities were summarized in a conceptual model (Fig. 8.10).

The relationships of desert lizard populations and rainfall can be complicated by interspecific competition or changes in vegetation cover. For example, the Australian netted dragon, *Ctenophorus nuchalis* was most abundant when the spinifex bunch grass (*Triodia basedowii*) was less then 10% cover. Another dragon, *C. isolepis*, reached numerical dominance following heavy rains that increased spinifex cover to greater than 20% (Dickman *et al.*, 1999). It was sug-gested that drought-wet cycles periodically reverse the dominance of the two species of *Ctenophorus* and probably of other species. This pattern of dominance shift is similar to that reported for two species of kangaroo rats, *Dipodomys* spp.,

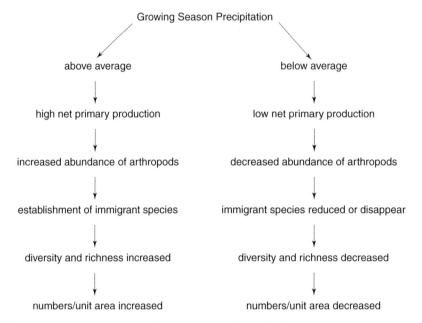

Figure 8.10   A conceptual model showing the relationship between growing season rainfall and density and diversity of Chihuahuan Desert lizard communities. (Modified from Whitford and Creusere 1977.)

in the Chihuahuan Desert of North America (Whitford and Steinberger, 1989). Although the community dominance may change, the biomass responses of the lizard assemblage remains the same as proposed by Whitford and Creusere, 1977).

Although the division of lizard communities into two 'foraging guilds' has been useful in distinguishing between those species that feed predominantly upon termites and those species that take whatever comes by, it does not account for species that are not 'sit and wait' but move slowly and deliberately from patch to patch. This is the foraging behavior of lizards that are ant specialists. Although ants are the most abundant surface-active arthropods, most of the biomass of an ant is indigestible chitin. Lizards that specialize on ants as a food source share some important behavioral traits. Several studies of the North American horned lizards (genus *Phrynosoma*) have examined the foraging behaviors of a deliberate slow-moving predator. Horned lizards are ant specialists. The juvenile and adults of a relatively large species, *Phrynosoma cornutum* (Fig. 8.11), preys almost exclusively on large harvester ants of the genus *Pogonomyrmex* (Whitford and Bryant, 1979). A small congener, *Phrynosoma modestum* (Fig. 8.12), feeds almost exclusively on honey-pot ants of the genus *Myrmecocystus* (Shaffer and Whitford, 1981). *P. cornutum* remains in one place as long as the time span

Figure 8.11   Texas horned lizard, *Phrynosoma cornutum*, Chihuahuan Desert, New Mexico.

Figure 8.12   Small round-tailed horned lizard, *Phrynosoma modestum*, Chihuahuan Desert, New Mexico.

Figure 8.13   Thorny devil, *Moloch horridus*, central Australian Desert.

between individual ants appearing within the capture space does not exceed some threshold value (Munger, 1984). *P. cornutum* seeks the canopy of subshrubs as a resting place during the heat of the day when ants cease foraging. By climbing into the canopy of a small shrub, this lizard can maintain its body temperature at 40°C, which allows it to digest the nonchitinous parts of its prey. There is morphological and behavioral convergence in North American horned lizards and a species of Australian agamid lizard, *Moloch horridus* (Fig. 8.13), which is also an ant specialist.

8.9 BELOW-GROUND FOOD WEBS

The soil fauna in deserts is important as detritus consumers and as regulators of decomposition and nutrient cycling. The soil macrofauna in warm deserts is dominated by subterranean termites. However, not all warm deserts have subterranean termites (Crawford *et al.*, 1993). At the southern margin of the Chilean arid zone, there is evidence that earthworms dominate the soil macrofauna. Earthworms have also been reported from the wetter margins of the Sahara. Earthworms are probably limited to areas with silty soils where soil moisture remains relatively high either from rainfall or from fog condensation (Crawford *et al.*, 1993). In most arid regions termites dominate the macrofauna. The termite faunas of Africa and Australia are the most diverse in the world and include a number of mound building and subterranean species. High biomass of subterranean and epigeic termites have been reported from Saudi Arabia and high species richness has been reported from the arid regions of Africa and the Arabian Peninsula (Johnson and Wood, 1980; Badawi *et al.*, 1984). Although the termite fauna of North American deserts is depauperate, the dominant species are generalists that consume most dead plant materials and animal dung (Whitford, 1991)

Desert soil mesofauna is dominated by mites (Acari) of the suborder Prostigmata with some mites in the suborder Cryptostigmata in microsites with sufficient litter layers. In mesic environments and semiarid woodlands, cryptostigmatid mites dominate the soil microarthropod assemblages (Kinnear, 1991). The prostigmatid mites in desert soils tend to be generalists that feed on fungi or on nematodes or both (Whitford, 1996). Other microarthropods that are abundant in some microhabitats include springtails, Collembola, book-lice, Psocoptera, and a variety of insect larvae (Wallwork, 1982). Most of these taxa feed directly on dead plant material and/or fungi. There are many similarities in the microarthropod faunas of the deserts of the world (Noble *et al.*, 1994). Many genera of soil mites are shared among North American deserts, Australian deserts, South American deserts and deserts of southern Africa (Crawford *et al.*, 1993; Noble *et al.*, 1994). The diversity and abundance of microarthropods varies with soil type and vegetation. In shrub deserts, microarthropods are abundant under shrub canopies where litter accumulates (Santos *et al.*, 1978). Soils under shrubs in more mesic microhabitats support higher densities and diversities of microarthropods (Wallwork *et al.*, 1985; Crawford *et al.*, 1993).

Desert soils also support an abundant and diverse microfauna: nematodes and protozoans. The desert nematode fauna is dominated by bacteria-feeding nematodes. Fungiphagous nematodes make up the second most abundant group of nematodes with omnivore-predators accounting for a small fraction of the nematode community (Freckman and Mankau, 1986; Freckman *et al.*, 1987). Nematode abundance and diversity is positively correlated with the quantity and quality of litter layers under shrubs and with the abundance of fine roots under grasses. In desert soils nematodes may be in an inactive, anhydrobiotic state when soils are dry. Nematodes require water films on the surface of soil particles in order to remain active (Whitford, 1989). Protozoans are the other abundant microfaunal element in desert soils. The most abundant protozoans in desert soils are naked amoebae with small numbers of testate amoebae, ciliates, and flagellates in the assemblage (Parker *et al.*, 1984). Protozoans also require soil films for their activity. When soils dry, protozoans encyst and remain in an inactive state until the soil is re-wetted (Whitford, 1989).

Desert soils support a varied fauna that are important consumers in the detritus-based food webs. The feeding relationships of these animals in desert soils are directly affected by soil water content (Fig. 8.14).

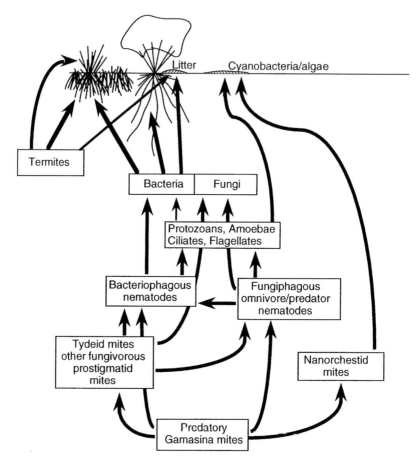

Figure 8.14   Diagramatic representation of a desert soil food web.

8.10 EPHEMERAL PONDS AND LAKES

Ephemeral ponds or lakes also known as playas, pans, chottes, and sebkas are common features of arid and semiarid regions (Crawford, 1981). Ephemeral ponds are sporadic aquatic systems that are inundated by large rain events. Ephemeral lake basins may be vegetated or completely devoid of vegetation. Vegetated basins frequently have a cover of grasses that are palatable to livestock and are therefore economically important features of arid landscapes. For example, some playas in the Chihuahuan Desert have a cover of *Panicum obtusum*, a grass which is considered an 'ice cream' plant by pastoralists. In arid rangelands, portions of ephemeral lakes are frequently deepened to form stock watering tanks. These tanks continue to hold water long after the flood waters that cover the lake basin have evaporated. The stock tanks provide a longer duration habitat for aquatic organisms and ephemeral lakes with stock tanks probably support a more complex aquatic food web than unmodified lake basins. For example in a study of a small ephemeral lake that is part of the Jornada Long-Term Ecological Research watershed, the lake basin decreased from maximum flood (average depth 30 cm) to completely dry in two weeks whereas the 2.5 m deep stock tank retained water for 60 days (Loring *et al.*, 1988). Water loss from flooded ephemeral lakes is mostly by evaporation in those lake basins where salts accumulate in the sediments. In lake basins with fine textured cracking clays sediments or silty clays that swell and shrink, deep cracks and sink holes provide avenues for water to infiltrate into the deep sediments. In these basins water loss by infiltration may exceed losses to evaporation.

Ephemeral waters are essential for the long-term survival of desert amphibians. An ephemeral lake in the Chihuahuan Desert was used as a breeding site by five species of anuran amphibians. Each of the species utilized a different microhabitat for calling sites and for breeding. Some species utilized the partially submerged clumps of grass in different areas of the flooded basin. Two species utilized the open, deep-water sites. Less than 1% of the metamorphosed frogs, survived to adult size (Creusere and Whitford, 1976). Tadpole survivorship is adversely affected by an increase in density as ephemeral lake size decreases with evaporation (Woodward, 1982). Pond drying can also cause death of tadpoles by desiccation. In an experimental study, spadefoot toad (*Scaphiopus couchi*) tadpoles developed rapidly and metamorphosed in low density ponds but exhibited slow growth and development in high density ponds. Tadpoles in the high density ponds died from desiccation before developing to the metamorphosis stage (Newman, 1987).

The aquatic fauna of ephemeral lakes should respond to flooding as predicted by the pulse-reserve model. The aquatic fauna of ephemeral lakes has been classified into three life cycle types: (1) fast onset, immediate development; (2) fast onset, prolonged development; and (3) delayed onset, rapid development (Loring *et al.*, 1988b). Based on data for a playa in the northern Chihuahuan Desert, fast onset, immediate development fauna included mosquitoes (*Aedes* sp.), clam shrimp (*Eulimnadia texana* (Conchostraca)) and tadpole shrimp (*Triops longicaudatus* (Notostraca)) (Fig. 8.15). These species complete their life cycle in two weeks or less. This allows them to exploit the high concentration of dissolved and suspended organic matter present immediately after flooding without competition from many other species. This strategy also allows such species to avoid predation from slower-developing species that become predators in the later stages of development.

Slower development is characteristic of the life histories of water fleas such as *Moina wierzejski* (Copepoda), fairy shrimp (*Streptocephalus texanus* and *Thamnocephalus playtyurus* (Ancostraca)) and spadefoot toad tadpoles (*Scaphiopus multiplicatus* (Amphibia)). These organisms grow, use and store resources as long as possible before reproducing (the Crustacea) or leaving the

Figure 8.15 Desiccated tadpole shrimps, *Triops longicaudatus*, that were caught in a rapidly drying pool at the edge of an ephemeral lake in the Chihuahuan Desert, New Mexico.

water (Amphibia). A percentage of the spadefoot toad tadpoles develop into predatory morphs while most of their siblings continue as filter feeders. The delayed onset–short duration development strategy was documented for rotifers in Chihuahuan Desert ephemeral lakes (Loring *et al.*, 1988b). Rotifers appeared 20 days after flooding and three species showed nonoverlapping peak population numbers separated by approximately 15 days (Fig. 8.16). Delayed onset of development was also exhibited by *Bufo cognatus* tadpoles which did not appear in the samples until day 50 after flooding and remained until the lake dried. This strategy allowed the bufonid tadpoles and rotifers to avoid competition with the crustaceans and to avoid predation by the *S. multiplicatus* tadpoles that developed the mouth parts of a carnivore (MacKay *et al.*, 1990).

Ephemeral ponds and lakes are essential breeding sites for desert amphibians. Even small ephemeral lakes can attract large numbers of breeding amphibians. In a 5 ha playa lake in southern New Mexico, the estimated numbers of breeding adult anurans (calling males and gravid females) were: 3000 *Scaphiopus multiplicatus*, 2500 *S. bombifrons*, 1000 *S. couchi*, 500 *Bufo debilis* and 150 *B. cognatus* (Creusere and Whitford, 1976). However, despite the large numbers and diversity of reproductive adults, the only tadpoles in the lake for the first 50 days after flooding were *S. multiplicatus* and *B. cognatus*. It is possible that these differences are the result of inadequate sampling. In a year when only the stock tank

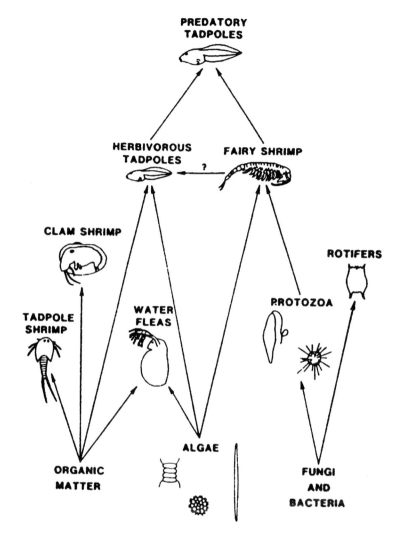

Figure 8.16   A typical food web of organisms found in flooded ephemeral ponds and lakes. (From Loring *et al.*, 1988b.) Reprinted with permission of the University of New Mexico Press.

in the lake basin held water the breeding population consisted of 700 *S. multiplicatus*, 200 *S. bombifrons* and less than 20 *S. couchi*. No bufonids were recorded in the breeding population that year.

In the Chihuahuan Desert playa lake, most of the aquatic species passed through a single generation during a flood cycle. The exception was the water flea (*Moina wierzejskii*) in which there were three generations during a single flood cycle (MacKay *et al.*, 1990).

The temporal patterns of development by the aquatic fauna of ephemeral lakes largely determines the structure of the food web. In the Chihuahuan Desert playa lake that was studied by Loring *et al.* (1988b), the fairy shrimp were filter-feeders and the tadpole shrimp were scavengers. The tadpole shrimp did not prey on the fairy shrimp although their well-developed mandibles suggest that they are capable of capturing and killing fairy shrimp. *Scaphiopus* spp. tadpoles are known predators on other tadpoles and were found to be important predators of fairy shrimp. The tadpoles did not actively hunt fairy shrimp but captured and ate fairy shrimp that were encountered in their slow movements along the bottom. However *Scaphiopus* tadpoles did not prey on tadpole shrimp possibly because the tadpole shrimp were nearly the size of the anuran tadpoles when both were present in the pond. The predator morphs of the *Scaphiopus* tadpoles were unable to kill the large herbivorous tadpoles but did remove large chunks from the tails of these tadpoles.

Ephemeral lake food webs probably begin as heterotrophic systems fueled by the allochthonous material washed into the basin from the surrounding drainage areas. Evidence for this is the changes in nitrate concentrations during a flooding cycle in the Chihuahuan Desert playa studied by Loring *et al.* (1988b). Within a few days, the tadpole shrimp, clam shrimp, bacterial and fungal populations peak and other populations such as protozoans, water fleas, and fairy shrimp begin to increase (Fig. 8.16). Diatoms and algae begin to reach high densities providing food for the water fleas, fairy shrimp and herbivorous tadpoles. Predaceous tadpoles do not appear in the systems until 2–4 weeks after flooding when the larger tadpoles have a suitable prey base. The high densities of the rapidly developing ephemeral fauna suggest that competition for food is more important than predation in reducing population size of the organisms of the food web.

The fauna of desert ephemeral ponds is essentially the same in the world's deserts. Many of the Anostraca, Conchostraca, Notostraca, and Cladocera in the rain pools in the Sahara are the same genera as found in North American ephemeral lakes (Rzoska, 1984, Maeda-Martinez, 1991). Because of these similarities, it is likely that the food web relationships and the temporal succession of the fauna will be similar to that documented in the Chihuahuan Desert in other deserts of the world.

REFERENCES

Abramsky, Z. (1983). Experiments on seed predation by rodents and ants in the Israeli desert. *Oecologia* **57**, 328–332.

Anderson, J. E. and Schumar, M. I. (1986). Impacts of black-tailed jackrabbits at peak population densities on sagebrush-steppe vegetation. *J. Range Mgmt* **39**, 152–156.

Anderson, R. A. and Karasov, W. H. (1988). Energetics of the lizard *Cnemidophorus tigris* and life history consequences of food-acquisition mode. *Ecol. Monogr.* **58**, 79–110.

Andrew, M. H. and Lange, R. T. (1986). Development of a new piosphere in arid chenopod shrubland grazed by sheep. 2. Changes to the vegetation. *Aust. J. Ecol.* **11**, 411–424.

Ayal, Y. (1994). Time-lags in insect response to plant productivity: significance for plant insect interactions in deserts. *Ecol. Entomol.* **19,** 207–214.

Badawi, A., Faragalla, A. A., and Dabbour, A. (1984). Population studies of some termites in Al-Kharji oasis, central region of Saudi Arabia. *Z. Angewandte Entomol.* **97,** 253–261.

Baharov, D. and Rosenzweig, M. L. (1985). Optimal foraging in Dorcas gazelles. *J. Arid Environ.* **9,** 167–171.

Belsky, A. J. (1986). Does herbivory benefit plants? A review of the evidence. *Am. Nat.* **127,** 870–892.

Belsky, A. J. (1987). The effects of grazing: confounding of ecosystem, community, and organism scales. *Am. Nat.* **129,** 777–783.

Belsky, A. J., Carson, W. P., Jensen, C. L., and Fox, G. A. (1993). Overcompensation by plants: herbivore optimization or red herring? *Evol. Ecol.* **7,** 109–121.

Bock, C. E. and Bock, J. H. (1994). Effects of predator exclusion on rodent abundance in an Arizona semidesert grassland. *Southwest. Nat.* **39,** 208–210.

Boldt, P. E. and Robbins, T. O. (1994). Phytophagous insect faunas of *Baccharis salicina, B. pteronioides,* and *B. bigelovii* (Asteraceae) in the southwestern United States and northern Mexico. *Environ. Entomol.* **23,** 47–57.

Bouskila, A. (1995). Interactions between predation risk and competition: a field study of kangaroo rats and snakes. *Ecology* **76,** 165–178.

Bowers, M. A. (1990). Exploitation of seed aggregates by Merriam's kangaroo-rat: harvesting rates and predatory risk. *Ecology* **71,** 2334–2344.

Briese, D. T. and Macauley, B. J. (1977). Physical structure of an ant community in semi-arid Australia. *Aust. J. Ecol.* **2,** 107–120.

Brown, J. H. and Davidson, D. W. (1977). Competition between seed-eating rodents and ants in desert ecosystems. *Science* **196,** 880–882.

Brown, J. H. and Heske, E. J. (1990). Control of a desert-grassland transition by a keystone rodent guild. *Science* **250,** 1705–1707.

Brown, J. H. and Ojeda, R. A. (1987). Granivory: patterns, processes, and consequences of seed consumption on two continents. *Revi. Chil. Hist. Nat.* **60,** 337–349.

Brown, J. H., Grover, J. J., Davidson, D. W., and Liebermann, G. A. (1975). A preliminary study of seed predation in desert and montane habitats. *Ecology* **56,** 987–992.

Bryant, J. P., Heitkonig, I., Kuropat, P., and Owen-Smith, N. (1991). Effects of severe defoliation on the long-term resistance to insect attack and on leaf chemistry in six woody species of the southern African savanna. *Am. Nat.* **137,** 50–63.

Campos, C. M. and Ojeda, R. A. (1997). Dispersal and germination of *Prosopis flexuosa* (Fabaceae) seeds by desert mammals in Argentina. *J. Arid Environ.* **35,** 707–714.

Cates, R. G. (1981). Host plant predictability and the feeding patterns of monophagous, oligophagous, and polyphagous insect herbivores. *Oecologia* **48,** 319–326.

Cates, R. G. and Rhoades, D. (1977a). Patterns in the production of antiherbivore chemical defenses in plant communities. *Biochem. Syst. Ecol.* **5,** 185–193.

Cates, R. G. and Rhoades, D. (1977b). Prosopis leaves as a resource for insects. In B. Simpson (ed.), *Mesquite, Its Biology in Two Desert Scrub Ecosystems,* pp. 61–83. US/IBP Synthesis Series 4. Dowden, Hutchinson, and Ross, Inc. Stroudsburg, PA.

Chew, R. M. (1974). Consumers as regulators of ecosystems: an alternative to energetics. *Ohio J. Sci.* **74,** 359–370.

Chew, R. M. (1976). Some ecological characteristics of the ants of a desert-shrub community in southeastern Arizona. *Am. Midl. Nat.* **98,** 33–49.

Chew, R. M. and Chew, A. E. (1970). Energy relationships of the mammals of a desert shrub (*Larrea tridentata*) community. *Ecol. Monogr.* **40,** 1–21.

Chew, R. M. and DeVita, J. (1980). Foraging characteristics of a desert ant assemblage: functional morphology and species separation. *J. Arid Environ.* **3,** 75–83.

Conley, M. R. (1985). Predation versus resource limitation in survival of adult burrowing spiders (Araneae: Lycosidae). *Oecologia* **67,** 71–75.

Cooper, S. M. and Owen-Smith, N. (1986). Effects of plant spinescence on large mammalian herbivores. *Oecologia* **68**, 446–455.

Coppock, D. L., Detling, J. K., Ellis, J. E., and Dyer, M. I. (1983a). Plant–herbivore interactions in a North American mixed-grass prairie. I. Effects of black-tailed prairie dogs on intraseasonal above-ground plant biomass and nutrient dynamics and plant species diversity. *Oecologia* **56**, 1–9.

Coppock, D. L., Ellis, J. E., Detling, J. K., and Dyer, M. I. (1983b). Plant–herbivore interactions in a North American mixed-grass prairie. II. Responses of bison to modification of vegetation by prairie dogs. *Oecologia* **56**, 10–15.

Crawford, C. S. (1981). *Biology of Desert Invertebrates*. Springer-Verlag, New York.

Crawford, C. S. (1988). Surface-active arthropods in a desert landscape: Influences of microclimate, vegetation, and soil texture on assemblage structure. *Pedobiologia* **32**, 373–385.

Crawford, C. S., MacKay, W. P., and Cepeda-Pizarro, J. G. (1993) Detritivores of the Chilean arid zone (27–32°S) and the Namib Desert: a preliminary comparison. *Rev. Chil. Hist. Nat.* **66**, 283–289.

Crawley, M. J. (1983). Herbivory, *The Dynamics of Animal–Plant Interactions*. University of California Press, Berkeley.

Creusere, F. M. and Whitford, W. G. (1976). Ecological relationships in a desert anuran community. *Herpetologica* **32**, 7–18.

De Casenave, J. L., Cueto, V. R., and Marone, L. (1998). Granivory in the Monte Desert, Argentina: is it less intense than in other arid zones of the world? *Global Ecol. Biogeogr. Lett.* **7**, 197–204.

Degen, A. A., Khokhlova, I. S., Kam, M., and Nagy, K. A. (1997). Body size, granivory and seasonal dietary shifts in desert gerbilline rodents. *Funct. Ecol.* **11**, 53–59.

Denny, M. J. S. (1982). Adaptations of the red kangaroo and euro (Macropodidae) to aridity. In W. R. Barker and P. J. M. Greenslade (eds), *Evolution of the Flora and Fauna of Arid Australia*, pp 179–183. Peacock Publications. Glen Osmond, South Australia.

De Vries, M. F. W. and Schippers, P. (1994). Foraging in a landscape mosaic: selection for energy and minerals in free-ranging cattle. *Oecologia* **100**, 107–117.

Dickman, C. R., Letnic, M., and Mahon, P. S. (1999). Population dynamics of two species of dragon lizards in arid Australia: the effects of rainfall. *Oecologia* **119**, 357–366.

Dyer, M. I. and Bokhari, U. G. (1976). Plant–animal interactions: studies of the effects of grasshopper grazing on blue grama grass. *Ecology* **57**, 762–772.

Edney, E. B., Haynes, S., and Gibo. D. (1974). Distribution and activity of the desert cockroach *Arenivaga investigata* (Polyphagidae) in relation to microclimate. *Ecology* **55**, 420–427.

Ernest, K. A. (1994). Resistance of creosotebush to mammalian herbivory: temporal consistency and browsing-induced changes. *Ecology* **75**, 1684–1692.

Estell, R. E., Fredrickson, E. L, and Havstad, K. M. (1996). Chemical composition of *Flourensia cernua* at four growth stages. *Grass Forage Sci.* **51**, 434–441.

Feeny, P. (1976). Plant apparency and chemical defense. *Recent Adv. Phytochem.* **10**, 1–40.

Floyd, T. (1996). Top-down impacts on creosotebush herbivores in a spatially and temporally complex environment. *Ecology* **77**, 1544–1555.

Foster, D. E., Ueckert, D. N., and Deloach, C. J. (1981). Insects associated with broom snakeweed (*Xanthocephalum sarothrae*) and threadleaf snakeweed (*Xanthocephalum microcephala*) in west Texas and eastern New Mexico. *J. Range Mgmt* **34**, 446–454.

Freckman, D. W. and Mankau, R. (1986). Abundance, distribution, biomass, and energetics of soil nematodes in a northern Mojave Desert ecosystem. *Pedobiologia* **29**, 137–139.

Freckman, D. W., Whitford, W. G., and Steinberger, Y. (1987). Effect of irrigation on nematode population dynamics and activity in desert soils. *Biology and Fertility of Soils* **3**, 3–10.

Gardener, C. J., McIvor, J. G., and Jansen, A. (1993). Passage of legume and grass seeds through the digestive tract of cattle and their survival in faeces. *J. Appl. Ecol.* **30**, 63–74.

Greenwood, K. L., MacLeod, D. A., and Hutchinson, K. J. (1997). Long-term stocking rate effects on soil physical properties. *Aust. J. Exp. Agric.* **37**, 413–419.

Grenot, C. and Vernet, R. (1972). Les reptiles dans l'ecosteme au Sahara occidental. *Soc. Biogeogr.* **1972**, 96–112.

Guo, Q., Thompson, D. B., Valone, T. J., and Brown, J. H. (1995). The effects of vertebrate granivores and folivores on plant community structure in the Chihuahuan Desert. *Oikos* **73**, 251–259.

Gutierrez, J. R., Meserve, P. L., Herrera, S., Contreras, L. C., and Jaksic, F. M. (1997). Effects of small mammals and vertebrate predators on vegetation in the Chilean semiarid zone. *Oecologia* **109**, 398–406.

Heatwole, H. (1996). *Energetics of Desert Organisms*. Springer, Berlin.

Heatwole, H. and Muir, R. (1991). Foraging, abundance and biomass of ants in the pre-Saharan steppe of Tunisia. *J. Arid Environ.* **21**, 337–350.

Heske, E. J., Brown, J. H., and Guo, Q. (1993). Effects of kangaroo rat exclusion on vegetation structure and plant species diversity in the Chihuahuan Desert. *Oecologia* **95**, 520–524.

Hilbert, D. W., Swift, D. M., Detling, J. K., and Dyer, M. I. (1981). Relative growth rates and the grazing optimization hypothesis. *Oecologia* **51**, 14–18.

Hoffman, M. T., James, C. D., Kerley, G. I. H., and Whitford, W. G. (1993). Rabbit herbivory and its effect on cladode, flower and fruit production of *Opuntia violacea* var. *macrocentra* (Cactaceae) in the northern Chihuahuan Desert, New Mexico. *Southwest. Nat.* **38**, 309–315.

Huey, R. B. and Pianka, E. R. (1981). Ecological consequences of foraging mode. *Ecology* **62**, 991–999.

Hunter, M. D. and Price, P. W. (1992). Playing chutes and ladders: heterogeneity and relative roles of bottom-up and top-down forces in natural communities. *Ecology* **73**, 377–406.

James, C. D. (1991). Temporal variation in diets and trophic partitioning by coexisting lizards (*Ctenotus*: Scincidae) in central Australia. *Oecologia* **85**, 553–561.

Johnson, H. B., Vasek, F. C., and Yonkers, T. (1978). Residual effects of summer irrigation on Mojave Desert annuals. *Bull. South. Calif. Acad. Sci.* **77**, 95–108.

Johnson, R. A. and Wood, T. G. (1980). Termites of the arid zones of Africa and the Arabian Peninsula. *Sociobiology* **5**, 279–293.

Jones, C. G., Lawton, J. H., and Shachak, M. (1994). Organisms as ecosystem engineers. *Oikos* **69**, 373–386.

Kelt, D. A., Brown, J. H., Heske, E. J., Marquet, P. O., Morton, S. R., Reid, J. R. W., Rogovin, K. A., and Shenbrot, G. (1996). Community structure of desert small mammals: comparisons across four continents. *Ecology* **77**, 746–761.

Kerley, G. I. H. (1991). Seed removal by rodents, birds and ants in the semi-arid Karoo, South Africa. *J. Arid Environ.* **20**, 63–69.

Kerley, G. I. H., and Whitford, W. G. (1994). Desert dwelling small mammals as granivores: intercontinental variations. *Aust. J. Zool.* **42**, 543–555.

Kerley, G. I. H. and Whitford, W. G. (2000). Impact of grazing and desertification in the Chihuahuan Desert: plant communities, granivores and granivory. *Am. Midl. Nat.* **144**, 78–91.

Kerley, G. I. H., Tiver, F., and Whitford, W. G. (1993). Herbivory of clonal populations: cattle browsing affect reproduction and population structure of *Yucca elata*. *Oecologia* **93**, 12–17.

Kerley, G. I. H., Whitford, W. G., and Kay, F. R. (1997). Mechanisms for the keystone status of kangaroo rats: graminivory rather than granivory? *Oecologia* **111**, 422–428.

Kinnear, A. (1991) Acarine communities of semi-arid soils from the Eastern Goldfields region of Western Australia. *Pedobiologia* **35**, 273–283.

Kotler, B. P. (1984). Risk of predation and the structure of desert rodent communities. *Ecology* **65**, 689–701.

Kramp, B. A., Ansley, R. J., and Tunnell, T. R. (1998). Survival of mesquite seedlings emerging from cattle and wildlife feces in a semi-arid grassland. *Southwest. Nat.* **43**, 300–312.

Lambert, M. R. K. (1984). Amphibians and reptiles. In J. L. Cloudsley-Thompson (ed.), *Sahara Desert*, pp. 205–227. Pergamon Press, Oxford.

Lawton, J. H. (1994). What do species do in ecosystems? *Oikos* **71**, 367–374.

Leman, C. A. (1978). Seed size selection in heteromyids: a second look. *Oecologia* **35**, 13–19.

Lightfoot, D. C. and Whitford, W. G. (1987). Variation in insect densities on desert creosotebush: is nitrogen a factor? *Ecology* **68**, 547–557

Lightfoot, D. C. and Whitford, W. G. (1989). Interplant variation in creosotebush foliage character-istics and canopy arthropods. *Oecologia* **81**, 166–175.

Lightfoot, D. C. and Whitford, W. G. (1991). Productivity of creosotebush foliage and associated canopy arthropods along a desert roadside. *Am. Midl. Nat.* **125**, 310–322.

Lobry de Bruyn, L. A. (1993). Defining soil macrofauna composition and activity for biopedologi-cal studies: a case study on two soils in the Western Australian wheatbelt. *Aust. J. Soil Res.* **31**, 83–95.

Loring, S. J., Weems, D. C., and Whitford, W. G. (1988a). Abundance and diversity of surface-active Collembola along a watershed in the northern Chihuahuan Desert. *Am. Midl. Nat.* **119**, 21–30.

Loring, S. J., MacKay, W. P., and Whitford, W. G. (1988b). Ecology of small desert playas. In J. L. Thames and C. D. Ziebell (eds), *Small Water Impoundments in Semi-arid Regions*, pp. 89–113. University of New Mexico Press, Albuquerque.

MacKay, W. P. (1991). The role of ants and termites in desert communities. In G. A. Polis (ed.), *The Ecology of Desert Communities*, pp. 113–150. The University of Arizona Press, Tucson.

MacKay, W. P., Loring, S. J., Frost, T. M., and Whitford, W. G. (1990). Population dynamics of a playa community in the Chihuahuan Desert. *Southwest. Nat.* **35**, 393–402.

Maeda-Martinez, A. M. (1991). Distribution of species of Anostraca, Notostraca, Spinicaudata, and Laevicaudata in Mexico. *Hydrobiologia* **212**, 209–219.

Main, B. Y. (1982). Adaptations to arid habitats by mygalomorph spiders. In W. R. Barker and P. J. M. Greenslade (eds), *Evolution of the Flora and Fauna of Arid Australia*, pp. 273–283. Peacock Publications, Glen Osmond, South Australia.

Mares, M. A. and Rosenzweig, M. L. (1978). Granivory in North and South American deserts: rodents, birds, and ants. *Ecology* **59**, 235–241.

Marone, L. and Horno, M. E. (1997). Seed reserves in the central Monte Desert, Argentina: impli-cations for granivory. *J. Arid Environ.* **36**, 661–670.

Marone, L., Rossi, B. E., and deCasenave, J. L. (1998). Granivore impact on soil-seed reserves in the Monte Desert, Argentina. *Funct. Ecol.* **12**, 640–645

Marsh, A. C. (1985). Forager abundance and dietary relationships in a Namib Desert ant community. *South Afr. J. Zool.* **20**, 197–203.

Matson, P. A. and Hunter, M. D. (1992). The relative contributions of top-down and bottom-up forces in population and community ecology. *Ecology* **73**, 723.

McNaughton, S. J. (1979). Grazing as an optimization process: grass-ungulate relationships in the Serengeti. *Am. Nat.* **113**, 691–703.

McNaughton, S. J. (1988). Mineral nutrition and spatial concentrations of ungulates. *Nature (London)* **334**, 343–345.

Meserve, P. L., Gutierrez, J. R., and Jaksic, F. M. (1993) Effects of vertebrate predation on a caviomorph rodent, the degu (*Octodon degus*) in a semiarid thorn scrub community in Chile. *Oecologia* **94**, 153–158.

Meserve, P. L., Gutierrez, J. R., Yunger, J. A., Contreras, L. C., and Jaksik, F. M. (1996). Role of biotic interactions in a small mammal assemblage in semiarid Chile. *Ecology* **77**, 133–148.

Milton, S. J. (1991). Plant spinescence in arid southern Africa: does moisture mediate selection by mammals? *Oecologia* **87**, 279–287.

Milton, S. J. (1993). Insects from the shrubs *Osteospermum sinuatum* and *Pteronia pallens* (Asteraceae) in the southern Karoo. *Afr. Entomol.* **1**, 257–262.

Mispagel, M. E. (1978). The ecology and bioenergetics of the acridid grasshopper, *Bootettix punctatus*, on creosotebush, *Larrea tridentata*, in the northern Mojave Desert. *Ecology* **59**, 779–788.

Montiel, S. and Montana, C. (2000). Vertebrate fugivory and seed dispersal of a Chihuahuan Desert cactus. *Plant Ecol.* **146**, 221–229.

Morton, S. R. (1985). Granivory in arid regions: comparison of Australia with North and South America. *Ecology* **66**, 1859–1866.

Morton, S. R. and Baynes, A. (1985). Small mammal assemblages in arid Australia: a reappraisal. *Aus. Mammal.* **8**, 159–169.

Morton, S. R. and James, C. D. (1988). The diversity and abundance of lizards in arid Australia: a new hypothesis. *Am. Nat.* **132,** 237–256.

Morton, S. R., Hinds, D. S., and MacMillen, R. E. (1980). Cheek pouch capacity in heteromyid rodents. *Oecologia* **46,** 143–146.

Munger, J. C. (1984). Optimal foraging? Patch use by horned lizards (Iguanidae: Phrynosoma) *Am. Nat.* **123,** 654–680.

Nagy, K. A., Huey, R. B., and Bennett, A. F. (1984). Field energetics and foraging mode of Kalahari lacertid lizards. *Ecology* **65,** 588–596.

Nash, M. S., Whitford, W. G., Van Zee, J., and Havstad, K. (1998). Monitoring changes in stressed ecosystems using spatial patterns of ant communities. *Environ. Monit. Assess.* **51,** 201–210.

Newman, R. A. (1987). Effects of density and predation on *Scaphiopus couchi* tadpoles in desert ponds. *Oecologia* **71,** 301–307.

Nilsen, E. T., Sharifi, M. R., Virginia, R. A., and Rundel, P. W. (1987). Phenology of warm desert phreatophytes: seasonal growth and herbivory in *Prosopis glandulosa* var. *torreyana* (honey mesquite). *J. Arid Environ.* **13,** 217–229.

Noble, J. C. (1975). The effects of emus (*Dromaius novaehollandiae* Latham) on the distribution of the nitre bush (*Nitraria billardieria* DC.) *J. Ecol.* **63,** 979–984.

Noble, J. C., Whitford, W. G., and Kaliszewski, M. (1994). Soil and litter microarthropod populations from two contrasting ecosystems in semi-arid eastern Australia. *J. Arid Environ.* **32,** 329–346.

Oesterheld, M. and McNaughton, S. J. (1991). Effect of stress and time for recovery on the amount of compensatory growth after grazing. *Oecologia* **85,** 305–313.

Otte, D. and Joern, A. (1977). On feeding patterns in desert grasshoppers and the evolution of specialized diets. *Proc. Acad. Nat. Sci. Philadelphia* **128,** 89–126.

Parker, L. W., Freckman, D. W., Steinberger, Y., Driggers, L., and Whitford W. G. (1984). Effects of simulated rainfall and litter quantities on desert soil biota: soil respiration, microflora, and protozoa. *Pedobiologia* **27,** 185–195.

Parmenter, R. R., MacMahon, J. A., and Vander Wall, S. B. (1984) The measurement of granivory by desert rodents, birds, and ants: a comparison of an energetics approach and a seed-dish technique. *J. Arid Environ.* **7,** 75–92.

Pianka, E. R. (1985). Some intercontinental comparisons of desert lizards. *Nat. Geogr. Res.* **1,** 490–504.

Polis, G. A. (1994). Food webs, trophic cascades and community structure. *Aust. J. Ecol.* **19,** 121–136.

Polis, G. A., Myers, C., and Quinlan, M. (1986). Burrowing biology and spatial distribution of desert scorpions. *J. Arid Environ.* **10,** 137–146.

Predavec, M. (1997). Seed removal by rodents, ants and birds in the Simpson Desert, central Australia. *J. Arid Environ.* **36,** 327–332.

Price, M. V. and Brown, J. H. (1983). Patterns of morphology and resource use in a North American desert rodent community. *Great Basin Nat. Mem.* **7,** 117–134.

Reichman, O. J. (1975). Relation of desert rodent diets to available resources. *J. Mammal.* **56,** 731–751.

Rhoades, D. F. (1977). The antiherbivore chemistry of *Larrea*. In T. J. Mabry, J. H. Hunziker and D. R. DiFeo Jr. (eds), *Creosotebush: Biology and Chemistry of Larrea in New World Deserts*, pp. 135–175. US/IBP Synthesis Series. Dowden Hutchinson and Ross Inc. Stroudsburg, PA.

Rhoades, D. and Cates, R. G. (1976). Toward a general theory of plant antiherbivore chemistry. *Recent Adv. Phytochem.* **10,** 168–213.

Rojas, P. and Fragoso, C. (2000). Composition, diversity, and distribution of a Chihuahuan Desert ant community (Mapimi, Mexico). *J. Arid Environ.* **44,** 213–227.

Roundy, B. A. and Ruyle, G. B. (1989). Effects of herbivory on twig dynamics of a Sonoran Desert shrub *Simmondsia chinensis* (Link) Schn. *J. Appli. Ecol.* **26,** 701–710.

Rzoska, J. (1984). Temporary and other waters. In J. L. Cloudsley-Thompson (ed.), *Sahara Desert*, pp. 105–114. Pergamon Press. Oxford.

Samson, D. A., Philippi, T. E., and Davidson, D. W. (1992). Granivory and competition as determinants of annual plant diversity in the Chihuahuan Desert. *Oikos* **65**, 61–80.

Santos, P. F., DePree, E., and Whitford, W. G. (1978). Spatial distribution of litter and microarthropods in a Chihuahuan Desert ecosystem. *J. Arid Environ.* **1**, 41–48.

Schowalter, T. D. (1996). Arthropod associates and herbivory on tarbush in southern New Mexico. *Southwest. Nat.* **41**, 140–144.

Schowalter, T. D., Lightfoot, D. C., and Whitford, W. G. (1999). Diversity of arthropod responses to host–plant water stress in a desert ecosystem in southern New Mexico. *Am. Midl. Nat.* **142**, 281–290.

Schultz, J. C., Otte, D., and Enders, F. (1977). Larrea as a habitat component for desert arthropods. In T. J. Mabry, J. H. Hunziker and D. R. DiFeo Jr (eds), *Creosotebush: Biology and Chemistry of Larrea in New World Deserts*, pp. 176–208. US/IBP Synthesis Series. Dowden Hutchinson and Ross Inc. Stroudsburg, PA.

Schumacher, A. and Whitford, W. G. (1976). Spatial and temporal variation in Chihuahuan Desert ant faunas. *Southwest. Nat.* **21**, 1–8.

Seely, M. K. and Louw, G. N. (1980). First approximation of the effects of rainfall on the ecology and energetics of a Namib Desert dune ecosystem. *J. Arid Environ.* **3**, 25–54.

Seely, M. K. and Mitchell, D. (1987). Is the subsurface environment a thermal haven for chthonic beetles? *South Afr. J. Zool.* **22**, 57–61.

Shachak, M. and Brand, S. (1983). The relationship between sit and wait foraging strategy and dispersal in the desert scorpion, *Scorpio marus palmatus*. *Oecologia* **60**, 371–377.

Shaffer, D. T. and Whitford, W. G. (1981). Behavioral responses of a predator, the round-tailed horned lizard, *Phrynosoma modestum*, and its prey, honey pot ants, *Myrmecocystus* spp. *Am. Midl. Nat.* **105**, 209–216.

Soholt, L. F. (1973). Consumption of primary production by a population of kangaroo rats (Dipodomys merriami) in the Mojave Desert. *Ecol. Monogr.* **43**, 357–376.

Steinberger, Y. and Whitford, W. G. (1983). The contribution of shrub pruning by jackrabbits to litter input in a Chihuahuan Desert ecosystem. *J. Arid Environ.* **6**, 177–181.

Stock, W. D., Le Roux, D., and Van der Heyden, F. (1993). Regrowth and tannin production in woody and succulent Karoo shrubs in response to simulated browsing. *Oecologia* **96**, 562–568.

Thompson, K. and Uttley, M. G. (1982). Do grasses benefit from grazing? *Oikos* **39**, 113–115.

Tigar, B. J. and Osborne, P. E. (1997). Patterns of arthropod abundance and diversity in an Arabian desert. *Ecography* **20**, 550–558.

Tigar, B. J. and Osborne, P. E. (1999). Patterns of biomass and diversity of aerial insects in Abu Dhabi's sandy deserts. *J. Arid Environ.* **43**, 159–170.

Turner, F. B. and Chew, R. M. (1981). Production by desert animals. In D. W. Goodall and R. A. Perry (eds), *Arid Lands Ecosystems*, pp. 199–269. Cambridge University Press, Cambridge, UK.

Valone, T. J., Brown, J. H., and Heske, E. J. (1994). Interactions between rodents and ants in the Chihuahuan Desert: an update. *Ecology* **75**, 252–255.

Vander Wall, S. B., Longland, W. S., Pyare, S., and Veech, J. A. (1998). Cheek pouch capacities and loading rates of heteromyid rodents. *Oecologia* **113**, 21–28.

Van Zee, J. W., Whitford, W. G., and Smith, W. E. (1997). Mutual exclusion by dolichoderine ants on a rich food source. *Southwest. Nat.* **42**, 229–231.

Vasquez, R. A., Bustamante, R. O., and Simonetti, J. A. (1995). Granivory in the Chilean matorral: extending the information on arid zones of South America. *Ecography* **18**, 403–409.

Wallwork, J. A. (1982). *Desert Soil Fauna*. Praeger, New York.

Wallwork, J. A., Kamill, B. W., and Whitford, W. G. (1985). Distribution and diversity patterns of soil mites and other microarthropods in a Chihuahuan Desert site. *J. Arid Environ.* **9**, 215–231.

Wandera, J. L., Richards, J. H., and Mueller, R. J. (1992). The relationships between relative growth rate, meristematic potential and compensatory growth of semiarid-land shrubs. *Oecologia* **90**, 391–398.

Watts, J. G., Huddleston, E. W., and Owens, J. C. (1982). Rangeland entomology. *Annu. Rev. Entomol.* **27**, 283–311.

White, T. C. R. (1984). The abundance of invertebrate herbivores in relation to the availability of nitrogen in stressed food plants. *Oecologia* **63,** 90–105.

Whitford, W. G. (1974). *Jornada Validation Site: Validation Studies.* US/IBP Desert Biome Research Memorandum 74–4. 110 p., Utah State University, Logan, UT.

Whitford, W. G. (1975). *Jornada Validation Site: Validation Studies.* US/IBP Desert Biome Research Memorandum 75–4. 104 p., Utah State University, Logan, UT.

Whitford, W. G. (1976). Temporal fluctuations in density and diversity of desert rodent populations. *J. Mammal.* **57,** 351–369.

Whitford, W. G. (1978). Structure and seasonal activity of Chihuahuan Desert ant communities. *Insectes Sociaux* **25,** 79–88.

Whitford, W. G. (1989). Abiotic controls of the functional structure of soil food webs. *Biol. Fertil. Soils* **8,** 1–6.

Whitford, W. G. (1991). Subterranean termites and long-term productivity of desert rangelands. *Sociobiology* **19,** 235–243.

Whitford, W. G. (1996). The importance of the biodiversity of soil biota in arid ecosystems. *Biodiversity Conserv.* **5,** 185–195.

Whitford, W. G. (1998). Contribution of pits dug by goannas (*Varanus gouldii*) to the dynamics of banded mulga landscapes in eastern Australia. *J. Arid Environ.* **40,** 453–457.

Whitford, W. G. (2000). Keystone arthropods as webmasters in desert ecosystems. In D. C. Coleman and P. F. Hendrix (eds), *Invertebrates as Webmasters in Ecosystems*, pp. 25–41. CABI Publishing, New York.

Whitford, W. G. and Bryant, M. (1979). Behavior of a predator and its prey: the horned lizard (*Phrynosoma cornutum*) and harvester ants (*Pogonomyrmex* spp.). *Ecology* **60,** 686–694.

Whitford, W. G. and Creusere, F. M. (1977). Seasonal and yearly fluctuations in Chihuahuan Desert lizard communities. *Herpetologica* **33,** 54–65.

Whitford, W. G. and Kay, F. R. (1999). Biopedturbation by mammals in deserts: a review. *J. Arid Environ.* **41,** 203–230.

Whitford, W. G. and Steinberger, Y. (1989). The long-term effects of habitat modification on a desert rodent community. In D. W. Morris, Z. Abramsky, B. J. Fox and M. R. Willig (eds), *Patterns in the Structure of Mammalian Communities*, pp 33–43. Texas Tech University Press, Lubbock, Texas.

Whitford, W. G., DePree, D. J., and Johnson, R. K. Jr. (1978). The effects of twig girdlers (Cerambycidae) and node borers (Bostrichidae) on primary production in mesquite (*Prosopis glandulosa*). *J. Arid Environ.* **1,** 345–350.

Whitford, W. G., Forbes, G. S., and Kerley, G. I. (1995). Diversity, spatial variability and functional roles of invertebrates in desert grassland ecosystems. In M. P. McClaran and T. R. Van Devender (eds), *The Desert Grassland*, pp. 152–195. University of Arizona Press. Tucson.

Whitford, W. G., Van Zee, J., Nash, M. S., Smith, W. E., and Herrick, J. E. (1999). Ants as indicators of exposure to environmental stressors in North American desert grasslands. *Environ. Monit. Assess.* **54,** 143–171.

Wisdom, C. S., Crawford, C. S., and Aldon, E. F. (1989). Influence of insect herbivory on photosynthetic area and reproduction in *Gutierrezia* species. *J. Ecol.* **77,** 685–692.

Wisdom, W. A. and Whitford, W. G. (1981). Effects of vegetation change on ant communities of arid rangelands. *Environ. Entomol.* **10,** 893–897.

Woodward, B. D. (1982). Tadpole interactions in the Chihuahuan Desert at two experimental densities. *Southwest. Nat.* **27,** 119–122.

Chapter 9 | Decomposition and Nutrient Cycling

9.1 NUTRIENT LIMITATIONS

There is no question that water availability limits productivity in deserts. That is intuitively obvious even to the casual observer. Because of the overriding importance of water, many investigators have tended to pass off potential nutrient limitations as unimportant. In arid regions nutrient limitations to productivity and general ecosystem functioning may be important only under low probability climates, e.g. a succession of years of above average rainfall. However, nutrient limitations may also occur at other times when water availability is high even if only for short time periods.

The nutrient most frequently cited as limiting productivity is nitrogen. Charley and Cowling (1968) were among the first ecologists to suggest that nitrogen and other nutrients may be limiting productivity in semiarid ecosystems. They reported that primary production was reduced during prolonged wet periods and attributed that reduction to less available nitrogen. Ludwig and Flavill (1979) reported a similar response in productivity in the northern Chihuahuan desert, and Floret *et al.* (1982) suggested that nitrogen limitation was the factor limiting production in the Tunisian desert during wet periods. In an extensive study in the Sahel, a group of scientists from Wageningen, the Netherlands, documented that at rainfall amounts above approximately 200 mm yr^{-1} nitrogen availability limited production (Penning de Vries and Djiteye, 1982).

In the northern Chihuahuan desert, Ettershank *et al.* (1978) reported large increases in net above ground production in the creosotebush, *Larrea tridentata* and a perennial grass, fluff-grass, *Erioneuron pulchellum*, in response to fertilization with ammonium nitrate. When fertilizer was applied at a low rate, only the shallow-rooted grass responded, probably because the grass sequestered the N before leaching carried it to the root zone of the shrubs. The nitrogen limitation–water relationships studied in a series of water supplementation–fertilization studies in the Chihuahuan desert has documented that nitrogen availability affects biomass production of perennial and annual plants and the species composition of annual plant communities (Gutierrez and Whitford, 1987; Fisher *et al.*, 1988; Gutierrez *et al.*, 1988). Nitrogen fertilization does not result in increased productivity of long-lived perennial desert grasses. Irrigation resulted in higher biomass production in black-grama, *Bouteloua eriopoda,* and also affected flowering, tiller mortality, and insect damage. Nitrogen fertilization reduced the root lengths of the grass clumps. Productivity in the *B. eriopoda* desert grassland appears to be closely linked to rainfall with no temporal lags due to nitrogen immobilization (Stephens and Whitford, 1993). These findings present a real dichotomy in the way that dominant plant species affect the ecosystem responses to rainfall.

Nitrogen fertilization and irrigation experiments demonstrated that nitrogen availability limits productivity if moisture availability is high for a complete plant growth cycle (James and Jurinak, 1978; Romney *et al.*, 1978). Nitrogen fertilization plus supplemental moisture increased in new shoot growth and/or leaf nitrogen content above that of unfertilized plants and in some but not all species of shrubs. Biomass production of several species of annual plants was increased by nitrogen fertilization (Romney *et al.*, 1978). Fertilization studies of native vegetation in the Sahel documented the importance of nitrogen and phosphorus in these semiarid systems. 'The low fertility of the soils, especially in nitrogen and phosphates, is often much more a limiting factor than the low irregular rainfall' (Penning de Vries and Djiteye, 1982). Soil N and P levels were relatively constant moving from north to south with the result that the quality of the vegetation decreased along a north to south gradient. The data of Penning de Vries and Djiteye (1982) suggested that productivity in the southern Sahara–Sahel was water limited at average precipitation of about 150 mm yr^{-1} and by nutrients at greater average rainfall amounts (Fig. 9.1).

Nutrient limitation of productivity is well documented in the semiarid and arid areas of Australia. Charley and Cowling (1968) found nutrient deficiencies, especially nitrogen, limited primary production saltbush desert areas of New South Wales (Charley and Cowling, 1968). They emphasized the importance of nitrogen availability in that system. Several nutrients were found to be potentially limiting in Central Australian ecosystems. In Mitchell grassland, open woodland, and mulga shrubland, nitrogen and phosphorus were found to limit grass production. Other nutrients were deficient and affected productivity in one or more of the ecosystems studied, e.g. sulfur (Freidel *et al.*, 1980). Freidel *et al.* (1980) pointed out two adaptations that are common in ecosystems of low fertility: (1) conservation of nutrients within living plant material and (2) rapid external cycling of nutrients through litter fall. It is important to keep these generalizations in mind as we examine the details of nutrient cycling processes in deserts. In addition to the direct evidence of increased production with application of fertilizer, other studies have invoked nutrient limitations to account for reduction in productivity despite adequate soil moisture availability (Ludwig and Flavill, 1979; Floret *et al.*, 1982). Of the potential long list of nutrients that could be limiting productivity in one or more arid ecosystems, nitrogen and phosphorus are most likely to be limiting in most desert ecosystems. Despite the differences in origin (atmospheric pool vs. rock weathering) both nutrients are processed by complex biological interactions in the soil. Because phosphorus is a product of rock weathering it is in lower concentration in the ancient shield-platform deserts than in the more recent basin and range deserts. Despite these differences, the biological processes and interactions with climate are likely to be similar in all subtropical deserts.

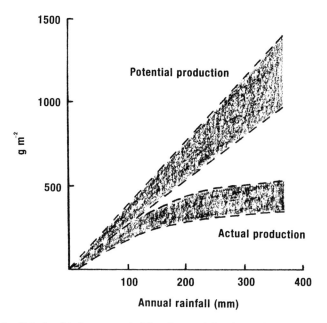

Figure 9.1 Relationship between rainfall and potential and actual primary production. The difference in the curves is due to nutrient limitations. Figure based on Penning de Vries, F. W. T. and Djiteye, M. A. (eds) 1982. La productivite des pasturages saheliens. Reproduced with permission of Backhuys Publishers B. V.

9.2 DECOMPOSITION

Decomposition is the process that releases nutrient materials held in the structure of plant and animal tissues to the soil solution where the nutrients are once again potentially available to plants. Since most nutrient elements cycle from soil solution to plant and back to the soil solution via decomposition, decomposition processes are key features of nutrient cycles. Studies in subtropical deserts have called attention to a number of striking differences in decomposition processes between moist temperate and subtropical arid and semiarid systems.

The decomposition process can be subdivided into two related processes: fragmentation and mineralization. Fragmentation results from abiotic forces that weaken and mechanically fragment large pieces of dead material or from biological activities such as comminution (chewing and passing through the gut) and lysis by enzymes. In mesic systems biological fragmentation is most important. However, in arid and semiarid ecosystems, fragmentation by abiotic means appears to be more important than biological activity. Mineralization processes involve the conversion of litter fragments into microbial biomass and the eventual death and breakdown of that microbial biomass. The breakdown of carbon compounds to CO_2 and H_2O is a mineralization process as is the deamination of protein and excretion of ammonia. Some mineralization occurs at the same time that biological fragmentation is occurring.

The spatial and temporal linkage between fragmentation and mineralization processes is tight in mesic systems: i.e. both processes occur simultaneously in these systems. In desert ecosystems, fragmentation and mineralization may be separated both temporally and spatially. This separation has profound implications for nutrient cycling processes and ultimately for soil characteristics in desert ecosystems. Both processes of decomposition are affected by biotic and abiotic parameters but to different extents and abiotic effects may be both direct or indirect. Wind, water, heat, and light may interact to effect mineralization of some plant compounds (Pauli, 1964).

In temperate mesic ecosystems there are a number of variables that have been shown to affect decomposition. In mesic systems decomposition rates vary directly as a function of actual evapotranspiration and inversely with lignin concentration (Meentemeyer, 1978). Plant materials with low C:N ratios decompose more rapidly than those with high C:N ratios. Soil arthropods accelerate decomposition by transporting spores and inoculating litter with microbial spores, and comminution (chewing into small fragments) (Swift *et al.*, 1979). Decomposition processes in deserts differ from mesic systems in terms of these variables. A general model for decomposition in deserts requires evaluation of the process with respect to each of the variables identified as important for mesic ecosystems.

9.2.1 Sources and Characteristics of Decomposable Material

Leaves of most desert plants are small and many have a waxy cuticle. Since most desert vegetation is composed of a mixture of drought deciduous, seasonally deciduous and evergreen shrubs or trees plus grasses and herbaceous plants, there is a continuous, but variable quantity of litter input in most desert ecosystems. The chemical and physical characteristics of the dead plant material are variables that affect the decomposition process. However, some of the structural correlates with decomposition rate like lignin content and carbon–nitrogen ratio of the plant material do not apply to deserts (Schaefer *et al.*, 1985). The leaves of desert plants frequently have high concentrations of waxes, volatiles, saponins and especially phenolics in the cuticular and epidermal cells. Lignin content of leaves of desert plants that have been assayed varied between 7.9% and 14.6% (Schaefer *et al.*, 1985; Steinberger and Whitford, 1988) which is low compared to oak leaves and conifer needles in which lignin content varied between 21% and 30% (Swift *et al.*, 1979). The leaves and stems of temperate mesic grasses had lignin contents between 11% and 14%. Carbon-nitrogen ratios of leaves of six desert plants varied between 16 and 52 (Schaefer *et al.*, 1985). Carbon–nitrogen ratios of the roots of two annual plants were 65 and 70 and for the shrubs, *L. tridentata* and *Zinnia acerosa* were 21 and 52, respectively (Whitford *et al.*, 1987). Carbon–nitrogen ratios of four standing dead desert grasses ranged from 71.5 to 115.4 (Montana *et al.*, 1988). These data indicate the considerable variablity in chemical and structural composition of desert plant structures.

The ultimate location of plant litter is probably of greater importance as a factor affecting decomposition than is the physical–chemical structure of the litter. Because of the generally small size and high surface to mass ratio of dead leaves, stem fragments, dead floral parts, etc., these materials are easily moved by the wind. As a consequence, litter tends to accumulate under shrubs, and in depressions (DeSoyza *et al.*, 1997). Because of wind transport, a significant fraction of the litter input is buried as small accumulations in shallow pits and depressions. The quantities buried vary seasonally as a function of digging activities of animals, wind, and litter fall (Steinberger and Whitford, 1983). Litter buried in depressions etc. is exposed to lower temperatures, protected from intense solar radiation and is in an evaporation-retarding environment. The decomposition of buried litter and dead roots, therefore, takes place in a more benign microenvironment than surface litter. Physical location thus has a dramatic effect on the mechanisms of decomposition.

Roots are the main source of carbon inputs into desert soils. Roots of annual plants enter the decomposable material pool in pulses following periods of annual plant production. The distribution of most annual plants is patchy, hence decomposing annual plant roots are concentrated in patches. The rates of turnover of roots of perennial plants are unknown in deserts. We can assume that during periods of rapid growth there is production of fine roots that die and decompose. Quantitative data on root turnover in perennials is difficult to obtain but may represent a significant fraction of the materials being cycled in an ecosystem in a year.

9.2.2 Rates and Mechanisms of Mass Loss (Surface Litter)

As pointed out above, the fate of dead plant material varies as a function of its physical–chemical composition and its physical location in the environment. Most studies of decomposition focus on litter on the soil surface. Conventional wisdom holds that surface litter in deserts should decompose slowly in brief pulses following rains. However, decomposition rates of a variety of litter types in several of the world's deserts demonstrate that this is not the case. Several investigators have suggested that decomposition rates are regulated by actual evapotranspiration (AET) and can be predicted from AET. Other factors that are included in general models of decomposition include C/N and C/P ratios, lignin concentrations, and lignin/N and lignin/P ratios. At a global scale (excluding data from desert regions) climate expressed as AET was the best predictor for decomposition constants (k values) (Aerts, 1997). In the semiarid Mediterranean region, the lignin/N ratio was the best chemical predictor of litter decomposition.

Decomposition rates in mesic environments can be predicted by a regression model that incorporates actual evapotranspiration and lignin content (Meentemeyer, 1978). That model has the form

$$Y_1 = 1.31369 + 0.05350X_1 + 0.18472X_2$$

where Y_1 is mass loss (%), X_1 is actual evapotranspiration (mm) and X_2 is actual evapotranspiration (mm) divided by lignin content in percent dry mass.

The AET model markedly underestimates mass loss in deserts. That underestimate was attributed to the relative independence of desert soil fauna from constraints imposed by dry, hot surface environments (Whitford et al., 1981). There was a slight positive not negative correlation between lignin content and mass loss and no relationship between AET and mass loss. High mid-day temperature and intense ultraviolet radiation characteristic of high elevation desert regions may have combined to produce photooxidation of lignins and other complex molecules (Schaefer et al., 1985).

The earliest suggestion that the high temperatures and intense ultraviolet light could be significant in the breakdown of organic matter in deserts was by Pauli (1964). Sunlight at wavelengths of 280–400 nm has sufficient energy to catalyze oxidations and reductions, especially of cyclic compounds like phenolics. Approximately 15% mass loss was measured in grass leaves and shrub leaves treated with mecuric chloride to eliminate all biota in short grass prairie and Chihuahuan desert respectively (Vossbrinck et al., 1979; MacKay et al., 1994). Photooxidation of structual materials which weakens leaves produces conditions where raindrop inpact will shatter pieces from the leaves. This scenario is consistent with the results of several studies of mass losses of materials on the soil surface in a desert (Moorhead and Reynolds, 1989). The contribution of abiotic factors such as high temperatures and sunlight is supported by data on grass decomposition rates which were relatively constant over time and independent of temperature and moisture. Litter C:N ratios remained constant during the study when mass losses were 40% and 80% for the grasses *Sporobolus airoides* and *Hilaria mutica* respectively (Montana et al., 1988).

Photodegradation and fragmentation by intermittent rainfall can produce the high decomposition rates reported for mass loss from surface litter in deserts (Table 9.1). There is a growing body of evidence that photodegradation can be a major factor in the decomposition of plant materials. A recent empirical study reported that UV-B decreased the proportion of lignin in plant residues, suppressed microbial respiration, and changed the decomposer fungal community structure (Gehrke et al., 1995). High amounts of UV-B radiation reach the soil surface in low- and mid-latitude deserts and that radiation combined with the high surface temperatures makes the abiotic breakdown hypothesis a reasonable explanation of the high rates of decomposition recorded in such deserts. This also explains the high *negative* correlation between decomposition rates and initial lignin content reported by Schaefer et al. (1985).

Correlations of surface litter disappearance and rainfall (Strojan et al. 1987) are partially the result of biological activity but also the result of raindrop impact fragmenting the material after the structural lignin has been broken down by photodecomposition. Rainfall causes fragmentation of structurally weakened material (Steinberger and Whitford, 1988). Thus the mechanisms involved in surface litter breakdown involve phothchemical lysis of lignin followed by physical

Table 9.1

Annual Rates of Decomposition of Plant Litter and Roots in Various Landscape Positions, and Deserts Around the World

Material	Location	Mass loss	Reference
Larrea tridentata (shrub) leaf litter	Buried, bajada, New Mexico	60%[b]	Santos and Whitford (1981)
Shrub and grass mixed	Flooded ephemeral lake, New Mexico watershed	55% 1 mo.[b]	MacKay *et al.* (1992)
Larrea tridentata leaf litter (buried)	Mojave Desert, southern Nevada	38% 6 mo.[b]	Santos *et al.* (1984)
Larrea tridentata leaf litter (buried)	Colorado Desert, California and Sonoran Desert, Arizona	43–48%[b]	Santos *et al.* (1984)
Larrea tridentata leaf litter (shrub)	Chihuahuan Desert watershed	47–58%[a]	Cepeda-Pizarro and Whitford (1990)
Larrea tridentata leaf litter (shrub)	Mojave Desert bajada	43%[a]	Strojan *et al.* (1987)
Ambrosia dumosa above-ground litter (shrub)	Mojave Desert bajada	58%[a]	Strojan *et al.* (1987)
Larrea tridentata leaf litter (shrub)	Chihuahuan Desert, basin	93%[a,c]	Elkins *et al.* (1982)
Quercus harvardii leaf litter (shrub)	Chihuahuan Desert coppice dunes	20%[a]	Elkins *et al.* (1982)
Salsola inermis whole plant (annual) buried	Negev Desert, Israel	85%[b]	Steinberger and Whitford (1988)
Hamada scoparia litter (shrub) buried	Negev Desert, Israel	40%[b]	Steinberger and Whitford (1988)
Salsola inermis whole plant – surface	Negev Desert, Israel	25%[a]	Steinberger and Whitford (1988)
Hamada scoparia litter	Negev Desert, Israel	35%[a]	Steinberger and Whitford (1988)
Stipa capensis litter (annual grass)	Irrigated – control various locations on watershed, Negev Desert	20–35%[a]	Steinberger and Whitford (1988)
Ambrosia dumosa litter (shrub)	Mojave Desert bajada	58%[a]	Strojan *et al.* (1987)
Larrea tridentata litter	Mojave Desert bajada	43%[a]	Strojan *et al.* (1987)
Lycium pallidum leaf litter (shrub)	Mojave Desert bajada	63%[a]	Strojan *et al.* (1987)
Tridens (*Erioneuron*) *pulchella* (grass) (foliage)	Upper bajada, New Mexico	62%[a]	Cepeda-Pizarro and Whitford (1990)
Baileya multiradiata (annual) stems and leaves	Mid-slope bajada, New Mexico	58%[a]	Cepeda-Pizarro and Whitford (1990)
Prosopis glandulosa (shrub) leaves	Lower-slope bajada, New Mexico	46%[a]	Cepeda-Pizarro and Whitford (1990)
Larrea tridentata litter (shrub)	Five locations – watershed, New Mexico	53–65%[a]	Cepeda-Pizarro and Whitford (1990)
Panicum obtusum litter (perennial grass)	Ephemeral lake basin, New Mexico	38%[a]	Cepeda-Pizarro and Whitford (1990)
Yucca elata leaves (perennial monocot)	Bajada, Chihuahuan Desert	64%[a]	Schaefer *et al.* (1985)
Chilopsis linearis litter (riparian shrub)	Ephemeral stream margin, New Mexico	77%[a]	Schaefer *et al.* (1985)

Table 9.1 – *continued*

Material	Location	Mass loss	Reference
Prosopis glandulosa litter (shrub)	Bajada, New Mexico	31%[a]	Schaefer *et al.* (1985)
Larrea tridentata litter	Bajada, New Mexico	35%[a]	Schaefer *et al.* (1985)
Flourensia cernua litter (shrub)	Bajada, New Mexico	41%[a]	Schaefer *et al.* (1985)
Annual plant litter	Bajada, New Mexico	62%[a]	Schaefer *et al.* (1985)
Zinnia acerosa (subshrub) roots	Bajada, New Mexico watershed	45%[b]	Whitford *et al.* (1987)
Larrea tridentata (shrub) roots	Bajada, New Mexico watershed	45%[b]	Whitford *et al.* (1987)
Baileya multiradiata (annual) roots	Bajada, New Mexico watershed	89%[b]	Whitford *et al.* (1987)
Dithyrea wislizenii (annual) roots	Bajada, New Mexico watershed	87%[b]	Whitford *et al.* (1987)
Panicum obtusum (grass) roots	Ephemeral lake, buried	20%[b]	Mun and Whitford (1998)
Bouteloua eriopoda (grass) roots	Piedmont toe slope, buried	38%[b]	Mun and Whitford (1998)
Tridens (Erioneuron) pulchella (grass) roots	Various locations on watershed	21–31%[b]	Mun and Whitford (1998)
Sporobolus airoides (grass)	Drainage basin	50%[a]	Montana *et al.* (1988)
Wheat straw	Desert plain, Judean Desert, Israel	16–22%[a]	Steinberger *et al.* (1990)
Gutierrezia sarothrae (subshrub) roots	Various locations on watershed, New Mexico	23–43%[b]	Mun and Whitford (1998)
Atriplex repanda (shrub)	Bajada – Chile	18%[a]	Cepeda-Pizarro (1993)
Atriplex semibaccata (shrub)	Bajada – Chile	11%[a]	Cepeda-Pizarro (1993)
Atriplex nummularia (shrub)	Bajada – Chile	16%[a]	Cepeda-Pizarro (1993)
Avena sterilis litter – buried (annual)	Judean Desert, Israel	60–65%[b]	Hamadi *et al.* (2000)
Avena sterilis litter	Judean Desert, Israel	20%–40%[a]	Hamadi *et al.* (2000)

[a] Decomposition rates of litter on the soil surface.
[b] Decomposition of buried litter.
[c] Litter consumed by termites during a 3-month period at the end of the growing season.

fragmentation by rainfall. Breakdown by microbial enzymes is obviously limited to those brief periods when litter and soil are moist and soil temperatures are moderate. Biological decomposition of surface materials is therefore sporadic and brief.

Measurement of mass loss cannot distinguish between fragmentation and mineralization. However, if mass loss is a function of biological activity, there should be a relationship between frequency and duration of litter wetting and mass loss since growth and activity of microorganisms requires moist environments. Experiments designed to examine this question by providing supplemental water by sprinkler irrigation have shown that there is no clear relationship between quantity and frequency of rainfall and mass loss (Whitford *et al.*, 1986; Mackay *et al.*, 1987; Steinberger and Whitford, 1988). Adding 305 mm yr^{-1} to the 230 mm annual total had no significant effect on mass losses nor did frequent pulses (Whitford *et al.*, 1986). When litter was placed in the field during an extended dry period, irrigated litter lost mass faster than nonirrigated litter. The rates, however, were equal thereafter (Fig. 9.2). Further insights into this process were gained in irrigation experiments in the Negev, Israel. In the Negev, the probability of rainfall in the summer months of May–September is essentially zero. Although

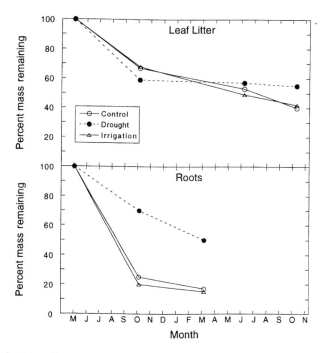

Figure 9.2 The effect of artificial drought and irrigation on decomposition of surface litter and buried roots in a cresotebush, *Larrea tridentata*, shrubland. (From Whitford *et al.*, 1995.) With kind permission from Kluwer Academic Publishers.

irrigation added considerable water and kept soils and litter wet for longer periods, there were no detectable irrigation effects during the normal winter–spring 'rainy period'. Irrigation increased mass losses during the rainless summer (Steinberger and Whitford, 1988).

Although rainfall supplementation experiments provide some insights into how climate variation affects decomposition, imposition of drought is essential to really understand climate–decomposition relationships. 'Rain-out' shelters constructed of steel frames with greenhouse plastic roofs were used in several studies in the Chihuahuan Desert. These studies demonstrated that the imposition of drought resulted in a marked decrease in average litter inputs (45 g per trap compared to 83 g per trap in year 1, and 72 g per trap compared to 112 g per trap in year 2 in droughted plots and controls, respectively). Imposition of drought had no effect on the rates of decomposition of creosotebush leaf litter on the soil surface (Fig. 9.2). However, decomposition rates of roots of a perennial herb were significantly reduced by imposed drought but not by irrigation (Fig. 9.2).

9.2.3 Spatial Variation

Variation in microclimate and physicochemical soil properties resulting from topographic position and vegetative cover can produce different patterns of decomposition, mineralization and soil organic matter accumulation (Seastedt, 1984). In a study of surface litter decomposition at several locations on a desert watershed, there were no differences in initial rates of mass loss (k_1) across the range of locations on the watershed. There were differences in average rates of mass loss among locations in the second phase of decomposition (k_2). Highest rates of decomposition were recorded for a run-on area at the base of the watershed. The lowest rates of decomposition were recorded in the dry lake basin on fine textured clay soils (Cepeda-Pizarro and Whitford, 1990). The high rates of decomposition in the run-on area were attributed to partial or complete burial of the litter bags by water-transported soil. The low rates of decomposition measured in the dry lake basin were attributed to the extremely low infiltration of water into those soils and shading by the dense grass cover.

In contrast to surface litter, decomposition of roots on the same watershed yielded no significant differences among sites, except for the dry lake basin, on the same watershed where the surface litter studies were conducted. There were no significant differences in decomposition rates of roots in nitrogen fertilized soils and in unfertilized soils (Mun and Whitford, 1998). The spatial differences in mass loss rates in the dry lake were attributed to the high clay content soils of the lake basin which reduced water availability and to the absence of termites in the lake basin. Therefore, it appears that topographic position and soil textural characteristics have little effect on decomposition processes except when landscape position results in litter burial or where fine texture soils affect the composition of the soil fauna.

9.2.4 Decomposition of Buried Litter and Roots

Litter that is buried in the soil decomposes more rapidly than surface litter and is primarily a biological process (Santos *et al.*, 1981; Elkins and Whitford, 1982; Parker *et al.*, 1984). Decomposition rates of buried litter varies inversely with initial lignin content and C:N ratio. The chemical changes in decomposing buried litter are the same as those reported for forest litter (Berg *et al.*, 1984). The early stages of buried litter decomposition involves losses of soluble materials and holocellulose but essentially no change in the lignin. Nitrogen immobilization occurs in the early stages of buried litter decomposition as the microbial biomass growing on the litter increases (Parker *et al.*, 1984).

The biotic regulation of decomposition of buried materials was clearly documented in studies in the hyperarid Namibian sand sea, southwestern Africa. Decomposition of cellulose (cotton cloth and filter paper) buried at 10 cm depth occurred in pulses following rain events. About 84% (range 64.7–97.2%) of the original mass was lost following rains of 9 mm or greater (Jacobson and Jacobson, 1998). Mass loss during a dry period was only 8.2% (range 0–16.7%). Most substrates were colonized by fungi. Termites and tenebrionid beetle larvae were recorded as feeding on the fungus-colonized substrates. In hyperarid systems where soils are too dry to support growth of fungi and activity of microarthropods, rainfall is the trigger that stimulates episodic periods of high rates of decomposition. The episodic pulses of decomposition result from rapid growth of fungi and activity of soil mesofauna and macrofauna that feed on fungi and fungal modified cellulose.

Considering that the recalcitrant materials in surface litter are rapidly destroyed by photodecomposition and that only a fraction of the litter is buried, it is obvious that the largest contribution to the recalcitrant soil organic carbon pool is that of decomposing roots. Roots vary considerably in initial lignin content (approximately 17% for roots of an annual plant, 24% for roots of a woody shrub, and 27–29% for roots of several grasses. Decomposition rates of roots were inversely related to the initial lignin content and also to the initial nitrogen content which were lower in the roots of grasses than in the other species examined (Mun and Whitford, 1998).

Decomposition rates of roots in the Chihuahuan Desert were equal to or higher than those reported for decomposing roots in mesic ecosystems (Whitford *et al.*, 1988). Rates of decomposition of roots of woody shrubs and of herbaceous annuals were not affected by supplemental irrigation. The absence of an irrigation effect was attributed to the relatively 'average' rainfall during the time that decomposition was measured. Average annual mass losses from woody roots (40%) were considerably lower than the average annual mass losses from roots of herbaceous annuals (85–90%) which was related to the lignin content of the roots and to the susceptibility of herbaceous roots to attack by subterranean termites. Simulated drought reduced the mass loss from roots of an herbaceous perennial from an average mass loss of 80% to an average mass loss of 50% (Whitford *et al.*, 1995; Fig. 9.3). The roots of the woody shrubs retained relatively constant C:N ratios through the experiment indicating little if any nitrogen immobilization. The C:N ratios of the roots of

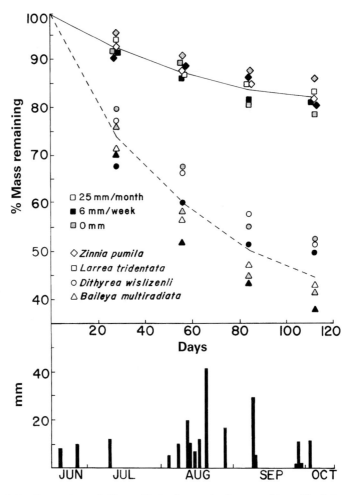

Figure 9.3 Comparison of effects of irrigation on the decomposition of buried roots (From Whitford *et al.*, 1988.) With permission of Springer-Verlag GmbH & Co KG.

annuals decreased by an average of 28 after 4 months indicating that microbial biomass on the roots of the annuals had immobilized a large amount of soil nitrogen.

The changes in carbon chemistry of decomposing roots in a desert are similar to those of decomposing roots in mesic forests and in mesic grasslands (McClaugherty *et al.*, 1984). The percentage composition of nonpolar compounds and water-soluble compounds in decomposing roots of all species dropped to approximately 20% of their original concentration within the first year (Mun and Whitford, 1998). The rapid decrease in these compounds is probably the result of rapid colonization of roots by bacteria and fungi and the subsequent rapid metabolism of these easily metabolizable compounds. The acid-soluble compounds are more recalcitrant than the nonpolar and water-soluble compounds and these decreased by only 10% of their original concentration at the end of the first year and to between 10% and 25% (depending on the species) after 42 months of decomposition. The relative concentration of lignin increased during the decomposition of roots and there were marked increases in the relative concentration of nitrogen in the roots of an annual but nitrogen immobilization in decomposing roots of woody plants did not occur until after 24 months of decomposition (Mun and Whitford, 1998). The evidence from changes in chemistry of decomposing roots and of buried litter supports the conclusion that decomposing roots are the principal source of recalcitrant carbon in desert soils.

9.2.5 The Role of Microfauna in Decomposition and Mineralization

The soil microfauna have been shown to play a major role in decomposition processes in mesic temperate ecosystem (Swift *et al.*, 1979; Seastedt, 1984). The soil microfauna in arid systems is nearly as diverse as the soil microfauna of mesic soils (Wallwork, 1982). In order to adequately address the role of soil biota in decomposition, it is necessary to distinguish between the plant litter, dung, wood, etc., exposed to the intense light and heat of the soil surface and the litter, litter fragments and dead roots at varying depths below the soil surface.

Organic materials on the soil surface have microclimatic conditions conducive to microbial growth and activity of soil microflora for brief periods at night and for slightly longer periods following rains (Whitford, 1989). However, the micro-mesofauna appear to have virtually no effect on the rates of mass loss from surface litter (Silva *et al.*, 1985; MacKay *et al.*, 1987).

Some plant litter is buried in pits produced by animals and an unknown fraction of the surface leaf litter is mixed into the surface soil as small fragments. The microclimate in the soil, even as shallow as 1 cm depth, is very much modified compared to surface conditions. The moderate soil temperatures and high humidity in the soil interstices, produces microclimates that allow some microarthropods to remain active even in soils with water potentials lower than −6.0 MPa (Whitford, 1989). In soils with water potentials lower than −3.0 MPa most of the soil protozoans are encysted, the nematodes are in a state of anhydrobiosis as are the collembolans and some of the soil acari. The active mesofauna in dry soils are soil acari. There are some species of soil acari that remain active in soils as dry as −6.0 MPa and some species that enter a cryptobiotic (inactive) state in dry soils (Whitford, 1989). In desert soils the complete soil micro- and mesofaunal community is involved in decomposition and mineralization only during periods when the soil is moist. Rainfall regulates decomposition and mineralization in arid ecosystems by stimulating rapid growth of fungi and activating the soil fauna. Rainfall-induced pulses of activity are observed most clearly in hyperarid systems where rainfall events are separated by sufficient time periods for soil to dry between events (Jacobson and Jacobson, 1998).

The importance of the soil microarthropod fauna in mass loss of buried litter in deserts has been well documented (Santos and Whitford 1981; Santos *et al.*,

1981; Elkins and Whitford, 1982; Parker *et al.*, 1984). In desert ecosystems, the activity of soil microarthropods (primarily soil acari), accounted for 30% of the initial mass losses. Despite the fact that the amount of soil respiration that can be attributed to all soil animals is 10% or less than the total amount (Reichle, 1971). The soil fauna directly or indirectly influence the soil microflora by their feeding activities. The effects of soil acari on decomposition in arid ecosystems appear to be indirect. The common soil acari that are active at high soil water potentials are small prostigmatid mites that feed on fungi and on nematodes. In the initial stages of decomposition of buried creosotebush leaf litter, microarthropods controlled the population densities of bacterial feeding nematodes. Where microarthropods were excluded, the high nematode populations reduced the bacterial biomass resulting in reduced rates of mass loss (Santos *et al.*, 1981) (Fig. 9.4).

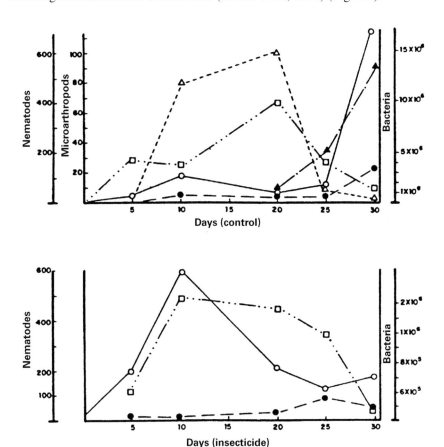

Figure 9.4   The effects of excluding soil microarthropods on the populations of soil nematodes and on decomposition rates in the Chihuahuan Desert. ○, Bacteriophagous nematodes; ●, fungivorous nematodes; Δ, Tydeidae; ▲, Pyemotidae; □, bacteria. (Data from Santos *et al.*, 1981.) With permission by The Ecological Society of America.

Whereas soil bacteria dominate the early stages of buried leaf litter decomposition, the later stages are dominated by fungi (Parker *et al.*, 1984). The microarthropods that are most abundant in buried decomposing litter or around decomposing roots are small fungal feeding mites. Although microarthropods have very little effect on mass losses during the later stages of decomposition, i.e. after the first 90 days they do directly affect rates of nitrogen mineralization.

9.2.6 The Role of Termites

Termites are abundant in the soils of most subtropical, arid, and semiarid regions of the world (Lee and Wood, 1971). Although large above-ground mounds are an impressive indication of the presence of these animals, the majority of termite species live below the soil surface. Based on the distribution of termites and the limited studies of the quantities of dead plant parts and other organic materials consumed by these animals, it is clear that these insects have a significant effect on nutrient cycling and nutrient availability in arid and semiarid systems. Termites are responsible for removing a large fraction of certain species of litter (Silva *et al.*, 1985; MacKay *et al.*, 1987). Soil moisture and temperature affect termite activity, hence mass loss from litter species consumed by termites shows a strong positive correlation to rainfall (MacKay *et al.*, 1987).

Materials such as grass leaves and annual plant leaves and stems that are preferred foods by Chihuahuan Desert termites are harvested during periods when soil moisture and temperature allows termites to forage on surface litter (Whitford *et al.*, 1982; MacKay *et al.*, 1986). For example, *Gnathamitermes tubiformans* cease foraging at soil moisture less than -5.4 ± 0.2 MPa and hence are capable of feeding in litter accumulations for much of the year when soil temperatures are higher than 5°C (MacKay *et al.*, 1986).

In southeastern Australia, termites were harvesting or directly consuming grass leaves and fragments, shrub and tree leaves and phyllodes, sheep and kangaroo dung, dead wood, fungal modified stem bark, and dead roots (Whitford *et al.*, 1992). In southern Somalia, litter decomposition was rapid (95% disappeared in one year) and not correlated with rainfall because termites were active throughout the dry season (Thurow, 1989). In the northern Chihuahuan Desert, mass loss from plant litter on plots from which termites had been chemically excluded were compared to mass loss from plant litter on plots with termites present. These studies demonstrated that a large fraction of dead plant parts and dung were consumed by termites (Whitford *et al.*, 1982, 1988) (Fig. 9.5). Materials ingested by termites are mineralized by the symbionts within the gut of the termite. The breakdown of materials within the termite gut includes not only the cellulosic compounds, but also lignins (Lee and Wood, 1971).

The small quantities of fecal material produced by termites are generally used in the cementing materials of gallery walls and nest chamber walls. The mineral nutrients in the food material are largely retained in the biomass of the colony and are returned to the ecosystem only via predators of the workers and reproductives

Livestock and rabbit dung
breakdown Δ40–50% yr^{-1}

Yucca inflorescence
decomposition Δ6% yr^{-1}

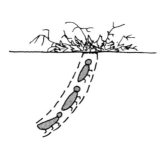

Dead leaves and annual
plant parts Δ10–50% yr^{-1}

Dead grasses Δ10–90% yr^{-1}

Figure 9.5   The keystone role of subterranean termites on decomposition processes and ecosystem function in Chihuahuan Desert ecosystems.

(Schaefer and Whitford, 1981). Because so much of the decomposition of organic matter in arid ecosystems occurs within the guts of termites, there appears to be a negative correlation between termite abundance and soil organic matter or soil nutrients. On a Chihuahuan Desert watershed where measurements of termite abundance and activity and soil organic matter were made at 30 m intervals over a 2 km transect, there was a highly significant correlation, $r = -0.97$ between termite abundance and soil organic matter (Nash and Whitford, 1995). The absence of a pattern of regularly increasing soil organic matter from the top to bottom of the watershed added further support to the contention that subterranean termites were primarily responsible for the variations in soil organic matter on that watershed. The most important macro-nutrient, nitrogen, tends to vary directly with soil organic matter (Whitford *et al.*, 1987). The negative relationship between termites and soil organic matter suggests that soil fertility varies inversely with termite abundance. However, the importance of termites as producers of macropores and their effects on infiltration of water is probably more important for sustaining the productivity of desert ecosystems than their effect on reducing the nitrogen content of desert soils.

9.2.7 Conceptual Model of Decomposition Processes in Deserts

The following conceptual model is proposed to account for the relationships among variables that have an effect on the decomposition processes in deserts. Initial high rates of mass losses from litter on the surface result from leaching of low-molecular-weight materials from leaf surfaces and by the breakdown of lignins by photochemical decomposition. This accounts for the direct correlation between initial lignin content and decomposition. At low light intensities and low sun angles, mass loss rates are low and occur primarily from low level activity of microorganisms. At high light intensities, photooxidation of complex structural molecules results in fragmentation by wind or water drops. The small fragments are transported into the soil by water where they are then decomposed by biological activity of the microflora during periods when soils are moist. This is the source of elevated CO_2 production measured when soil and litter are moist (Parker *et al.*, 1983).

A major component of the high rates of decomposition of surface litter is the consumption of many of these materials by termites. Termites consume more than 50% of the total surface litter and may consume up to 70–80% of the litter and most of the animal dung. Materials consumed by termites are virtually completely converted into carbon dioxide and water because the symbiotic gut flora of many termites is capable of processing recalcitrant organic molecules such as lignins. Thus litter falling to the surface of the soil in deserts contributes little to the soil organic carbon pool and minerals in litter are returned to the soil in a mineralized state largely through the feces of animals that prey on termites.

A variable quantity of the litter input in a desert ecosystem may be blown into soil pits and buried. The process of decomposition of buried litter and dead roots is very different from that of surface litter in arid environments. The moderate environment of the soil affects both decomposition rates and patterns of decomposition. Buried litter accumulates moisture and remains moist even in dry soils. The moderate temperature and moisture content of buried litter is conducive to growth of microflora and microfauna. Dead roots encounter the same moderate conditions in soil and serve as ready substrates for growth of the microflora. Increasing populations of microflora on decomposing materials attract microfloral grazers (protozoans, nematodes, and microarthropods), which in turn attract predators (predatory nematodes and nematophagous mites). Complex food webs are quickly established around moist roots and buried litter accumulations. The extracellular enzymes of the microflora rapidly decompose the water-soluble and nonpolar compounds (sugars, fats, waxes, starches, and cellulose). The complex acid-soluble compounds are slowly attacked by a small subset of the microbial heterotrophs. Decomposition of buried litter and roots is primarily a biological process.

9.3 LANDSCAPE PATTERNS OF NUTRIENT DISTRIBUTION

One of the best-documented spatial patterns of nutrient distributions in arid ecosystems are the 'islands of fertility' associated with shrubs and trees (Garcia-Moya and McKell, 1970; Charley and West, 1975, 1977; Barth and Klemmedson, 1978; Parker et al., 1982; Virginia and Jarrell, 1983; Schlesinger et al., 1996). The accumulation of nutrients in the surface soil beneath shrubs and the lower nutrient levels in soil between shrubs results from a whole suite of processes: plant-nutrient uptake, litter fall, dust interception, soil-erosion, soil deposition, soil biota-mineralization, decomposition, animal decomposition, fecal decomposition, and predation. The relationship between shrub morphology and spatial pattern on below-canopy litter accumulation has been documented for one species but undoubtedly applies to a number of other species of shrubs (DeSoyza et al., 1997). In landscapes composed of patches with a variable mix of shrubs, grasses, and perennial herbs, the nitrogen stores in the soil are related to the characteristics of the patch. Available nitrogen was considerably higher in soils from complex patches, intermediate in simple shrub–grass patches and lowest in patches of bare soil (Mazzarino et al., 1996).

Although the activity of subterranean termites may be enhanced by the presence of leaf litter and the moderating effect of the canopy on soil temperature, shrub and tree leaves rank low on the food preference scale of most termite species. Tree and shrub leaf litter, therefore, provides the energy and nutrient source for the soil microbiota, resulting in nutrient enrichment. Another factor affecting the nutrient enrichment of the subshrub soil is the activity of animals like lizards. Many lizards move quickly from shrub to shrub spending most of their time foraging or simply sitting in the more benign subcanopy environment (Creusere and Whitford, 1982; Peterson and Whitford, 1987). In the northern Chihuahuan Desert, lizards are the most important predators of subterranean termites. In other arid regions, e.g. Australia and the Kalahari of southern Africa, lizards are the most important termite predators. Many lizards feed almost exclusively on termites (Pianka, 1985; James, 1991). In these deserts, lizards use clumps of vegetation as shelter and as foci for their foraging activities thereby contributing to the 'islands of fertility'. Given the abundance of termites and lizards in arid and semiarid regions, it is likely that this linkage between detritivore–predator and spatial nutrient pattern is generally important in all such systems. The general pattern of nutrient distribution with high concentration in center to mid-canopy soils decreasing rapidly from mid-canopy to canopy edge and into the intershrub space, is characteristic not only of shrubs, but also of perennial grass clumps and small trees. Nutrients are accumulated where soil organic matter is elevated either by debris accumulation or turnover of fine roots.

Fossorial mammals, burrowing mammals, mammals and lizards that dig for invertebrate prey contribute to the nutrient heterogeneity of soils in arid regions. Foraging pits fill with litter, frass, dung, and seeds creating small nutrient-rich 'hot spots'. Burrow systems that are occupied by successive generations of central place foragers, e.g. banner-tail kangaroo rats, hairy-nosed wombats, harvester ants, and mound building termites) gradually accumulate nutrients in the mound soils and eventually develop into nutrient rich patches (Fig. 9.6). Fossorial mammals turn over soil and leave patches that are rich in organic matter in storage chambers and defecation sites (Whitford and Kay, 1999; Whitford, 2000). The abundance, species composition, and behavior of those animals that disturb soil or accumulate organic materials in burrows are important variables to consider when assessing the patterns of nutrient distribution across the landscape.

The spatial distribution patterns, morphologies, and physiological characteristics of dominant plant species clearly affect the richness of the below canopy 'islands of fertility'. However, when the frame of reference is moved from the local patch to the landscape scale there are many other variables that affect the nutrient status of soils and nutrient cycling in the ecosystems that make up a landscape. One of the most important variables that affect both nutrient status and the temporal characteristics of nutrient cycling is the spatial distribution of different root systems in a landscape. There are some generalizations about root systems

Figure 9.6   Vigorous growth of a herbaceous annual, *Erodium texanum*, in a corona around the nest of a central place forager – seed-harvesting ants, *Pogonomyrmex rugosus*. Nitrogen concentrations in the soils of the corona were double that of the surrounding soil.

that can be inferred from the above-ground life form of the vegetation. Perennial grasses tend to produce a dense but relatively shallow mat of roots below the basal area of the plant. The root system of some species may occupy an area somewhat larger than the canopy crown of the grass but few grasses have spatially extensive root systems. Some species of grasses have physiological characteristics that result in complex rhizosheaths of mucilages and mucigels, and microbes (Wullstein *et al.*, 1979). In some species the mucilages and mucigels secreted from the roots of grasses serve to attract and support free-living nitrogen fixing bacteria in the rhizosheath (Mandimba *et al.*, 1986; Reinhold *et al.*, 1986).

The root systems of most herbaceous annuals may consist of a main tap root with fine roots emerging from the bottom third or of three or more roots that extend from the root crown. Herbaceous annuals that germinate during the growing season may have shallow root systems and root:shoot ratios as low as 0.2. In herbaceous annuals that germinate during the nongrowing season (i.e. winter–spring annuals in North American deserts), the rooting depth and root biomass is dependent on the depth of the wetting front and the length of time that the deeper soil profile remains moist. During cool periods when above-ground growth does not occur, the roots of the rosettes continue to grow. This can produce root:shoot ratios as high as 1.2 at peak above-ground biomass production. The root biomass of annual plants can vary both temporally and among species.

Perennial herbaceous plants are generally but a small fraction of the cover and biomass of desert vegetation. Some perennial herbaceous species are clonal and may occur in dense patches. Usually there is more below-ground biomass of herbaceous perennials than above-ground biomass. The root systems are generally fleshy (storage organs) with fine roots emanating from the actively growing segments.

Most desert shrubs and small trees have diffuse, relatively deep (> 1 m) woody roots systems. Many desert trees and shrubs also have extensive shallow root systems in addition to the deep roots (Gile *et al.*, 1998). In most desert soils, nitrogen is concentrated in the upper 10 cm of soil. The growth responses of a nitrogen-limited desert shrub and small rain events suggest that the shallow roots contribute most of the nitrogen uptake and allow the shrub to respond rapidly to relatively small rain events (Fisher *et al.*, 1987).

Plants with symbiotic nitrogen fixers are common in arid landscapes. Many species of small tree and shrub legumes have nodules containing the nitrogen-fixing *Rhizobium* spp. Not all legumes host symbiotic nitrogen fixers on the root system. Several species of legumes have increased in abundance and cover in areas where domestic stock grazing has modified the environment, e.g. *Acacia caven* (in Chile), *Prosopis* spp. (in North America) and *Acacia* spp. (in Africa). These species produce leaves with lower C:N values than those of other non-leguminous species even in those plants with root systems devoid of nodules. Even when a legume is removed, there is a residual elevation in soil nitrogen levels (Klemmedson and Tiedeman, 1986). This has important implications for the patchiness of vegetation, species composition of the vegetation and the productivity of arid areas in which woody legumes dominate the landscape. The abundance and spatial distribution of such species affects soil nitrogen levels and productivity of grasses and other plants that grow under the canopies of woody legumes (Weltzin and Coughenour, 1990). Nutrient enrichment of soils under legume canopies can result in production values that are twice those measured in the between-canopy space (Weltzin and Coughenour, 1990).

Herbaceous nitrogen-fixing legumes are a relatively minor component of the vegetation. Some of the nitrogen-fixing legumes are ephemeral plants, e.g. species such as *Astragalus* spp. in North America. Most of the herbaceous nitrogen-fixing legumes are perennial. However, perennial herbaceous legumes accounted for only a fraction of one percent of the vegetative cover on most locations on a desert watershed (Whitford *et al.*, 1987). It is unlikely that herbaceous legumes contribute much to the nitrogen economy of desert ecosystems.

There is a clear correlation between soil organic matter and nutrient availability in desert ecosystems. This is not a surprising correlation because the primary source of nutrients in soils is from the decomposition and mineralization of organic materials. In a study using nitrogen mineralization potential as a measure of nutrient availability to plants, Whitford *et al.* (1987) found good correlations between mean soil organic matter and mineralized N (Fig. 9.7). The spatial distribution of soil organic matter in deserts is determined by (1) location and relative abundance of termites, (2) run-off, run-on patterns, (3) resistance of the vegetation to wind transport of litter and (4) structure of the vegetation mosaic. There is a positive relationship between soil organic matter, total nitrogen and nitrogen mineralization potential (the rate of nitrogen mineralization under defined conditions of soil incubation). This relationship is well documented in the northern Chihuahuan Desert (Fig. 9.7). As the data for the dry lake bed soils show, the direct positive relationship may be somewhat modified by soil texture.

Total N–organic matter relationships are also affected by the genetic physiological characteristics of the vegetation. For example, the total soil N is elevated in mesquite coppice dune soils because of the lower C:N ratio of *Prosopis glandulosa* leaves in comparison to other shrubs and grasses in the area. Another important consideration affecting the spatial variability of soil nutrients is the negative relationship between soil organic matter and subterranean termites. Soils that are unfavorable for termites will tend to have higher levels of soil organic matter and higher nutrient concentrations. The spatial pattern of termite colonies and their foraging patterns will certainly affect the variability in nutrient availability across an arid landscape. The morphology and physiology of the dominant plants also affect the variability and the relative importance of various processes. For example, in plants with canopies shaped as inverted cones, accumulation of dust on the leaves may contribute greatly to nutrient concentration around the root crown because of effective stem flow (DeSoyza *et al.*, 1997). Perennial species with symbiotic N fixers may produce nitrogen rich patches, but the concentrations of other nutrients may not be proportioned to the N concentration.

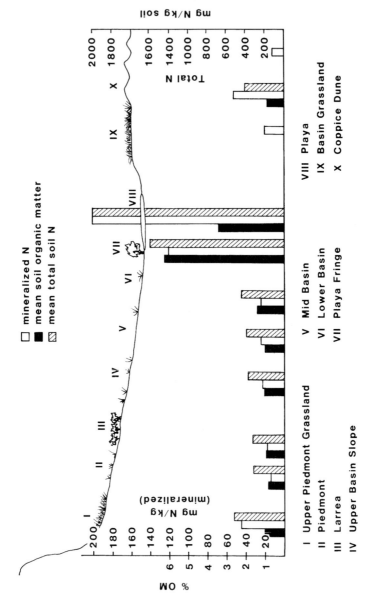

Figure 9.7 The distribution of nitrogen, soil organic matter (OM), and nitrogen mineralization potential across a Chihuahuan Desert landscape. (From Whitford *et al.*, 1987).

The patterns of nitrogen mineralization and soil organic matter across a Chihuahuan Desert landscape reflect these variables. The inverse correlation between termite relative abundance and soil organic matter has been well established for a Chihuahuan Desert watershed. In deserts where termites are more abundant, soil nutrient levels may be considerably lower than the concentrations recorded in North American deserts. The pattern of nitrogen mineralization on a watershed also reflected the run-off, run-on patterns of organic materials. The dry lake basin where the highest nitrogen mineralization and highest total soil N were recorded is not only enriched by organic matter run-on but is also an area where the fine textured soils and infrequent inundation eliminate termites. The relationships between soil organic matter and landscape position, soil characteristics, and nitrogen mineralization are probably applicable to most arid and semiarid landscapes.

9.4 RESORPTION

An important variable that affects decomposition and mineralization is the resorption of nutrients from foliage prior to leaf abscission. In the process of resorption, nutrients are mobilized from senescing leaves and transported to other plant tissues. Nutrients that are resorbed are conserved by the plant for use in primary production in the next growing season. Nutrient resorption is especially important for plants growing in nutrient poor environments (Aerts, 1996). In a review of nutrient resorption in perennial plants, Aerts (1996) reported that mean nitrogen resorption efficiency was significantly lower in evergreen trees and shrubs (47%) than in deciduous trees and shrubs (54%). Phosphorus resorption efficiency was essentially the same for evergreen (51%) and deciduous (50%) shrubs and trees.

Nutrient resorption patterns in desert plants differ from the general patterns for mesic species. Resorption of phosphorus in creosotebush (*L. tridentata*) was very high (72–86%) (Lajtha, 1987). However, the nitrogen resorption values for creosotebush fell within the range of values for evergreen trees and shrubs (Lajtha and Whitford, 1989). Mean resorption efficiencies calculated on the data for desert shrubs for nitrogen and phosphorus were 57% and 53%, respectively. There was a large intersite and interyear variation in resorption efficiencies of ocotillo, *Foquieria splendens*, suggesting that the regulation of resorption efficiencies in desert shrubs may be quite complex.

Winter-deciduous shrubs growing on the margins of ephemeral stream channels exhibited nutrient resorption patterns related to their status as facultative or riparian species. Nitrogen and phosphorus resorption efficiencies of obligate riparian species (*Brickellia laciniata* and *Chilopsis linearis*) (N = 63%, P = 58%) were substantially higher than in facultative riparian species (*Fallugia paradoxa*, *Rhus microphylla*, *Prosopis glandulosa*, and *Flourensia cernua*) (N = 46%, P = 46%). The species resorbing the least nitrogen from the leaves was mesquite, *P. glandulosa*; a species known to harbor nitrogen-fixing symbionts on the roots

(Killingbeck and Whitford, 2001). It has been hypothesized that plant species with nitrogen-fixing root symbionts will resorb less foliar nitrogen than other species in the same environment. In the Sonoran Desert, no resorption of nitrogen was reported in senescing leaves of *P. glandulosa* (Rundel *et al.*, 1982).

9.5 NITROGEN CYCLE

9.5.1 Nitrogen Fixation

Of all of the soil nutrients required for plant growth, nitrogen appears to be the most important limiting nutrient in semiarid and arid ecosystems. Despite the huge atmospheric pool of nitrogen, that pool represents a nonusable form of nitrogen for use in photosynthesis. The biologically active forms of nitrogen are ammonium and nitrate which can be used to produce amino acids and other nitrogen-containing organic molecules. The fate of fixed nitrogen is complex (Fig. 9.8) and involves a variety of microorganisms, soil microfauna, plus plants and animals.

The fixation of atmospheric nitrogen (N_2) is an energy-expensive reduction to ammonium (NH_4) that occurs in some cyanobacteria, in nodules on the roots of legumes by *Rhizobium* spp. bacteria, and in some free-living N-fixing bacteria

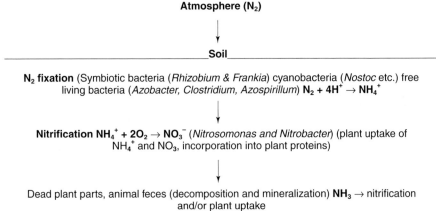

Atmosphere (N_2)

↓

_____**Soil**_____

N_2 **fixation** (Symbiotic bacteria (*Rhizobium & Frankia*) cyanobacteria (*Nostoc* etc.) free living bacteria (*Azobacter, Clostridium, Azospirillum*) $N_2 + 4H^+ \rightarrow NH_4^+$

↓

Nitrification $NH_4^+ + 2O_2 \rightarrow NO_3^-$ (*Nitrosomonas and Nitrobacter*) (plant uptake of NH_4^+ and NO_3, incorporation into plant proteins)

↓

Dead plant parts, animal feces (decomposition and mineralization) $NH_3 \rightarrow$ nitrification and/or plant uptake

Nitrogen Losses from Organic Pool

Ammonia Volatilization $NH_4^+ + OH^- \leftrightarrow NH_3\uparrow + H_2O$

Denitrification $NO_3 \rightarrow N_2O \rightarrow N_2\uparrow$ (*Pseudomonas, Achromobacter*)(requires anaerobic or extremely low soil oxygen concentration)

Figure 9.8   A general nitrogen cycle applicable to desert ecosystems.

such as *Azobacter* spp. Some other genera of free-living bacteria such as *Azospirillum* sp. may fix atmospheric nitrogen but there is conflicting evidence concerning these organisms as N-fixers.

Cyanobacteria are widely distributed as components of the algal–lichen (cryptogamic) crusts that are common in many deserts (Fig. 9.9). Cyanobacteria also occur in some desert soils where microphytic crusts appear to be absent. In areas with well-developed cyanobacterial crusts, atmospheric N-fixation may contribute significant amounts of nitrogen to the soils. Soils with cyanobacterial and lichen crusts had nitrogen contents 4 to 7 times higher than soils with no lichen or algal crusts (Shields, 1957). Nitrogen fixation in desert algal crusts, measured by uptake of isotopically labeled nitrogen, was rapid after initial wetting of the crusts. Cyanobacterial and lichen crusts leak the fixed nitrogen to the soil, usually in the form of ammonium, thereby increasing soil nitrogen levels (Mayland and McIntosh, 1966).

Nitrogen fixation is not the only influence of cyanobaterial and lichen crusts on soil nutrient enrichment. The concentrations of the elements N, P, K, Ca, Mg, and Fe, were significantly higher in a grass, *Festuca* spp. growing on cryptogamic (algal and lichen) crusted soil than on soil with no cryptogamic crust (Belnap and Harper, 1995). They reported that cryptogamic crusts fixed considerable nitrogen, which becomes available to vascular plants via decomposition and mineralization of the cryptogams or by cellular secretion in the living cryptogams. In deserts, the

Figure 9.9   Cryptogamic crusts covering bare patches in a tabosa grass, *Pleuraphis mutica*, swale in the northern Chihuahuan Desert, New Mexico.

nutrient enrichment of soils by cyanobacterial and lichen crusts is rainfall dependent. In an area with 15% surface cover of algal crusts, up to 1.3 g N ha^{-1} h^{-1} was added to the soil during the daylight hours for the first 24 hours after a rainfall (Loftis and Kurtz, 1980).

Symbiotic nitrogen fixation in desert plants may be by *Rhizobium* nodules or nodules containing actinomycetes. One of the important tree species in the temperate and tropical semiarid zone of Australia is *Casuarina cunninghamiana,* which harbors actinomycete symbionts (*Frankia* spp.) in nodules on its roots. *Frankia* spp. on the roots of this tree fixed considerable nitrogen at soil temperatures between 20°C and 30°C (Reddell *et al.,* 1985). Actinomycete symbionts may be important in nitrogen fixation in other woody plant species in arid regions but there is insufficient information on most species to evaluate this possibility.

Tree and shrub legumes are abundant in most arid and semiarid regions of the world. Many of the most common genera harbor *Rhizobium* spp. or *Bradyrhizobium* spp., symbiotic, nodule-forming, nitrogen-fixing bacteria. Among the common genera confirmed to have nodulation and nitrogenase activity are *Acacia, Prosopis, Pterocarpus,* and *Pericopsis* (Shearer *et al.,* 1983; Hogberg, 1986). Studies using nitrogen abundance ratios (N^{15}/N^{14}) confirmed symbiotic nitrogen fixation in *Prosopis* at six of seven sites in the Sonoran Desert. Nitrogen fixation was also reported in other Sonoran Desert legumes of the subfamily Papilionoideae: *Lupinus, Dalea, Astragalus* and *Lotus* (Shearer *et al.,* 1983). An abundant shrub (*Psorothamnus spinosus*) growing in the low nitrogen environment at the margins of ephemeral streams obtained approximately one-third of its nitrogen from symbiotically fixed N. From the limited available data, it appears that most of the woody legumes and a large portion of the herbaceous legumes obtain fixed atmospheric nitrogen from symbiotic *Rhizobium* or *Bradyrhizobium* species. Nitrogen fixation by root symbionts is probably the primary source of nitrogen in the fertile islands that develop under the canopies of leguminous trees and shrubs.

Although nitrogen fixation in *Rhizobium* nodules benefits the host plant with the nitrogen enrichment most of the nitrogen gain does not enter the ecosystem until the N-rich leaves decompose. The published estimates of quantities of fixed-N relative to the quantities of N incorporated into growing plant tissues are not particularly reliable. In desert ecosystems, the relative importance of the various N fixation processes can probably be assessed by the presence and cover of cyanobacterial crusts, cover of legumes known to support *Rhizobium* nodules, and distribution and cover of vegetation life forms that provide rhizospheres conducive to the growth of free-living N-fixers.

Some desert grasses and herbaceous perennials produce rhizosheaths of soil particles bound together and to the root surface by mucilages and other complex carbohydrates (Fig. 9.10). Rhizosheaths surrounding root surfaces are several times the diameter of the root that secreted the complex carbohydrates (Rougier and Chaboud, 1985). Rhizosheaths provide an ideal habitat for a variety of soil microorganisms. Some of the microorganisms inhabiting rhizosheaths are species

Figure 9.10   The rhizosheath on the roots of a porcupine grass, *Triodia pungens*, northwestern New South Wales, Australia.

of bacteria that are free-living nitrogen fixers (*Clostridium* spp., *Azotobacter* spp., and *Azospirillum* spp.) (El Shahaby, 1988). In plants that produce rhizosheaths, there appears to be stimulation of nitrogen fixation. The relationship between root secretion and nitrogen fixation by rhizosheath bacteria is an example of associative nitrogen fixation. Associative nitrogen fixation is a loose symbiosis in which the energy supplied by the plant supports nitrogen fixing bacteria that supply nitrogen to the plant. Studies of associative free-living nitrogen fixation in the rhizospheres of grasses from a desert watershed showed peak nitrogen fixation during the vegetative growth stage and seed production phenophase of the grasses (El Shahaby, 1988). Grasses grown in sterile quartz sand inoculated with bacterial cultures of species isolated from grass rhizosheaths in the field, exhibited growth related to the inoculum. Grasses grown in sterile quartz sand inoculated with either *Azotobacter* spp. or *Azospirillum* spp. exhibited higher growth rates than grasses grown with a balanced nutrient solution minus nitrogen. Grasses grown with an inoculum of both species exhibited considerably higher growth than grasses grown with a single species of N-fixer (El Shahaby, 1988). The growth stimulation produced by associative rhizosheath bacteria may be produced by growth factors (hormones or hormone-like compounds) released in addition to nitrogen.

In addition to N_2 that is fixed by biological fixation, there are atmospheric inputs of nitrogen in the form of dust. Dryfall inputs to desert ecosystems vary considerably depending on the proximity of the desert area to industry and population centers. In the Mediterranean region of southern Europe, long-distance transport of polluted air masses increased atmospheric wet nitrogen input by at least a factor of 1.6 (Loye-Pilot *et al.*, 1990). In the northern Chihuahuan Desert there were high values of calcium, sulfate, nitrate, and total nitrogen in dryfall during the growing season. The growing season values of sulfate and nitrogen were nearly double the annual average values. These values were attributed to the differences between directional movement of air masses during the winter months and variable direction of air masses during the season of convectional storms (growing season). Air masses passing over metropolitan areas pick up nitrogen, sulfur, and other minerals that are deposited in rainfall or as dryfall. The concentrations of nitrate and total nitrogen in dry fall ranged between 4.4 ± 2.7 and 1.4 ± 0.9 ppm in dry fall and averaged 0.23 ppm in bulk precipitation (Whitford *et al.*, 1997). Rainfall inputs of NO_3–N that averaged 39 g ha^{-1} cm^{-1} rainfall were measured in the Chihuahuan Desert in west Texas (Loftis and Kurtz, 1980). In the semiarid chaparral of southern California, precipitation and dryfall resulted in mean annual deposition of Ca, K, NH_4^+–N, NO_3^-–N, Na, Mg, and SO_4^{2-}–S equal to or more than equal to annual losses reported in run-off from these ecosystems (Schlesinger *et al.*, 1982).

9.5.2 Nitrogen Losses

Ammonia volatilization and denitrification are mechanisms of loss of nitrogen from soil pools back to the atmosphere. Ammonia volatilization occurs in soils with a pH of 7.0 or greater. Since many desert soils are basic, there is a large potential for ammonia volatilization from desert landscapes. Ammonia is formed by the loss of a proton from NH_4 by the following reaction:

$$NH_4 + OH^- \Rightarrow NH_3 \Uparrow + H_2O$$

Ammonia is very soluble in water, therefore volatilization from moist soils is minimal. The loss of ammonia is greatest from coarse, dry soils with limited cation exchange capacity (Fleisher *et al.*, 1987). Ammonia volatilization measured in three habitats on a Chihuahuan Desert watershed demonstrated that wetting soils increased the rates of nitrogen mineralization and of ammonia volatilization (Schlesinger and Peterjohn, 1991). Another study found that irrigation decreased nitrogen mineralization which would have the effect of decreasing ammonia volatilization (Fisher *et al.*, 1987). The differences in results of these studies emphasizes the importance of long-term studies for understanding ecosystem processes in arid ecosystems. The studies by Schlesinger and Peterjohn were short-term incubations of soils enclosed in PVC pipe into which water and/or chemicals were added. The studies by Fisher *et al.* (1987) were conducted in the field over several years and incorporated 'droughted plots' under 'rain-out shelters', irrigated, and unmanipulated plots. The relationship between ammonia volatilization and nitrogen mineralization is affected by the effects of rainfall on development of the ephemeral plant community. In shrub-dominated systems, irrigation stimulates growth of dense stands of ephemerals. The decomposition of the dead roots of those plants results in nitrogen immobilization and reduction in mineralization rates with irrigation in the following year (Fisher and Whitford, 1995). Thus the relative importance of ammonia volatilization in desert ecosystems is dependent on climate patterns as well as spatial variability.

Denitrification is another avenue of loss of nitrogen from the soil pool back to the atmospheric pool. Denitrification results from the microbial reduction of nitrate (NO_3) or nitrite (NO_2) to either N_2O or dinitrogen (N_2). There are many species of heterotrophic soil bacteria that are capable of denitrification. Factors that have been shown to affect the rate of denitrification include: pH, temperature, available carbon, available nitrogen and the partial pressure of oxygen in the soil (Firestone, 1982). Desert soils are notoriously low in carbon and nitrogen, are frequently basic, and subject to extremes in temperature. Most desert upland soils are coarse textured and well aerated thus not good environments for denitrification.

In most desert watersheds, losses of nutrients in litter and plant debris transported into ephemeral streams is a spatial phenomenon, i.e. concentration of organic materials at the watershed terminus. Nutrient losses from desert watersheds only occur in those few watersheds that are drained by perennial streams. Thus in most desert landscapes, denitrification is the most important means of loss of this nutrient. In a nitrogen budget developed for the Great Basin Desert in the western US, it was estimated that denitrification accounted for 95% of the annual nitrogen losses and 65% of the total N inputs (West and Skujins, 1977). Estimates of denitrification rates in the cold Great Basin Desert in North America (19 kg N ha^{-1} yr^{-1}) were approximately equal to 65% of the nitrogen inputs. Other studies in the Great Basin estimated denitrification losses equal to approximately 80% of the nitrogen fixed by cryptogamic crusts and of added ^{15}N-labeled ammonium (Klubek and Skujins, 1981). In the Sonoran Desert, high rates of denitrification were measured in wet soil under nitrogen-fixing mesquite tree canopies (Virginia *et al.*, 1982). Denitrification was estimated to account for > 77% of the N inputs in the desert regions of the US (Peterjohn and Schlesinger, 1990). The content of denitrifying enzymes in upland desert soils was associated with indices of N and C availability. However, wet desert soils provided optimal conditions for the several variables that affect denitrification (Peterjohn and Schlesinger, 1991). In the Chihuahuan Desert denitrification rates were affected by water availability and the C/N ratio of the soil. Highest rates were measured in desert grassland and there were no 'fertile island' effects on denitrification rates in shrubland soils. Annual rates of denitrification on a Chihuahuan Desert watershed were estimated at 2.3 kg N ha^{-1} yr^{-1} on piedmont grassland soils, 4.3 kg N ha^{-1} yr^{-1} in creosotebush (*L. tridentata*) shrubland soils, 10.5 kg N ha^{-1} yr^{-1} in basin grassland soils, and 3.5 kg N ha^{-1} yr^{-1} in ephemeral lake soils. The differences in rates of denitrification in soils from various locations on the watershed were attributed to soil textural differences that affect soil water potentials (Peterjohn and Schlesinger, 1991).

These studies suggest that denitrification is probably an important process in the ecosystems of most desert regions. However, caution must be applied when extrapolating the results of studies of denitrification, based on irrigated soil cores, to desert regions of the world. In incubations used to measure denitrification of soils, cores were flooded with the equivalent of a 31.5 mm rain (Peterjohn and Schlesinger, 1990). Rains of that intensity have a return time of less than one storm per 20 years in the area where these studies were conducted (Fisher and Whitford, 1995). Since low oxygen tensions are characteristic of supersaturated soils, measures of denitrification under such conditions clearly optimize denitrification. Although the spatial variation and ratio of soil carbon and nitrogen are important variables affecting desertification, rates of denitrification are probably considerably lower, with water inputs equivalent to an average rainfall event of 10–15 mm, than those reported in the literature based on saturated soil incubations.

Most of the available nitrogen in the soil comes from the mineralization of organic nitrogen. Available soil nitrogen is affected by the relative rates of mineralization and immobilization. Immobilization of nitrogen occurs when ammonia or nitrate are incorporated into microbial biomass. Microbial biomass can change very rapidly under suitable conditions of substrate (available organic matter) and soil moisture. In desert ecosystems there appears to be an increase in microbial biomass, an increase in soil carbohydrates, and a decrease in extractable soil nitrogen in response to irrigation (Gallardo and Schlesinger, 1995). In desert grassland soils, fertilization with nitrogen increased microbial biomass but addition of carbohydrate increased microbial biomass only when accompanied by nitrogen fertilization. In shrub-dominated ecosystems, addition of carbohydrate resulted in increased microbial biomass and decreased extractable nitrogen and phosphorus (Gallardo and Schlesinger, 1995). Thus increases in microbial biomass requires different conditions in desert grasslands in comparison to desert shrublands.

Mineralization of nitrogen involves the soil microfauna as well as the soil microflora. As pointed out earlier, much of the organic matter that serves as a source of nutrients for plant growth is in the form of high C:N ratio dead plant roots (especially roots of ephemeral plants). As soon as the roots of annual plants die, they are colonized by heterotrophic bacteria and fungi. In soils where available N is very low, the growth of microbes on the roots is limited by available N. Microbes scavange the available N and convert it into microbial biomass N. The N in microbial biomass is not available to plant roots until the microbial biomass dies and that microbial biomass N is released as inorganic N. When desert soils are wet (less than –0.3 MPa) there is a complex of soil microfauna that graze on the microbial biomass and release mineral N as excretory products. That soil microfauna is composed of protozoans, nematodes, and microarthropods. However, in dry soils, most of the soil microfauna is encysted or anhydrobiotic, leaving only a few taxa of soil mites to graze on the fungal hyphae (Whitford, 1989). During dry periods, some fungi grow slowly on the organic substrates in the soil and there is limited mineralization of N via the grazing activities of the few soil mites. The relative rates of N mineralization and N immobilization, therefore, are determined by the pulses of rainfall and pulses of inputs of organic substrates. Periodic drought results in the death of microbial biomass. Rainfall that ends such drought produces conditions for rapid mineralization of the dead microbial biomass and competition of the growing microbial biomass and plant roots for available nutrients (Fisher and Whitford, 1995).

The differences in responses of shrubland and desert grassland to nitrogen fertilization appear to be related to the spatial and temporal distribution of ephemeral plants and to the growth characteristics of the dominant plants. In desert grasses the gradual turnover of root biomass and location of nitrogen-fixing organisms in the rhizosheaths of roots of some species produce conditions where nitrogen mineralization rates always exceed rates of nitrogen immobilization (Fig. 9.11) (Whitford and Herrick, 1996). The temporal patterns of mineralization–immobilization on the dead roots of ephemeral plants also explains the temporal variation in productivity of annual plants in shrublands where that productivity is decoupled from rainfall (Guo and Brown, 1997). In the northern Chihuahuan Desert, high production of winter–spring annuals is linked to reductions in production of summer annuals even in years with above-average summer rainfall.

The relationships between rainfall, plant productivity, and the relative rates of nitrogen mineralization to immobilization are complex. Rainfall that initiates growth and establishment of ephemeral plants sets up the potential for an input pulse of decomposable organic matter. Spatial distribution and densities of ephemerals are a function of the morphologies of the dominant vegetation. With C_3 shrubs, a large residual pool of dead roots from the previous growing season

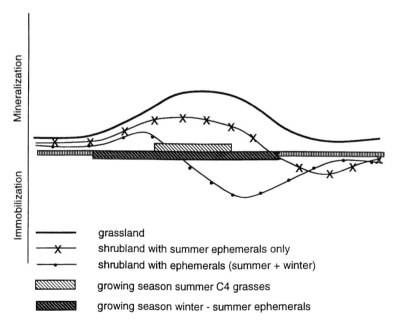

Figure 9.11   The hypothesized relationships between nitrogen mineralization and nitrogen mobilization in perennial desert grasslands and in desert shrublands. (From Whitford and Herrick, 1995.) With permission from The Society for Range Management.

or a pool of dead roots from the winter–spring ephemerals, establishes conditions for nutrient immobilization and reduction in growth rates of the shrubs. In desert grassland dominated by C_4 grasses, spatial distribution of ephemerals is random and pulses of dead roots do not result in rates of immobilization exceeding rates of mineralization (Whitford and Herrick, 1995) (Fig. 9.11).

9.6 OTHER POTENTIALLY LIMITING NUTRIENTS

Phosphorus is frequently a limiting element in cultivated agricultural systems. In studies on a northern Chihuahuan desert watershed, there was no evidence for P limitation of primary production (Lajtha and Schlesinger, 1988). However, phosphorus may be limiting in deserts with very ancient soils, e.g. Australian, South African arid regions, but there are no quantitative data available to support this supposition. There are also suggestions that other elements may be limiting factors in primary production in arid and semiarid ecosystems but there are no empirical studies that examine this suggestion. Thus for most materials the processes of decomposition and mineralization turn over sufficient quantities of elements for the water-limited plant growth.

9.7 ROLE OF MYCORRHIZAE

Mycorrhizae (literally fungal roots) are fungi that develop mutualistic symbioses with roots of many species of plants. Mycorrhizal fungi are known to enhance the nutrition of the host plant (especially with respect to phosphorus nutrition) and receive benefit from the host plant in the form of useable energy: carbohydrates (Read *et al.*, 1985). Mycorrhizae produce nondestructive interfaces with root cells as intercellular hyphae, intracellular coils, arbuscules and vesicals (Fig. 9.12). The external micorrhizal hyphae extend from the root surface into the soil where the hyphae serve as functional extensions of the roots. Some studies have documented mycorrhizal mycelial links between different species of plant (Watkins *et al.*, 1996). Most of the research on mycorrhizal associations and the ecological significance of these symbioses has been conducted with agricultural plants or with naturally occurring host plants in mesic environments. Approximately 95% of the world's plant species belong to families that are typically mycorrhizal and most of these families form vesicular–arbuscular endomycorrhizae associations (Trappe, 1981). A survey of the literature on mycorrhizal colonization of roots of arid and semiarid plants revealed that some species in most of the plant families surveyed were colonized by mycorrhizae. Most of the reports were based on single observations but where multiple samples were examined, some samples were mycorrhizal and some were not, e.g. *Atriplex canescens* (Chenopodiaceae) (Trappe, 1981). Arid zone plants that were usually mycorrhizal were members of the Compositae, Gramineae, Leguminosae,

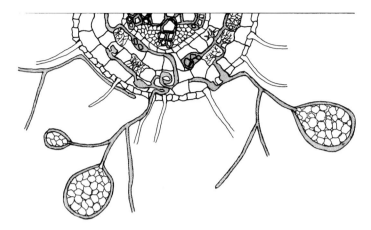

Figure 9.12   Vesicular–arbuscular mycorrhizal structures in a root.

Rosaceae, Salicaceae, and Solonaceae. The Cruciferae and Zygophyllaceae were reported to be largely nonmycorrhizal. Out of 93 species of plants from 37 families in semiarid, open mallee (*Eucalyptus* spp.) 85 were mycorrhizal and most of the mycorrhizae were vesicular–arbuscular (McGee, 1986). Several genera that were not previously known to harbor mycorrhizae were found to be mycorrhizal. Some species of Cactaceae and Chenopodiaceae were mycorrhizal at some times and nonmycorrhizal at other times. Ephemeral species in families that are typically mycorrhizal appear to develop mycorrhizal symbionts as readily as the perennial species in those families.

Vesicular–arbuscular mycorrhizae fruit as single spores in the soil. The concentrations of vesicular–arbuscular spores in the soil vary seasonally, with soil moisture, and probably with the phenological stage of the host plant. Mycorrhizal spores are dispersed by anything that moves soil (wind, water, and animals) (Allen *et al.*, 1989). A variety of arthropods (crickets, grasshoppers, ants, wasps, and other insects) and some small mammals have been found to be spore dispersers (Trappe, 1981). Mycophagous arthropods may adversely affect the symbiosis between vesicular–arbuscular mycorrhizae and plants by grazing on the spores and external hyphae. Some microarthropods may also act as symbiotic mutualists by grazing and disseminating fungal propagules or by stimulating fungal growth and activity by grazing (Moore, 1988). Harvester ants, *Pogonomyrmex occidentalis*, leave high densities of vesicular–arbuscular mycorrhizal fungi in the soils of the nest mounds that assist in the establishment of mutualistic mychorrizal associations after the ant colonies have abandoned the mounds (Friese and Allen, 1993).

The mutualism between mychorrizal fungi and plant roots is one in which the fungus receives carbon from the plant and the plant receives nutrients from the fungus. Mycorrhizal fungi differ in their ability to take-up and metabolize various carbon sources (Smith and Read, 1998). Vesicular–arbuscular mycorrhizae are totally dependent on photosynthetically derived glucose. The external mycelia of other mycorrhizae can use a greater variety of carbon sources. Root exudation may trigger mycorrhizal colonization but the quantities of exudate are apparently insufficient to support continuous mycorrhizal development. Mycorrhizal colonization may reduce and change the quality of root exudates (Ingham and Molina, 1991).

There is some evidence that mycorrhizal plants are more drought tolerant than non-mycorrhizal species. This is thought to result from the increased soil volume exploited by mycorrhizal plants and not to an increase in uptake of water by the fungi (Ingham and Molina, 1991). The enhancement of growth in mycorrhizal plants is generally attributed to the exploitation of a larger volume of soil by the mycorrhizal mycelia that are external to the root. Increased phosphorus and ammonium uptake and inflow to plant roots has been documented (Read et al., 1985). Direct connection of plant roots of different individuals of the same species or roots of different species by mycorrhizal bridges has also been documented (Robinson and Fitter, 1999). Also, there is evidence for the movement of nutrients from dying roots to living plants via mycorrhizal links between dying and living roots (Newman and Eason, 1989). Experimental evidence for the direct interplant transfer of carbon suggests that mycorrhizal associations may be extremely important for productivity of patches of plants with different rooting depths and phenologies. An immediate supply of carbohydrate could allow shallow-rooted plant species to initiate growth of photosynthetic tissue as soon as soil moisture becomes available.

Mycorrhizal associations may not be the only root–fungus mutualism that improves the growth and survival of desert plants. The roots of dominant grass and shrub species of the arid southwestern United States, were extensively colonized by dark septate fungal endophytes (Barrow et al., 1997). Dark septate fungi were either the only fungi in the root or were found in combination with vesicular–arbuscular mycorrhizae in the same root. The internal morphology of dark septate fungi is different from other mycorrhizal associations. The dark septate hyphae penetrate the root and grow inter- and intracellularly in the root cortex. Dark septate fungi form microsclerotia of various structural configurations over time. These microsclerotia take the form of coils and vesicles. Dark septate fungi are widespread and are abundant in stressed ecosystems suggesting that they are ecologically important (Jumpponen and Trappe, 1998). Dark septate fungal associations with the roots of a desert shrub, *Atriplex canescens*, were found to contain large quantities of lipid during the period of leaf production by the shrub in the early spring. It was hypothesized that the storage of carbon in the dark septate fungi during peak growth periods of the plant could serve as a carbon sink and enhance water retention in the fungal hyphae. From the fungal reservoir, sufficient carbon, water, and nutrients could be supplied to the plant to insure survival during periods of climatic stress (Barrow, 2001). This fascinating fungus–plant association should be the focus of future research on desert plants and nutrient cycling processes.

REFERENCES

Aerts, R. (1996). Nutrient resorption from senescing leaves of perennials: are there general patterns. *J. Ecol.* **84,** 597–608.

Aerts, R. (1997). Climate, leaf litter chemistry and leaf litter decomposition in terrestrial ecosystems: a triangular relationship. *Oikos* **79,** 439–449.

Allen, M. F., Hipps, L. E., and Wooldrige, G. L. (1989). Wind dispersal and subsequent establishment of VA mycorrhizal fungi across a successional arid landscape. *Land. Ecol.* **2,** 165–171.

Barrow, J. R. (2001). The transfer of lipids to fourwing saltbush, *Atriplex canescens* (Pursh) Nutt. by a dark septate fungal root endophyte. *J. Arid Environ.* (In Press)

Barrow, J. R., Havstad, K. M., and McCaslin, B. D. (1997). Fungal root endophytes in fourwing saltbush, *Atriplex canescens,* on arid rangelands of southwestern USA. *Arid Soil Res. Rehabil.* **11,** 177–185.

Barth, R. C. and Klemmedson, J. O. (1978). Shrub-induced spatial patterns of dry matter, nitrogen and organic carbon. *Soil Sci. Soc. Am. J.* **42,** 804–809.

Belnap, J. and Harper, K. T. (1995). Influence of cryptobiotic soil crusts on elemental content of tissues of two desert seed plants. *Arid Soil Res. Rehabil.* **9,** 107–115.

Berg, B., Wickbohm, G., and McClaugherty, C. (1984). Lignin and holocellulose relations during the long-term decomposition of some forest litters. Long-term decomposition in a Scots pine forest. IV. *Can. J. Bot.* **62,** 2540–2550.

Cepeda-Pizaro, J. G. (1993) Litter decomposition in deserts: an overview with an example from coastal arid Chile. *Rev. Chil. de Hist. Nat.* **66,** 323–336.

Cepeda-Pizaro, J. G. and Whitford, W. G. (1990). Decomposition patterns of surface leaf litter of six plant species along a Chihuahuan Desert watershed. *Am. Midl. Nat.* **123,** 319–330.

Charley, J. L. and Cowling, S. W. (1968). Changes in soil nutrient status resulting from overgrazing and their consequences in plant communities of semi-arid areas. *Proc. Ecol. Soc. Aust.* **3,** 23–38.

Charley, J. L. and West, N. E. (1975). Plant-induced chemical patterns in some shrub-dominated semi-desert ecosystems of Utah. *J. Ecol.* **63,** 945–964.

Charley, J. L. and West, N. E. (1977). Micro-patterns of nitrogen mineralization activity in soils of some shrub-dominated semi-desert ecosystems of Utah. *Soil Biol. Biochem.* **9,** 357–365.

Cox, G. W. and Gakahu, C. G. (1985). Mima mound microtopography and vegetation pattern in Kenyan savannas. *J. Trop. Ecol.* **1,** 23–36.

Cox, G. W. and Gakahu, C. G. (1986). A latitudinal test of the fossorial rodent hypothesis of Mima mound origin. *Z. Geomophol.* **30,** 485–501.

Creusere, F. M. and Whitford, W. G. (1982). Temporal and spatial resource partitioning in a Chihuahuan Desert lizard community. In N. J. Scott Jr. (ed.), *Herpetological Communities*, pp. 121–127. U S Fish and Wildlife Service, Wildlife Research Report 13, U.S. Department of the Interior, Washington, DC.

deSoyza, A. G., Whitford, W. G., Martinez-Meza, E., and Van Zee, J. W. (1997). Variation in creosotebush (*Larrea tridentata*) canopy morphology in relation to habitat, soil fertility and associated annual plant communities. *Am. Midl. Nat.* **137,** 13–26.

Elkins, N. Z. and Whitford, W. G. (1982). The role of microarthropods and nematodes in decomposition on a semi-arid ecosystem. *Oecologia* **55,** 303–310.

Elkins, N. Z., Steinberger, Y., and Whitford, W. G. (1982). Factors affecting the applicability of the AET model for decomposition in arid environments. *Ecology* **63,** 579–580.

El Shahaby, A. F. (1988). Associative nitrogen fixation with C_4 grasses of the northern Chihuahan Desert. PhD. Dissertation. New Mexico State University, Las Cruces, NM. 134 pp.

Ettershank, G., Ettershank, J., Bryant, M., and Whitford, W. G. (1978). Effects of nitrogen fertilization on primary production in a Chihuahan Desert ecosystem. *J. Arid Environ.* **1,** 135–139.

Firestone, M. K. (1982) Biological denitrification. In F. J. Stevenson (ed.) *Nitrogen in Agricultural Soils*, pp. 289–326. Soil Science Society of America. Madison. Wisconsin.

Fisher, F. M. and Whitford, W. G. (1995). Field simulation of wet and dry years in the Chihahuan Desert: soil moisture, N mineralization and ion-exchange resin bags. *Biol. Fertil. Soils* **20,** 137–146.

Fisher, F. M., Parker, L. W., Anderson, J. P., and Whitford, W. G. (1987). Nitrogen mineralization in a desert soil: interacting effects of soil moisture and nitrogen fertilizer. *Soil Sci. Soc. Am. J.* **51,** 1033–1041,

Fisher, F. M., Zak, J. C., Cunningham, G. L., and Whitford, W. G. (1988). Water and nitrogen effects in growth allocation patterns of creosotebush in the northern Chihuahuan Desert. *J. Range Mgmt* **41,** 387–391.

Fleisher, Z., Ravina, A. K. I., and Hagin, J. (1987). Model of ammonia volatilization from calcareous soils. *Plant Soil* **103,** 205–212.

Floret, C., Pontanier, R., and Rambal, S. (1982). Measurements and modelling of primary production and water use for a south Tunisian steppe. *J. Arid Environ.* **5,** 77–90.

Friedel, M. H., Cellier, K. M., and Nicolson, K. P. (1980). Nutrient deficiencies in central Australian semi-desert rangelands, with reference to decline in range condition. *Aust. Rangeland J.* **2,** 151–161.

Friese, C. F. and Allen, M. F. (1993) The interaction of harvester ants and vesicular–arbuscular mycorrhizal fungi in a patchy semi-arid environment: the effects of mound structure on fungal dispersion and establishment. *Funct. Ecol.* **7,** 13–20.

Gallardo, A. and Schlesinger, W. H. (1995). Factors determining soil microbial biomass and nutrient immobilization in desert soils. *Biogeochemistry* **28,** 55–68.

Garcia-Moya, E. and McKell, C. M. (1970). Contribution of shrubs to the nitrogen economy of a desert-wash plant community. *Ecology* **51,** 81–88.

Gehrke, C., Johanson, U., Callaghan, T. V., Chadwick, D., and Robinson, C. H. (1995). The impact of enhanced ultraviolet-B radiation on litter quality and decomposition processes in *Vaccinium* leaves from the subarctic. *Oikos* **72,** 213–222.

Gile, L. H., Gibbens, R. P., and Lenz, J. M. (1998). Soil-induced variability in root systems of creosotebush (*Larrea tridentata*) and tarbush (*Flourensia cernua*). *J. Arid Environ.* **39,** 57–78.

Guo, Q. and Brown, J. H. (1997). Interactions between winter and summer annuals in the Chihuahuan Desert. *Oecologia* **111,** 123–128.

Gutierrez, J. R. and Whitford, W. G. (1987). Chihuahuan Desert annuals: importance of water and nitrogen. *Ecology* **68,** 2032–2045,

Gutierrez, J. R., DaSilva, O. A., Pagani, M. I., Weems, D., and Whitford, W. G. (1988). Effects of different patterns of supplemental water and nitrogen fertilization on productivity and composition of Chihuahuan Desert annual plants. *Am. Midl. Nat.* **119,** 336–343.

Hamadi, Z., Steinberger, Y., Kutiel, P., Lavee, H., and Barness, G. (2000) Decomposition of *Avena sterilis* litter under arid conditions. *J. Arid Environ.* **46,** 281–293.

Hogberg, P. (1986). Nitrogen-fixation and nutrient relations in savanna woodland trees (Tanzania). *J. Appl. Ecol.* **23,** 675–688.

Ingham, E. R. and Molina, R. (1991). Interactions among mycorrhizal fungi, rhizosphere organisms, and plants. In P. Barbosa, V. A. Krischik, and C. G. Jones (eds), *Microbial Mediation of Plant–Herbivore Interactions*, pp. 169–197. John Wiley, New York.

Jacobson, K. M. and Jacobson, P. J. (1998). Rainfall regulates decomposition of buried cellulose in the Namib Desert. *J. Arid Environ.* **38,** 571–583.

James, C. D. (1991). Temporal variation in diets and trophic partitioning by coexisting lizards (*Ctenotus*: Scincidae) in central Australia. *Oecologia* **85,** 553–561.

James, D. W. and Jurinak, J. J. (1978). Nitrogen fertilization of dominant plants in the northeastern Great Basin Desert. In N. E. West and J. Skujins (eds), *Nitrogen in Desert Ecosystems*, pp. 219–243. Dowden, Hutchinson and Ross, Stroudsburg, PA.

Johnson, H. B. and Mayeux, H. S. Jr.(1988). Mesquite nodulation and nitrogen fixation on rangelands. *Proceedings 41st Annual Meeting of Society for Range Management*. Abstract 296.

Johnson, H. B. and Mayeux, H. S. Jr (1990). *Prosopis glandulosa* and the nitrogen balance of rangelands: extent and occurrence of nodulation. *Oecologia* **84,** 176–185.

Jumpponen, A., and Trappe, J. M. (1998) Dark spetate endophytes: a review of facultative biotrophic root-colonizing fungi. *New Phytol.* **140,** 295–310.

Killingbeck, K. T. (1993) Nutrient resorption in desert shrubs. *Rev. Chil. Hist. Nat.* **66,** 345–355.

Killingbeck, K. T. and Whitford, W. G. (2001). Nutrient resorption in shrubs growing by design and by default in Chihuahuan Desert arroyos. *Oecologia* (in press)

Klemmedson, J. O. and Tiedeman, A. R. (1986). Long term effects of mesquite removal on soil characteristics: II. nutrient availability. *Soil Sci. Soc. Am. J.* **50,** 476–480.

Klubek, B. and Skujins, J. (1981). Gaseous nitrogen losses from ^{15}N-ammonium and plant material amended with labeled nitrogen in Great Basin Desert surface soils. *Geomicrobiol. J.* **2,** 225–236.

Lajtha, K. (1987). Nutrient resorption efficiency and the response to phosphorus fertilization in the desert shrub *Larrea tridentata* (DC.) Cov. *Biogeochemistry* **4,** 265–276.

Lajtha, K. and Schlesinger, W. H. (1988). The biogeochemistry of phosphorus cycling and phosphorus availability along a desert soil chronosequence. *Ecology* **69,** 24–39.

Lajtha, K. and Whitford, W. G. (1989). The effect of water and nitrogen amendments on photosynthesis, leaf demography, and resource-use efficiency in *Larrea tridentata*, a desert evergreen shrub. *Oecologia* **80,** 341–348.

Lee, K. E. and Wood, T. G. (1971). *Termites and Soils.* Academic Press, London.

Lobry de Bruyn, L. A. and Conacher, A. J. (1990). The role of termites and ants in soil modification: a review. *Aust. J. Soil Res.* **28,** 55–93.

Loftis, S. G. and Kurtz, E. B. (1980). Field studies of inorganic nitrogen added to semiarid soils by rainfall and blue–green algae. *Soil Sci.* **129,** 150–155.

Loye-Pilot, M. D., Martin, J. M., and Morelli, J. (1990). Atmospheric input of inorganic nitrogen to the western Mediterranean. *Biogeochemistry* **9,** 117–134.

Ludwig, J. A. and Flavill, P. (1979). Productivity patterns of *Larrea* in the northern Chihuahuan Desert. In E. C. Lopez, T. J. Mabry, and S. F. Tavizon (eds), *Larrea*, pp. 139–150. Centro de Investigacion en Quimica Aplicada, Saltillo, Mexico.

MacKay, W. P., Silva, S., Lightfoot, D. C., Pagani, M. I., and Whitford, W. G. (1986). Effect of increased soil moisture and reduced soil temperature on a desert soil arthropod community. *Am. Midl. Nat.* **116,** 45–56.

MacKay, W. P., Silva, S., Loring, S. J., and Whitford, W. G. (1987). The role of subterranean termites in the decomposition of above-ground creosotebush litter. *Sociobiology* **13,** 235–239.

MacKay, W. P., Zak, J., and Whitford, W. G. (1992). Litter decomposition in a Chihuahuan Desert playa. *Am. Midl. Nat.* **128,** 89–94.

MacKay, W. P., Loring, S. J., Zak, J. C., Silva, S. I., Fisher, F. M., and Whitford, W. G. (1994). Factors affecting loss in mass of creosotebush leaf-litter on the soil surface in the northern Chihuahuan Desert. *Southwest. Nat.* **39,** 78–82.

Mandimba, G., Heulin, T., Bally, R., Guckert, A., and Balandreau, J. (1986). Chemotaxis of free-living nitrogen-fixing bacteria towards maize mucilage. *Plant Soil* **90,** 129–139.

Mayland, H. F. and McIntosh, T. H. (1966). Distribution of nitrogen fixed in desert algal-crusts. *Soil Sci. Soc. Am. Proc.* **30,** 605–609.

Mazzarino, M. J., Bertiller, M. B., Spain, C. L., Laos, F., and Coronato, F. R. (1996). Spatial patterns of nitrogen availability, mineralization, and immobilization in northern Patagonia, Argentina. *Arid Soil Res. Rehabil.* **10,** 295–309.

McClaugherty, C. A., Aber, J. D., and Melillo, J. M. (1984) Decomposition dynamics of fine roots in forested ecosystems. *Oikos* **42,** 378–386.

McGee, P. (1986). Mycorrhizal associations of plant species in a semiarid community. *Aust. J. Bot.* **34,** 585–593.

Meentemeyer, V. (1978). Macroclimate and lignin control of litter decomposition rates. *Ecology* **59,** 465–472.

Montana, C., Ezcurra, E., Carrillo, A., and Delhoume, J. P. (1988). The decomposition of litter in grasslands in northern Mexico: a comparison between arid and non-arid environments. *J. Arid Environ.* **14,** 55–60.

Moore, J. C. (1988). The influence of microarthropods on symbiotic and non-symbiotic mutualism in detrital-based below-ground food webs. *Agric. Ecosyst. Environ.* **24,** 147–159.

Moorhead, D. L. and Reynolds, J. F. (1989). Mechanisms of surface litter mass loss in the northern Chihuahuan Desert: a reinterpretation. *J. Arid Environ.* **16,** 157–163.

Moorhead, D. L., Fisher, F. M., and Whitford, W. G. (1988). Cover of spring annuals on nitrogen-rich kangaroo rat mounds in a Chihuahuan Desert grassland. *Am. Midl. Nat.* **20**, 443–447.

Mun, H. T. and Whitford, W. G. (1998). Changes in mass and chemistry of plant roots during long-term decomposition on a Chihuahuan Desert watershed. *Biol. Fertil. Soils* **26**, 16–22.

Nash, M. H. and Whitford, W. G. (1995). Subterranean termites: regulators of soil organic matter in the Chihuahuan Desert. *Biol. Fertil. Soils* **19**, 15–18.

Newman, E. I. and Eason, W. R. (1989). Cycling of nutrients from dying roots to living plants, including the role of mycorrhizas. In M. Clarholm and L. Bergstrom (eds), *Ecology of Arable Land*, pp. 133–137. Kluwer Academic Publishers, New York.

Parker, L. W., Fowler, H., Ettershank, G., and Whitford, W. G. (1982). The effects of subterranean termite removal on desert soil-nitrogen and ephemeral flora. *J. Arid Environ.* **5**, 53–59.

Parker, L. W., Miller, J., Steinberger, Y., and Whitford, W. G. (1983). Soil respiration in a Chihuahuan Desert rangeland. *Soil Biology and Biochemistry* **15**, 303–309.

Parker, L. W., Santos, P. F., Phillips, J., and Whitford, W. G. (1984). Carbon and nitrogen dynamics during decomposition of litter and roots of a Chihuahuan Desert annual, *Lepidium lasiocarpum. Ecol. Monogr.* **54**, 339–360.

Pauli, F. (1964). Soil fertility problem in arid and semi-arid lands. *Nature* **204**, 1286–1288.

Penning de Vries, F. W. T. and Djiteye, M. A. (1982). *La Productivite des Paturages Saheliens.* Centre for Agricultural Publishing and Documentation. Wageningen, Netherlands.

Peterjohn, W. T. and Schlesinger, W. H. (1990). Nitrogen loss from deserts in the southwestern United States. *Biogeochemistry* **10**, 67–79.

Peterjohn, W. T. and Schlesinger, W. H. (1991). Factors controlling denitrification in a Chihuahuan Desert ecosystem. *Soil Sci. Soc. Am. J.* **55**, 1694–1701.

Peterson, D. K. and Whitford, W. G. (1987). Foraging behavior of *Uta stansburiana* and *Cnemidophorus tigris* in two different habitats. *Southwest. Nat.* **32**, 427–433.

Pianka, E. R. (1985). Some intercontinental comparisons of desert lizards. *Nat. Geogr. Res.* **1**, 490–504.

Read, D. J., Francis, R., and Finlay, R. D. (1985). Mycorrhizal mycelia and nutrient cycling in plant communities. In D. Atkinson, D. J. Read, and M. B. Usher (eds), *Ecological Interactions in Soil: Plants, Microbes, and Animals*, pp. 193–217. Blackwell Scientific Publications, Oxford, England.

Reddell, P., Bowen, G. D., and Robson, A. D. (1985). The effects of soil temperature on plant growth, nodulation, and nitrogen fixation in *Casuarina cunninghamiana* Miq. *New Phytol.* **101**, 1–10.

Reichle, D. E. (1971). Energy and nutrient metabolism of soil and litter invertebrates. In P. Duvigneaud (ed.), *Productivity of Forest Ecosystems*, pp. 465–477. UNESCO, Paris.

Reinhold, B., Hurek, T., Niemann, E., and Fendrik, I. (1986). Close association of *Azospirillum* and diazotrophic rods within different root zones of Kollar grass. *Appl. Environ. Microbiol.* **52**, 520–526.

Robinson, D., and Fitter, A. (1999). The magnitude and control of carbon transfer between plants linked by a common mycorrhizal network. *J. Exp. Bot.* **50**, 9–13.

Romney, E. M., Wallace, A., and Hunter, R. B. (1978). Plant responses to nitrogen fertilization in the northern Mojave Desert and its relationship to water manipulation. In N. E. West and J. Skujins (eds), *Nitrogen in Desert Ecosystems*, pp. 232–243. Dowden, Hutchinson and Ross, Stroudsburg, PA.

Rougier, M. and Chaboud, A. (1985). Mucilages secreted by roots and their biological functions. *Isr. J. Bot.* **34**, 129–136.

Rundel, P. W., Nilsen, E. T., Sharifi, M. R., Virginia, R. A., Jarrel, W. M., Kohl, D. H., and Shearer, G. B. (1982). Seasonal dynamics of nitrogen cycling for a *Prosopis* woodland in the Sonoran Desert. *Plant and Soil* **67**, 343–353.

Santos, P. F. and Whitford, W. G. (1981). The effects of microarthropods on litter decomposition in a Chihuahuan Desert ecosystem. *Ecology* **62**, 654–663.

Santos, P. F. and Whitford. W. G. (1983). The influence of soil biota on decomposition of plant material in a gypsum sand dune habitat. *Southwest. Nat.* **28**, 423–427.

Santos, P. F., Phillips, J., and Whitford, W. G. (1981). The role of mites and nematodes in early stages of buried litter decomposition in a desert. *Ecology* **62,** 664–669.

Santos, P. F., Elkins, N. Z., Steinberger, Y., and Whitford, W. G. (1984). A comparison of surface and buried *Larrea tridentata* leaf litter decomposition in North American hot deserts. *Ecology* **65,** 278–284.

Schaefer, D. A. and Whitford, W. G. (1981). Nutrient cycling by subterranean termites in a Chihuahuan Desert ecosystem. *Oecologia* **65,** 382–386.

Schaefer, D., Steinberger, Y., and Whitford, W. G. (1985). The failure of nitrogen and lignin control of decomposition in a North American desert. *Oecologia* **65,** 382–386.

Schlesinger, W. H. and Peterjohn, W. T. (1991). Processes controlling ammonia volatilization from Chihuahuan Desert soils. *Soil Biol. Biochem.* **23,** 637–642.

Schlesinger, W. H., Gray, J. T., and Gilliam, F. S. (1982). Atmospheric deposition processes and their importance as sources of nutrients in chaparral ecosystems of southern California. *Water Resources Res.* **18,** 623–629.

Schlesinger, W. H., Raikes, J. A., Hartley, A. E., and Cross, A. F. (1996). On the spatial pattern of soil nutrients in desert ecosystems. *Ecology* **77,** 364–374.

Seastedt, T. R. (1984). The role of microarthropods in decomposition and mineralization processes. *Ann. Rev. Entomol.* **29,** 25–46.

Shearer, G., Kohl, D. H., Virginia, R. A., Bryan, B. A., Skeeters, J. L., Nilsen, E. T., Sharifi, M. R., and Rundel, P. W. (1983). Estimates of N_2-fixation from variation in the natural abundance of ^{15}N in Sonoran Desert ecosystems. *Oecologia* **56,** 365–373.

Shields, L. M. (1957). Algal and lichen floras in relation to nitrogen content of certain volcanic and arid range soils. *Ecology* **38,** 661–663.

Silva, S. I., MacKay, W. P., and Whitford, W. G. (1985). The relative contributions of termites and microarthropods to fluff grass litter disappearance in the Chihuahuan Desert. *Oecologia* **67,** 31–34.

Smith, S. E. and Read, D. J. (1998). *Mycorrhizal Symbiosis,* 2nd edn. Academic Press, London.

Steinberger, Y. and Whitford, W. G. (1983). The contribution of rodents to decomposition processes in a desert ecosystem. *J. Arid Environ.* 6, 177–181.

Steinberger, Y. and Whitford, W. G. (1988). Decomposition process in Negev ecosystems. *Oecologia* **75,** 61–66.

Steinberger, Y., Shmida, A., and Whitford, W. G. (1990). Decomposition along a rainfall gradient in the Judean Desert, Israel. *Oecologia* **82,** 322–324.

Stephens, G. and Whitford, W. G. (1993). Responses of *Bouteloua eriopoda* to irrigation and nitrogen fertilization in a Chihuahuan Desert grassland. *J. Arid Environ.* **24,** 415–421.

Strojan, C. L., Randall, D. C., and Turner, F. B. (1987). Relationship of litter decomposition rates to rainfall in the Mojave desert. *Ecology* **68,** 741–744.

Swift, M. J., Heal, O. W., and Anderson, J. M. (1979). *Decomposition in Terrestrial Ecosystems.* Blackwell Scientific, Oxford .

Thurow, T. L. (1989). Decomposition of grasses and forbs in coastal savanna of southern Somalia. *Afr. J. Ecol.* **27,** 201–206.

Trappe, J. M. (1981) Mycorrhizae and productivity of arid and semiarid rangelands. In *Advances in Food Producing Systems for Arid and Semi-arid Lands,* pp. 581–599. Academic Press, New York.

Virginia, R. A. and Jarrell, W. M.. (1983). Soil properties in a mesquite-dominated Sonoran Desert ecosystem. *Soil Sci. Soc. Am. J.* **47,** 138–144.

Virginia, R. A., Jarrell, W. M., and Franco-Vizcaino, E. (1982). Direct measurement of denitrification in a *Prosopis* (mesquite) dominated Sonoran Desert ecosystem. *Oecologia* **53,** 120–122.

Virginia, R. A., Jenkins, M. B., and Jarrell, W. M. (1986). Depth of root symbiont occurrence in soil. *Biol. Fertil. Soils* **2,** 273–284.

Vossbrinck, C. R., Coleman, D. C., and Wooley, T. A. (1979). Abiotic and biotic factors in litter decomposition in semiarid grassland. *Ecology* **60,** 256–271.

Wallwork, J. A. (1982). *Desert Soil Fauna.* Praeger, New York.

Watkins, N. K., Fitter, A. H., Graves, J. D., and Robinson, D. (1996). Carbon transfer between C_3 and C_4 plants linked by a common mycorrhizal network, quantified using stable carbon isotopes. *Soil Biol. Biochem.* **28,** 471–477.

Weltzin, J. F. and Coughenour, M. B. (1990). Savanna tree influence on understory vegetation and soil nutrients in northwestern Kenya. *J. Veget. Sci.* **1,** 325–334.

West, N. E. and Skujins, J. (1977). The nitrogen cycle in North American cold-winter, semi-desert ecosystems. *Oecol. Plant.* **12,** 45–53.

Whitford, W. G. (1989). Abiotic controls on the functional structure of soil food webs. *Biol. Fertil. Soils* **8,** 1–6.

Whitford, W. G. (2000). Keystone arthropods as webmasters in desert ecosystems. In D. C. Coleman and P. F. Hendrix (eds), *Invertebrates as Webmasters in Ecosystems*, pp. 25–41. CABI Publishing, New York.

Whitford, W. G. and DiMarco, R. (1995). Variability in soils and vegetation associated with harvester ant (*Pogonomyrmex rugosus*) nests on a Chihuahuan Desert watershed. *Biol. Fertil. Soils* **20,** 169–173.

Whitford, W. G. and Herrick, J. E. (1995). Maintaining soil processes for plant productivity and community dynamics. In N. E. West (ed.), *Proceedings of the Fifth International Rangeland Congress* vol. II, pp. 33–37. Society for Range Management, Denver, CO.

Whitford, W. G. and Kay, F. R. (1999). Biopedturbation by mammals in deserts: a review. *J. Arid Environ.* **41,** 203–230.

Whitford, W. G., Meentemeyer, V., Seastedt, T. R., Cromack, K. Jr, Crossley, D. A. Jr., Santos, P. F., Todd, R. L., and Waide, J. B. (1981). Exceptions to the AET model: deserts and clear-cut forests. *Ecology* **62,** 257–277.

Whitford, W. G., Repass, R., Parker, L. W., and Elkins, N. Z. (1982). Effects of initial litter accumulation and climate on litter disappearance in a desert ecosystem. *Am. Midl. Nat.* **108,** 105–110.

Whitford, W. G., Steinberger, Y., MacKay, W. P., Parker, L. W., Freckman, D. F., Wallwork, J. A., and Weems, D. (1986) Rainfall and decomposition in the Chihuahuan Desert. *Oecologia* **68,** 512–515.

Whitford, W. G., Reynolds, J. F., and Cunningham, G. L. (1987). How desertification affects nitrogen limitation of primary production of Chihuahuan Desert watersheds. In E. F. Aldon, C. E. Gonzales-Vincent and W. H. Moir (eds), *Strategies for Classification and Management of Native Vegetation for Food Production in Arid Zones,* pp. 143–153. General Technical Report RM 150. USDA Forest Service, Rocky Mountain Forest and Range Experiment Station, Ft. Collins, CO.

Whitford, W. G., Stinnett, K., and Anderson, J. (1988). Decomposition of roots in a Chihuahuan Desert ecosystem. *Oecologia* **75,** 147–155.

Whitford, W. G., Ludwig, J. A., and Noble, J. C. (1992). The importance of subterranean termites in semi-arid ecosystems in south-eastern Australia. *J. Arid Environ.* **22,** 82–91.

Whitford, W. G., Martinez-Turanzas, G., and Martinez-Meza, E. (1995). Persistence of desertified ecosystems: explanations and implications. *Environ. Monit. Assess.* **37,** 319–332.

Whitford, W.G., Anderson, J., and Rice, P. M. (1997). Stemflow contribution to the 'fertile island' effect in creosotebush, *Larrea tridentata*. *J. Arid Environ.* **35,** 451–457.

Wullstein, L. H., Bruening, M. L., and Bollen, W. B. (1979) Nitrogen fixation associated with sand grain root sheaths (rhizosheaths) of certain xeric grasses. *Plant Physiol.* **46,** 1–4.

Chapter 10 | Desertification

Desertification has been described as

> the diminution or destruction of the biological potential of the land, and can lead ultimately to desert-like conditions. It is an aspect of the widespread deterioration of ecosystems under the combined pressure of adverse and fluctuating climate and excessive exploitation. Such pressure has diminished or destroyed the biological potential, i.e. plant and animal production, for multiple purposes at a time when increased productivity is needed to support growing populations in quest of development. (Verstraete, 1986).

The 1992 UN Conference on Environment and Development that was held in Rio de Janeiro, defined desertification as land degradation in arid, semiarid, and dry subhumid areas resulting from various factors including climatic variations and human activities. These definitions of desertification focus on degradation and economic decline in arid and semiarid regions of the world. The United Nations conferences focused attention on the human populations affected by desertification '250 million people around the world and an indirect threat to a further 750 million people' (Williams and Balling, 1996). The economic implications of desertification result primarily from a reduction in management options as a result of the degradation of land resources. The United Nations conference recognized that the immediate cause of degradation of the ecosystems was the interaction of variation in climate (drought) and overuse of the natural resources during these periods. When evaluating desertification, as with any widespread environmental degradation, it is necessary to separate the effects of *disturbance*, i.e. environmental factors that are within the range of the evolutionary history of the ecosystems, i.e. drought, flood, fire, and wind storms, from those effects that are anthropogenic and which impose *stress* on the ecosystems. The magnitude of the changes resulting from the combination of drought disturbance and stress by over-use of natural resources is the result of management decisions made by the land managers and indigenous people. Some aspects of desertification such as the harvesting of woody plants for fuel have adverse effects on the ecosystems that are independent of climatic disturbance.

Desertification produces many changes in the ecosystems of a region. Desertification is recognized primarily by the physical changes in the environment: reduction of plant cover, soil loss, loss of soil organic matter, deposition of sand bodies, increased run-off, etc. (Kassas, 1977). In extreme cases, desertification can result in marked reduction in vegetative cover and loss of soil by water and wind erosion. In such areas there is a loss in primary productivity and in potential productivity. The impacts of desertification are not evenly distributed among the arid and semiarid regions of the world. Drought combined with high human population growth and poor land management practices had devastating effects on the humans in the Sahel and Ethiopia during periods in the 1970s and 1080s. More than 1.3 million people starved to death in this region from the combined effects of desertification and civil war. The result of the recurring droughts interacting with farming of marginal lands has resulted in alarming drops in agricultural production and loss of productive capacity in the northern margins of the Sahel (Darkoh, 1998). In most other parts of the world, desertification has adversely affected livestock production capacity and the economic status of the residents (Grove, 1986; Arshad and Rao, 1994; Omar and Abdal, 1994; Fullen and Mitchell, 1994; Dean *et al.*, 1995; Noble, 1997) but has not caused human death due to starvation.

Not all desertified regions suffer a loss of primary productivity. In many regions, the loss of productivity is measured as reduction in the production of harvestable animals (livestock) which results from changes in species composition of the vegetation. In southern Africa recognition that desertification is occurring begins when desirable forage species are replaced by species that are largely inedible by livestock (Milton *et al.*, 1994). In much of the southwestern US and in eastern Australia, desertification is recognized as the loss of forage species (primarily perennial grasses) and replacement of these species by woody shrubs that are largely inedible by livestock (Buffington and Herbel, 1965; Archer et al., 1988; Archer, 1990; Grover and Musick, 1990; Noble, 1997). This is well documented in the case study of the Jornada Experimental Range in southern New Mexico (Buffington and Herbel, 1965). The percentage of land area classified as grassland (with few or no shrubs) decreased from 60% in 1858 to 0% by 1970. In 1858, the land area dominated by mesquite, *Prosopis glandulosa*, accounted for 20% of the land area and this increased to 60% by 1970. At the present time, more than half of the land area dominated by mesquite is in the form of coppice dunes. Coppice dunes form when blowing sand accumulates around the stems of multi-stemmed mesquite plants. The plants continue to grow, producing adventitious roots into the accreting dune. Mature coppice dunes may exceed 4 m in height and be stabilized by a coppice of 20–50 cm stems that form a crown of more than 4 m in diameter. The conversion of grassland to coppice dunes reduces the livestock forage to nearly zero and effectively eliminates livestock production as a use of that land area.

In the Australian arid and semiarid zones, desertification is referred to as the woody weed problem. In Australia the problem is primarily one of loss of palatable species and replacement with nonpalatable largely woody species (Noble, 1997). In the semiarid region at the southern edge of the Atacama, desertification resulted from the loss of perennial grasses and the increased cover of nonpalatable small tree, *Acacia caven* (Fabaceae:Mimosoideae). The resulting landscape is a pseudo-savanna consisting of the trees and a large variety of native and exotic annuals (Ovalle et al., 1993). Similar changes have occurred as a result of degradation in the semiarid and arid regions of Argentina, i.e. invasion of the small tree, *Geoffroea decorticans* (Fabaceae).

10.1 HISTORY OF DESERTIFICATION

Desertification is a problem with ancient roots. Cutting of forests, overgrazing and salt accumulation in irrigated lands led to desertification in Mesopotamia and the lands bordering the Mediterranean more than 2000 years ago. There is evidence that the climate of the Mediterranean region has been in a drying phase for much of the past 5000 years (El Baz, 1983). However, in the Mediterranean region, the climatic drying was coincident with developing agricultural technology and the increased demand for resources such as timber. The development of city-states was dependent upon the grain fields, timber, and animal products from the semiarid lands fringing the Mediterranean. As these resources were overexploited, the processes of desertification such as soil loss, reduced the productivity of the region. The declining productivity resulting from desertification contributed to the collapse of major civilizations such as Carthage, Greece, and the Roman Empire. The long history of degradation of the lands surrounding the Mediterranean had parallels in India and southwestern Asia and western China associated with technological advances in the cultures of those regions. However, it was not until the 20th century and the industrial revolution that technology allowed development of groundwater and opened up vast areas of the arid and semiarid lands of the world to commercial livestock production and to irrigation of lands not adjacent to rivers. The technological advances of the 20th century also allowed exponential growth of human populations. The rapidly expanding human populations increased the pressure on the natural resources of the arid and semiarid regions. This has resulted in the spread of desertification to every continent except Antarctica. In some areas such as the Sahel, drought combined with intensification of resource use has had devastating effects on the ecosystems and the human inhabitants. Desertification is a continuing process in many areas of the world.

The productivity of semiarid lands around the world continues to be reduced by the increase in unpalatable woody plants. This coupled with reduction in vegetative cover leads to further deterioration of soil structure and soil hydraulic properties. Now more than 75% of the area that can potentially be desertified has already been affected to some degree by desertification processes (Karrar and Stiles, 1984). The history of desertification in the 20th century may best be illustrated by a case study of the southwestern US. By examining the progress of desertification in a region with a well-documented history, we may identify some of the underlying causes of degradation and gain some insights into the degradation process.

Bahre and Shelton (1993) have documented the vegetation changes that have occurred in southeastern Arizona since 1870. These include: decline in grass, increases of woody xerophytes, decline in riparian wetlands, decline in desert scrub and cactus communities, and expansion of exotics. Most but not all of these changes have occurred in New Mexico and west Texas since 1870 (Hastings and Turner, 1965; Grover and Musick, 1990). Not only has vegetation changed but there have been less well-documented changes in the soils (erosion, deposition) and drainages (change from perennial flow to ephemeral flooding, stream bed cutting, and head cutting). The degradation of western rangelands was well recognized by the turn of the 20th century (Griffiths, 1901). Griffiths (1901) described decreased rangeland productivity because of the destruction of grasslands by overstocking. For example 'There were fully 50,000 head of stock at the head of Sulphur Spring Valley and the valley of the Aravipa in 1890. In 1900 there was no more than one-half that number and they were doing poorly' (Griffiths, 1901). Changes in vegetation were accompanied by soil erosion, arroyo cutting and changes in soil structure (Gibbens *et al.*, 1983; Grover and Musick, 1990). Although there is ample evidence that desertification has occurred and is continuing on these desert rangelands there is no agreement among scientists and land managers about the causes of these changes. Overgrazing and other land-use practices are frequently cited as primary desertification by some groups while climatic drought and competition between native grasses and shrubs are invoked as causal agents by other groups (Bahre and Shelton, 1993). However, responses to questionnaires sent by Griffiths (1901) to ranchers in southeastern Arizona consistently attributed the degradation to overstocking. Degradation of southeastern Arizona grasslands was documented in a letter from C. H. Bayless of Oracle, Arizona to Griffiths:

> . . . the San Pedro Valley consisted of a narrow strip of subirrigated and very fertile lands. Beaver dams checked the flow of water and prevented cutting of a channel. Trappers exterminated the beavers and less grass on the hillsides permitted greater erosion, so that within four to five years a channel varying in depth from 3 to 20 feet (approximately 1 m to 6 m) was cut almost the whole length of the river. . . . Of the rich grama grasses that originally covered the country, so little now remains that no account can be taken of them. . . . Where stock water is far removed, some remnant of perennial grasses can be found. Grasses that grow only from seed sprouted by summer rains are of small and transitory value. The foliage of mesquite and catclaw bushes is eaten by most animals and even the various cacti are attempted by starving cattle. . . . No better pasture was ever found in any country than that furnished by our native grama grasses, now almost extinct. . . . The present unproductive conditions are due entirely to overstocking. . . . Twelve years ago, 40,000 cattle grew fat along a certain portion of the San Pedro where now 3,000 cannot find sufficient forage for proper growth and development. If instead of 40,000 head, 10,000 had been kept on this range, it would in all probability be furnishing good pasture for the same number today. Very few of these cattle were sold or removed from the range. They were simply left there until the pasture was destroyed and the stock then perished by starvation.

Griffiths (1901) included letters from a number of other ranchers that basically told the same story. The degradation of rangelands in southeastern Arizona was not confined to two decades before the beginning of the 20th century. Recent studies in the San Pedro drainage show that desert grassland fragments have continued to diminish in size and become more isolated as small islands surrounded by a sea of shrubland (Kepner *et al.*, 1999). Their analysis showed that the number of grassland patches increased by 61% and the average patch size decreased by 60% between 1974 and 1987. This fragmentation involved the loss of grass cover, replacement of palatable grasses with unpalatable shrubs, and soil loss due to the decrease in vegetation cover. The increase in areas dominated by woody species is frequently called brush encroachment because most of the shrub–tree species have been components of the vegetation for thousands of years. The driving force for the increase in cover and density of woody shrubs in desert grasslands 'seems to be chronic, high levels of herbivory by domestic animals' (Van Auken, 2000).

The history of rapid degradation of arid rangelands in the southwestern US is paralleled by the history of degradation of eastern Australian rangelands. Degradation in the eastern Australian rangelands in the 1890s was exacerbated by the 'rabbit problem' which led to accelerated erosion of topsoil. It took only one season to destroy the plant cover and replace it with 'great patches of scalded clay, as bare as on the day when the last wavelet of some receding ocean lapped over them and left them to evolve a covering for their nakedness' (Noble, 1997). In 1900 a Royal Commission was established to examine the conditions of the rangelands in the Western Division of New South Wales. That commission listed the causes of the degradation of the rangelands that led to the economic depression in the region: low rainfall, rabbits, overstocking, sandstorms, and growth of nonedible scrub and fall in prices. According to the pastoralist who testified at the Royal Commission hearings

> the system of stocking which has been practised during the last twenty years, with the result that the majority is eaten out, leaving only isolated specimens to propagate the species upon the return of more favoured conditions . . . at present is producing vegetation capable of adapting itself to its altered surroundings by its unpalatableness and protective spiny growths . . . is apparent to every casual observer who notes the entire absence of grasses or herbage amongst the dense growth of injurious scrub on the red lands, which fifteen years ago were open well-grassed country (Noble, 1997).

The changes in vegetation and soils in the eastern Australian rangelands resulted from the interaction between the natural variability in rainfall and overuse of the palatable vegetation during periods when low rainfall limited productivity. Many of the perennial herbaceous and grass species were unable to cope with the grazing stress imposed during climatic conditions that placed the plants at their tolerance limits.

Change in species composition and the spatial distribution of vegetative cover plus soil loss are not the only impacts of historical episodes of desertification.

Dramatic changes in the structure of landscapes can occur in a short period of time. One of the major consequences of the loss of grass cover on the watersheds in southeastern Arizona was the shift from low-energy overland flow of run-off water through broad grass covered swales, to high-energy, flash-flood, overland flow. Historical sources indicate that the gullying and arroyo head cutting that characterizes the present-day landscapes of southeastern Arizona occurred in the period 1865–1915 with a peak in the 1880s (Cooke and Reeves, 1976). Today the San Simon and San Pedro valleys are deeply dissected landscapes with steep-sided ephemeral channels draining the sparsely vegetated uplands (Fig. 10.1).

There is evidence that the changes described as desertification occur even in the absence of grazing. There was no increase in perennial grass cover in a grazing exclosure established in 1933 and on an area partially occupied by mesquite, *P. glandulosa*, when re-measured in 1980. There were no differences between the grazing exclosure and grazed area transects that were grassland in 1935 and that had degraded to mesquite dunes when re-measured in 1980 (Hennessy *et al.*, 1983). Similar results were obtained in grazing exclosure studies in Arizona where the densities of mesquite plants doubled between 1932 and 1949 in both cattle-excluded and grazed areas (Glendening, 1952). During that same period, there was a 20-fold decrease in densities of perennial bunch grasses. In a comparison of grazed and ungrazed paddocks established in 1941, Smith and Schmutz (1975) recorded some recovery of grasses and reduction in subshrubs in the

Figure 10.1   The deeply incised channel of the San Simon River in southeastern Arizona. The sparse creosotebush shrubland of the uplands replaced the desert grassland in the 1880s.

ungrazed paddock but a large increase in mesquite was recorded in both paddocks. These authors also reported that an exotic, Lehmann lovegrass (*Eragrostis lehmanniana*) was establishing on both sites thereby contributing to the differences between paddocks. Vegetation change and degradation may continue in areas where grazing is excluded (Gibbens and Beck, 1988). Animal impact is not absolutely essential for ecosystem deterioration to continue. The results of these and other studies suggest that once shrub establishment and grass cover reduction has passed some threshold, the degradation trajectory is followed even when the environmental stress (grazing) is removed.

The presence of some shrubs provides a seed source, which can then be distributed over varying distances by animals such as rodents. Seed dispersal distances and relative success of plant establishment are dependent on the size and behavior of the seed feeder. Rodents have been suggested as important dispersers of shrub seeds into grasslands even in the absence of grazing (Cox *et al.*, 1993). The problem with the interpretation of these studies is that the areas excluded from grazing are rarely much larger than 2.3 km^2 and most are considerably smaller. Although rodents may be capable of effectively dispersing seeds in small exclosures, it is less likely that they can account for shrub dispersal into areas that are hundreds of square kilometers in size. Another factor affecting continued change in exclosures even in the absence of grazing is that the surrounding landscape continues to be impacted by grazing which can lead to the development of erosion cells (Pickup, 1985; Pickup and Chewings, 1986). The growth of erosion cells eventually envelop exclosures, destroying grasses by burial in wind-blown sand, abrasion, and by the loss of topsoil by sheetflow from up-slope disturbed areas. The end result is that either with or without continued grazing, the areas change sufficiently to be identified as desertified.

The most important historical development in the desertification story has been the development of technology that allows exploitation of deep, groundwater supplies. Deep wells, windmills and pumps have made all potential grazing land available to livestock. The development of pastoral industries in the semiarid regions of the world in the 20th century has centered on establishing reliable supplies of water for livestock. Established water points result in concentric areas of decreasing degradation radiating out from the water point reflecting the concentric nature of animal activity around water (Andrew and Lange, 1986a). Within 2.5 years after the establishment of a well and trough in a near pristine chenopod shrubland in South Australia, there were measurable changes due to the activity of sheep. At this location sheep were stocked at the lowest levels in the Australian arid zone. Soil changes included marked reduction in surface lichen crusts, increased soil compaction and increased bulk density, especially within 40 m of the trough. Vegetation changes included reduction in biomass of *Atriplex vesicaria*, a palatable shrub, and short-lived perennial grasses. Changes also included establishment of an exotic species, *Marsubium vulgare*, hore hound, within a 13 m radius from the trough. There was substantial mortality of *A. vesicaria* close to the trough after 8 years of grazing. Andrew and Lange (1986a) emphasize that these changes occurred despite the low stocking rate in the paddock. In addition, water points tend to concentrate the activities of native herbivores thereby exacerbating the grazing and browsing effects. Andrew and Lange (1968b) cite an extreme case where 6000 sheep denuded an area of saltbush up to 100 m from water in the first 3 months of stocking. Reductions of vegetative cover in the disturbance gradients around water points produce the initial erosion cells, which expand by abrasion and burial of vegetation plus soil loss by erosion.

Human activities contributing to desertification are not limited to overgrazing or overutilization of plants that are palatable for livestock, harvesting trees, shrubs and dung for fuel decreases productivity of the land, opens areas to wind and water erosion and contributes to the loss of soil fertility (Fernandez and Busso, 1999).

10.2 ECOSYSTEM AND LANDSCAPE CONSEQUENCES OF DESERTIFICATION

Most attention focused on the process of desertification of arid and semiarid lands has been on reduction in cover and loss of desirable plant species, invasion by woody plants, and soil erosion. The conceptual framework that has served as the working model for scientists and managers attempting to deal with desertification has been an autogenic succession model. Encroachment by shrubs has frequently been viewed as 'invasion' by species that were superior competitors with the more desirable plant species. Experience with grazing exclosures not only caused questioning of overgrazing as a causal agent of desertification but also caused us to question the applicability of autogenic succession models. If the ecosystem does not recover or even move in a trajectory toward the prestressed state when the environmental stressor is removed, is an autogenic succession model applicable? For semiarid rangelands, a conceptual model of desertification must allow for ecosystems to change structurally and functionally to alternate stable states, to experience discontinuous and irreversible transitions, and/or to establish nonequilibrium communities. Westoby *et al.* (1989) proposed a nonequilibrium conceptual model for rangelands which allows for rangelands to transition to one or more states that are not in equilibrium. The resulting rangeland communities may assume more than one configuration (Fig. 10.2) depending on the interaction of a number of physical and biological variables. Rangeland ecosystem dynamics may best be described by catastrophe theory (Lockwood and Lockwood, 1993). Rangelands exhibit five characteristics of catastrophe systems: (1) distinct conditions or states of existence; (2) conditions that are very unstable; (3) relatively rapid movement between states; (4) hysteresis, i.e. processes associated with degradation or recover are not readily reversible by simply inverting the sequence of events; and (5) relatively small changes in initial conditions can result in dramatically different outcomes with time. A nonequilibrium, state and transition model is consistent with the structural changes in

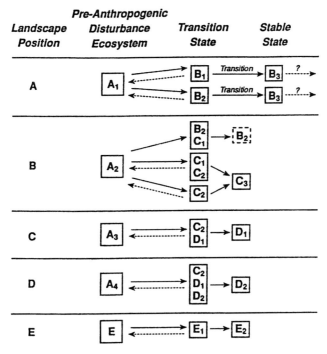

Figure 10.2 A conceptualization of the dynamics of arid and semiarid ecosystems in response to environmental stressors. Transition state may be either similar to or dissimilar to the eventual stable state, which is generally structurally and functionally different from the pre-stressed ecosystem.

desert rangelands that have been recorded in many areas of the world but there are systems in which the succession conceptual model seems to apply. Although the catastrophe model captures both the continuous (succession) and discontinuous (state and transition) dynamics, it is most applicable to the discontinuous dynamics that characterize most arid and semiarid ecosystems that are inhabited by humans.

Change from one stable state to an alternate stable state requires a conceptual model for desertification that accounts for the resulting changes in spatial and temporal patterns of limiting resources. The limiting resources are water and nutrients. The environmental stressors that initiate the trajectory of change is the interaction between overgrazing by domestic livestock and rainfall amounts that are close to the threshold for survival of the dominant vegetation. Large numbers of domestic livestock not only reduce the cover of palatable species, soil-trampling effects are severe especially in the vicinity of water points. Reduced vigor of perennial grasses and herbaceous plants exacerbates the effects of trampling by livestock. Even short duration, high-intensity grazing by cattle can result in significant changes in the microtopography of landscape units. Reduction in vegetation cover increases the susceptibility to surface erosion by either wind or water. Reduction in cover of grasses and herbaceous perennials produces scattered patches of unvegetated soil where shrub seedlings can establish with no competition between the seedling roots and the roots of perennial grasses and herbaceous plants. Evidence for the importance of bare patches in heavily grazed grassland as establishment sites for shrubs was provided by a study where seedling creosotebushes (*Larrea tridentata*) were transplanted into nitrogen fertilized–irrigated, irrigated, and unmodified plots in an ungrazed black-grama (*Bouteloua eriopoda*) grassland, a creosotebush shrubland, and a grazed grassland with a mixture of grasses. After two years, the only site where cresotebush seedlings survived was in the grassland that had been intensively grazed prior to the initiation of the study (Whitford *et al.*, 2001).

A conceptual model for desertification in the Chihuahuan Desert proposed that the reduction in vegetative cover resulting from excess grazing pressure plus drought resulted in a redistribution of soil resources. In desert grasslands, the distribution of soil nutrients is relatively uniform as is the distribution of soil water following rain events (Schlesinger *et al.*, 1990). Loss of grass cover produced open patches that were subject to overland flow of run-off on sloping terrain, and/or movement of materials by wind irrespective of slope. Wind and water redistributed organic materials such as rabbit dung, grass fragments and leaves, and surface soil particles. The loss of grasses eliminated efficient transfer of water from the leaves via stemflow to the root mat thus reducing infiltration rates. Erosion of the soil surface also eliminates biopores, which further reduces infiltration rates. Thus reduction in grass cover quickly resulted in the spatial redistribution of the critical water and nutrient resources. Unvegetated patches within such a landscape provided germination–establishment sites for shrubs. In the northern Chihuahuan Desert, the 'increaser' shrubs are creosotebush and mesquite (*P. glandulosa*). The distribution patterns of these species prior to the development of a commercial livestock industry was creosotebushes on steep, well-drained, gravelly or sandy slopes and mesquite along drainage channels, edges of ephemeral lakes and Indian encampments (York and Dick-Peddie, 1969). Episodes of mesquite and creosotebush invasion and establishment frequently followed reduction in grass cover resulting from drought and grazing (Grover and Musick, 1990). Creosotebush seed dispersal is primarily by overland water flow. Mesquite seed dispersal is primarily by animals that feed on the seedpods but do not destroy the seeds. Seedling mesquite are frequently seen emerging from cattle dung pats during summer rains in the Chihuahuan Desert. Seedlings of both mesquite and creosotebush rapidly produce deep taproots and long lateral roots near the surface of the soil (Hennessy *et al.*, 1983) (Figs 10.3 and 10.4).

Replacement of grasses by shrubs set the stage for concentration of litter under shrubs and within coppices of shrubs thereby creating 'islands of fertility' or 'resource islands' (Fig. 10.5). Organic fragments are trapped in a very different way in shrub-dominated ecosystems than they are in grassland ecosystems. The

Figure 10.3   A remnant patch of black grama (*Bouteloua eriopoda*) desert grassland on the Jornada Experimental Range in southern New Mexico.

Figure 10.4   Mesquite (*Prosopis glanduosa*) coppice dunes that replaced black grama (*Bouteloua eriopoda*) desert grassland on the Jornada Experimental Range in Southern New Mexico.

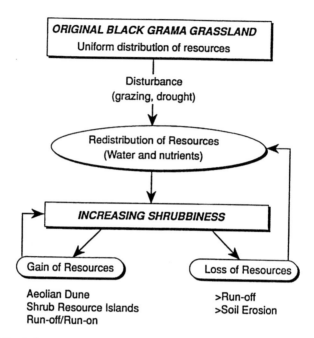

Figure 10.5   A diagram of the Jornada conceptual model for desertification. Redistribution of water and nutrients involves both spatial and temporal changes in availability of these resources.

architecture of the shrubs determines their characteristics with respect to creating surface roughness and eddy currents in wind and with respect to their efficacy as obstructions to sheet flow of water (DeSoyza *et al.*, 1997a). Mesquite (*P. glandulosa*), which grew as a single-stemmed plant in most areas prior to the introduction of large herds of livestock, assumed a multiple-stem morphology in the degrading grassland This multistemmed mesquite trapped blowing sand thus forming the nucleus for a coppice dune. As the stems were buried by blowing sand, buried stems developed adventitious roots and the meristematic stem tips formed the coppice.

Although the Jornada conceptual model is useful and is obviously applicable to desertification processes in the Chihuahuan Desert, it does not fit all cases of desertification. There are examples of desertification in which the system moves from a heterogeneous spatial structure to a more-homogeneous spatial structure and as a consequence, temporal variability may be reduced. For example, in regions where the vegetation is structured into a mosaic of patches of bunch grasses and shrubs, loss of bunch grasses may actually increase the homogeneity of the desertified ecosystems. The common operational feature in the process is a *change* in the spatial and temporal heterogeneity of the ecosystems not simply an increase in heterogeneity. The change in ecosystem characteristics both temporally and spatially should lead to an increase in the variability of ecosystem processes and rates of processes.

The processes involved in the spread of mesquite in the desert grasslands of North America differ in several ways from shrub invasion in the South African Karoo. In the Karoo, *Acacia karoo*, is the most important woody invader of grassland. Germination of *A. karoo* did not occur in livestock dung nor did reduction of grass cover enhance the establishment of this shrub (O'Conner, 1995). However, many of the consequences of the establishment of *A. karoo* in grasslands of southern Africa are similar to those experienced in North America with the spread of mesquite.

Desertification in the Australian semiarid and arid regions also parallels the changes in ecosystem structure and function reported for North American rangelands. The impacts of domestic livestock combined with limited rainfall contribute to changes in the spatial heterogeneity of rangeland landscape units. In the Australian rangelands fine-scale heterogeneity is replaced by large-scale heterogeneity associated with the changes in abundance and spatial distribution of woody species (Tongway and Ludwig, 1990; Ludwig and Tongway, 1995). Changes in vegetation species composition and the spatial patterns of distribution of those species modify run-off patterns and the abundance and distribution of unvegetated patches that are subject to wind erosion. The conceptual model for degradation of Australian rangelands is very similar to the Jornada model.

The ecosystem changes resulting from the spread of invasive shrubs in the South American arid zone also fit the Jornada model. Invasive shrubs, such as *Acacia caven*, result in the redistribution of soil and water resources. The changes in spatial and temporal availability of these critical resources result in positive feedback loops that exacerbate the degradation of the rangelands (Aronson *et al.*, 1993). Degradation of semiarid and arid rangelands in shrubs and trees that are unpalatable to livestock replace much of the grass and herbaceous cover produces changes in the spatial and temporal distribution of water and nutrients. When the modified spatial and temporal patterns of water and nutrient distribution have been in place for a sufficient period of time, one or more of the 'thresholds of irreversibility' have been crossed. The configuration of the degraded ecosystem will remain in the 'degraded' condition unless massive intervention is imposed (Aronson *et al.*, 1993).

Degradation of arid and semiarid ecosystems does not always involve the replacement of desirable forage species by less desirable species. Degradation can result from the spread of nonnative species and the environmental consequences of abundant exotics. The Great Basin Desert in North America has been colonized by a Eurasian annual grass, *Bromus tectorum*, colloquially known as cheat grass along with tumbleweeds, *Salsola* spp., and a succulent halophyte, *Halogeton glomeratus*. In years with adequate precipitation, the intershrub spaces in the sagebrush-dominated rangelands develop a nearly complete cover of alien annual grasses and forbs. Although cover and abundance of alien species are greater in heavily grazed areas than in ungrazed or lightly grazed areas, they appear to be increasing in all Great Basin shrubland ecosystems. These alien species increase fuel loads sufficiently to greatly shorten the fire return cycles. Following fires, recolonization of sites by chenopods and sagebrush may be compromised by changes in disease organisms in the soil and inhibition of establishment by the presence of the alien species (Harper *et al.*, 1996). Intense fire in these shrublands can leave the hillsides devoid of vegetation cover and susceptible to wind and water erosion.

Fuel loads by alien species can result in fire in more arid regions where sparse vegetation could not carry fire. For example, red brome (*Bromus rubens*) has spread rapidly in the Mojave Desert and the upper Sonoran Desert of southern Nevada and Arizona. Wet winters since 1976 have encouraged the spread of *B. rubens* and the buildup of fine fuels in the native shrublands. Large fires have been frequent in desert communities that rarely burned in the past. Desert plants such as saguaros (*Carnegia gigantea*) that grow slowly and recruit episodically are essentially elim- inated by these fires (Betancourt, 1996). Another alien grass that is rapidly expanding in the Sonoran Desert is *Cenchrus ciliare*. *C. ciliare* is providing fine fuel in areas of the Sonoran Desert where fire was absent or extremely rare. The long- term effects of expansion of alien annual grasses in desert rangelands can only be surmised because this is a very recent phenomenon. The increasing frequency of fire will undoubtedly change the spatial and temporal patterns of distribution of water and nutrients. These changes will undoubtedly affect the structure and processes of the ecosystems that develop in desert areas exposed to frequent fire.

Desertification effects on temporal and spatial distribution of soil resources produces variability that changes coefficients of variation in biomass production and vegetation cover. For example, temporal changes in the chemical profiles (both quantitative and qualitative) of shrub-dominated ecosystems results in increased variability in foliage as a source of food for herbivores.

The Jornada conceptual model focused on the physical processes that redis- tributed soil and nutrients but did not address any of the implications of changes in the composition of the animal communities nor changes in the spatial and tem- poral redistribution of animals in degrading grasslands. The establishment of commercial livestock production was accompanied by a number of activities designed to 'improve' the desert grasslands for livestock production. These activ- ities included extirpation of prairie dogs (*Cynomys ludovicianus*), trapping and poisoning rodents such as the banner-tail kangaroo rats (*Dipodomys spectabilis*), and trapping, shooting, and poisoning predators such as the lobo wolf (*Canis lupus*), mountain lion (*Felis concolor*), and coyote (*Canis latrans*). The poison- ing campaigns resulted in killing large numbers of small predators such as grey fox (*Urocyon cinereoargenatus*), skunks (*Mephitis* spp.) and badger (*Taxidea taxus*). The reduction in small predators probably led to an increase in the abun- dance of desert rodents that was noted by early workers at research stations such as the Jornada Experimental Range (unpublished reports, 1912–1918) despite the rodent control efforts. Recent studies have documented the effects of rodents on productivity, species composition, and landscape patchiness (Moorhead *et al.*,

1988; Brown and Heske, 1990; Mun and Whitford, 1990; Chew and Whitford, 1992; Guo, 1996; Kerley and Whitford, 2000). The complex relationships between ecosystem function and desert rodents have led to a consideration of these animals as keystone species (Brown and Heske, 1990). Since the cessation of rodent control efforts, the rodent populations have reached high levels. Population densities of rodents are highest in the creosotebush shrublands on the Jornada and lowest in the desert grasslands. There are feedbacks between rodent abundance and impacts on the ecosystem. In ecosystems where bunch grasses provide as much as 10% of the vegetative cover, some species of rodents destroy a large fraction of the flowering tillers thereby affecting seed production and the seed bank (Kerley *et al.*, 1997). Because there are no quantitative data on the relative abundance and species composition of the desert grassland rodent communities prior to the onset of the desertification processes, we have no baseline against which to assess the effects of changes in desert rodent communities that were concomitant with changes in vegetation and soils. However, there are large differences in abundance and species composition of rodent communities in remnant patches of desert grassland and rodent communities in shrublands (Whitford, 1993). Because rodents affect soil processes by their burrow systems, seed-caching behavior, and by granivory and gramnivory, they contribute a positive feedback in the maintenance of shrub communities once these are established (Whitford, 1993). In North American desert grasslands, the extirpation of the black-tail prairie dog (*Cynomys ludovicianus*) may have contributed to the spread of shrubs into the grassland and contributed to the extinction of other desert grassland species (Miller *et al.*, 1994).

The consequences of desertification are not limited to biological impacts and reduction of productivity. Desertification affects a number of biophysical variables: albedo, temperature, precipitation, soil moisture, wind erosion, and water erosion. In semiarid regions where vegetation cover protects soil from wind erosion, degradation processes that reduce plant cover and increase fetch lengths of unvegetated patches result in an increase in wind erosion. Dust generated from erosion of recently exposed surfaces contributes to the global deposition of dust over the oceans and in land areas not subject to wind erosion (Arimoto *et al.*, 1989; Derbyshire *et al.*, 1998).

10.3 RESISTANCE AND RESILIENCE OF DESERTIFIED LANDSCAPES

Evidence is accumulating from experimental studies that, once established, shrub-dominated ecosystems are both resistant to disturbance and resilient following perturbation. Field experiments using 'rain-out' shelters to impose summer drought on plots in creosotebush shrubland showed that the annual growth increments of plants subjected to drought were not significantly different from the growth of controls. Summer drought was imposed on the plots for five consecutive years.

There were no differences in growth increments in the fifth year of the experiment. Leaf litter fall patterns were affected by the drought but the annual quantity of leaf litter production was the same for shrubs subjected to drought and control shrubs (Whitford *et al.*, 1995). There were similar responses to imposed drought in mesquite shrubs. Small mesquite shrubs exhibited greater decline in photosynthesis when subjected to drought than the large shrubs. The results of this study suggested that large, established plants were better able to withstand drought than were the small, establishing shrubs (DeSoyza *et al.*, 1997b). However, after five consecutive years of growing-season drought, perennial herbaceous plants and perennial grasses disappeared on the drought plots (Whitford *et al.*, 1995). In creosotebush, growth was suppressed during the growing-season drought. The shrubs exposed to drought exhibited rapid growth after the first rainfall after 'rain-out' canopies were removed. This growth spurt compensated for the lack of growth during the growing season. The rapid growth spurt was attributed to the higher rates of nitrogen mineralization in the soils under 'rain-out' shelters. The resilience of creosotebush to long-term, repetitive drought is attributed to its ability to utilize deep soil moisture and the soil system's nitrogen mineralization responses to drought (Whitford *et al.*, 1995).

Fifteen years after creosotebush seedlings were planted in a grazed desert grassland community, there was essentially no other perennial vegetation under the canopies of the shrubs and very low cover of annual plants. The soils under the recently established creosotebushes had lower aggregate stability than soils in adjacent plots where creosotebush had not established (Whitford *et al.*, 2001).

These studies suggest that a number of mechanisms account for the resistance and resilience of shrubs that establish in degrading rangelands. Shrubs modify the water infiltration patterns by stemflow, throughfall fractions. Stemflow water may be channeled to deep soil layers along root surfaces. The shallow roots of shrubs probably compete with the root systems of perennial grasses and herbaceous plants for water and nutrients. Many shrubs are resistant to repetitive drought whereas grasses and herbaceous plants are not. In addition, the capacity of shrubs to respond rapidly to rainfall allows them to take advantage of the water and soil nutrient availability more rapidly than grasses and herbaceous plants. This provides resilience to disturbance in shrub dominated ecosystems.

10.4 DESERTIFICATION AND CLIMATE CHANGE

Human activities that alter the ocean–atmosphere interactions, that change atmospheric and stratospheric chemistry and that alter land surface – atmosphere interactions directly affect regional climates around the world. Although it is uncertain how global warming will affect the thermal and chemical properties of the oceans, it is clear that changes in oceanic surface temperatures will impact the weather patterns of arid and semiarid regions. The linkages between ocean surface temperatures and precipitation patterns in the world's arid regions have been described in general terms but there is insufficient information to allow robust predictions of the effects of global warming on the climates of the worlds arid regions (Williams and Balling, 1996).

Overgrazing that reduces vegetation cover initiates biogeophysical feedbacks that modify the climate of an arid or semiarid region. Reduced plant cover results in increased albedo. Increased albedo decreases the net radiation from the soil surface. Decreased net radiation from the soil surface results in surface cooling. Surface cooling promotes subsidence of upper atmosphere air masses. Subsidence decreases convection and cloud formation which leads to reduced precipitation (Charney, 1975; Charney *et al.*, 1975, 1977). Humans have affected the earth surface–atmospheric processes primarily by reducing vegetation cover by overgrazing, cultivation, and deforestation. Changes in vegetation cover alters surface roughness, thereby modifying wind speeds. Changes in vegetation cover also alter soil moisture patterns. Reduction in vegetation cover exposes soils to erosion and increases atmospheric dust loads. Atmospheric dust affects surface temperatures and provides raindrop nuclei. Burning of fuel woods and of savannas contributes to atmospheric particulates and trace gasses that impact the climate of arid and semiarid regions. There are no quantitative studies of these effects; therefore the net effect on the climate arid regions cannot be evaluated nor can generalizations be made about the impact of these on future climate change (Williams and Balling, 1996).

Global circulation models provide predictions of global and regional-scale climate changes. The model simulations are generally considered to be uncertain at regional geographic scales. The models predict increases in annual average temperature of 4–5°C for most of the worlds desert regions. For North American deserts, precipitation and soil moisture changes are 'highly variable by season and model' (Williams and Balling, 1996). Predictions for precipitation in South American deserts are variable but 'summer-time soil moisture levels are generally expected to decrease' (Williams and Balling, 1996). Lower precipitation is predicted for deserts in the Middle East and northeastern Africa and southern Africa. Increased winter rainfall is expected in the Asian arid lands with decreased soil moisture in summer. Increased rainfall in both summer and winter is predicted for the central Australian deserts. There are also predictions of increased frequency of drought in most of these regions. Global climate changes are likely to increase climate variability of the arid and semiarid regions of the planet. Whatever the direction and magnitude of climate changes are, the end result is that these changes will simply exacerbate desertification processes resulting from human activities in the future.

10.5 SOCIAL AND ECONOMIC CONSEQUENCES OF DESERTIFICATON

The most severe social and economic impacts of desertification are seen in sub-Saharan Africa, the Sahel. Beginning with the severe drought in the 1970s and continuing today, overgrazing, fuelwood harvesting, and cultivation of marginal lands have drastically reduced per-capita food production in the region and increased the severity of poverty in an already impoverished area. Increasing poverty, food shortage, problems with land ownership and land tenure patterns interact with the rapid increase in human population to create social upheaval. Desertification forces large numbers of people to abandon subsistence agriculture and to emigrate to urban areas in hopes of finding employment and economic stability. In many areas of the Sahel, this migration is primarily of young adult males which leaves an inadequate labor force to sustain the subsistence agriculture (Darkoh, 1998). In some of the countries in the region government policies encourage permanent immigration into the semiarid and arid lands from urban areas. The immigration of culturally nonadapted peoples into these regions further exacerbates desertification. These migrants settle into these fragile environments with inappropriate technologies and no knowledge of sustainable management practices. The intrusions of row-crop agriculture into traditional pastoral lands creates civil strife in many parts of this region. In much of the region there have been repeated episodes of breakdown of law and order in the last few decades. These breakdowns have frequently resulted in mass exodus of people who are held in refugee camps in neighboring countries (Darkoh, 1998). Other aspects of the economies of the countries of the region lead to increasing the severity of desertification. A prime example is the demand for charcoal in the urban areas. The value of charcoal makes cutting small trees attractive to entrepreneurs. In most of the semiarid and arid lands in Africa, the economic costs of soil erosion resulting from desertification have been estimated by a World Bank study. Average US dollar losses by soil erosion were estimated at $80 to $150 per hectare for grazing lands. In much of Africa desertification 'translates into a spiral of declining production, increasing poverty and diminished potential productivity. It exacerbates poverty, which, in turn, exacerbates desertification because, as the pressure increases, people are forced to exploit their land to survive. . . . Entire societies and cultures are now threatened. The pastoralists of the Sahel are a case in point. For most, the loss of their livelihood means a life in relief camps or shanty towns mushrooming around Sahelian cities and those of the countries to the south' (Darkoh, 1998). Although the social and economic costs of desertification are less severe in Eurasia than in Africa, the loss of production potential and increasing human populations produce a positive feedback system that impacts the human communities especially in Pakistan, western India, Kazakstan, Turkmenistan, Uzbekistan, and Afganistan.

There are social and economic consequences of desertification even in countries with robust economies. In the western US, the social and economic consequences of continuing degradation of rangelands are the cause of intense conflicts among the various user groups of the public rangelands. Environmental organizations claim that ranching is destroying the production base and adversely affecting plant and animal diversity and abundance on the rangelands. Ranchers and pastoralists are trying to protect a 'life-style' and see themselves as stewards of the land and its resources. The economic realities are frequently overlooked by both groups. For example, ranchers in the desert grassland region of the southwestern United States realize a return on investment of only 1–3%. 'From an investment standpoint, Chihuahuan Desert cattle ranching would be considered unprofitable and risky by any Wall Street analyst' (Holachek, 1992).

Changes in wholesale cattle prices, government policies such as drought subsidies, and climate fluctuations combine to influence management decisions by ranchers that frequently result in further degradation. Ranches on rangeland that is considered in fair to poor condition are frequently 'hobby ranches' where the desirable 'cowboy lifestyle' is subsidized by employment in another profession (Holachek, 1992). Ranches that are primarily poor to fair condition rangeland have to be managed on the basis of short-term financial returns rather than on sustaining the long-term health of the resource. Economic necessity (repayment of bank loans, feed a family, maintain fences, water points and other infrastructure etc.) forces ranchers to maximize herd size. This leads to conflicts with government land management agencies and environmental groups. When the conflicts cannot be resolved by compromise, they are moved to the court system where the resolution is rarely satisfactory to any of the litigants.

10.6 EFFECTS OF DESERTIFICATION ON BIODIVERSITY

One of the major biological challenges of the 21st century is the resolution of the relationship between the diversity of organisms in ecosystems and the long-term behavior of those ecosystems. There are four prevalent hypotheses concerning this relationship: the diversity–stability hypothesis, the rivet hypothesis, the redundancy hypothesis, and the idiosyncratic response hypothesis (Johnson *et al.*, 1996). The diversity–stability hypothesis predicts that ecosystems will increase in productivity and the ability to recover from disturbance as the number of species in the system increases. According to this hypothesis, deletion of species from the complex trophic webs will result in an increase in susceptibility of the system to disruption in response to a perturbation or environmental stress. The rivet hypothesis views species in ecosystems as the rivets holding an airplane together. The removal of rivets beyond some threshold number may cause the plane or ecosystem to crash. Implicit in this hypothesis is the view that a few extinctions may have little effect on ecosystem stability and resilience but as extinctions increase, the redundant components are lost and the ecosystem fails to resist disturbance and fails to rebound after experiencing disturbance. The redundancy hypothesis views ecosystem structure as species associations in functional groups. Extinction of a redundant species in a functional group will result in a neighbor species assuming the 'work' of the extinct species in the ecosystem. The idiosyncratic hypothesis views the relationship as intermediate between species composition and ecosystem function because of higher-order interactions among the biotic communities.

Species richness in deserts is moderate to high in semiarid regions (> 200 mm rainfall per year) and declines with increasing aridity for most groups of organisms (Huenneke and Noble, 1996). Species richness appears not to be a good measure of biodiversity with respect to ecosystem function in arid and semiarid regions because species richness is frequently higher in the 'degraded' ecosystems than in the relatively unchanged systems. There is some evidence for 'keystone' species or species complexes that have multiple direct and indirect effects on ecosystem processes and structure. Loss of a 'keystone' species may compromise the ability of the system to resist environmental stress and the ability to recover structure and function after changes resulting from stress. The importance of 'keystone' species is not considered in species richness or species diversity measures. Loss of keystone species may be the critical feature of desertification that results in the ecosystem remaining in an altered stable state.

Empirical data suggest that changes resulting from desertification have little effect or at least little negative effect on species richness. The alternate, stable configuration of shrubland ecosystems that replaced grassland systems, supported higher species richness and diversity of small mammals and birds in the desertified systems than in the relatively undisturbed desert grasslands (Whitford, 1997). However, in desertified systems dominated by an alien species (an introduced South African grass, *Eragrostis lehmanniana*) species richness of birds was significantly lower than in desertified systems with no alien species. In the systems dominated by *E. lehmanniana*, there were fewer taxa of 'ecosystem engineers' that affect soil processes and significantly less soil turnover than in systems without the alien grass (Kulas, 2001).

Taxa or functional groups that are most important in maintaining the integrity of arid ecosystems are those species that directly or indirectly affect soil processes (Whitford, 1996). The species richness of the soil biota in arid ecosystems appears to be more related to fine temporal and spatial scale patterns of organic materials than to the dominant vegetation. Soil compaction may affect the composition of the soil biotic community by reducing pore neck size in the pedon but soil compaction is generally a feature of intense impact of large herbivores in the immediate vicinity of a water point. Subterranean termites are a keystone component of many arid ecosystems. Termites appear to be very resistant to ecosystem degradation and structural change. There was no change in the spatial and temporal abundance and feeding rates of the 'keystone' subterranean termites in desertified and intact Chihuahuan Desert ecosystems (Nash *et al.*, 1999). The importance of termites in nutrient cycling and energy flow is maintained even in ecosystems in which vegetation and soils have experienced dramatic change within less than a century.

As desertification proceeds, relatively intact 'pre-anthropogenic stress' ecosystems are reduced in size and become fragments or 'islands' in the sea of changed landscape. This fragmentation frequently leads to local extinction of species, particularly top predators. For example a desert grassland specialist, the Aplomodo falcon (*Falco femoralis*), has disappeared from the Chihuahuan Desert grasslands in the United States and survives as a sparse population in Mexico.

10.7 REHABILITATION OF DESERTIFIED LANDSCAPES

Efforts to rehabilitate desertified landscapes have been focused on recovery of some if not all of the production potential of the rangelands. In some desertified regions, rehabilitation has focused on stabilizing mobile sands in order to protect transportation routes and human settlements from being buried under sand dunes. In desert rangelands where small trees and shrubs have invaded grasslands, rehabilitation efforts have focused on reducing shrub cover and re-establishing grasses.

In North America, efforts to rehabilitate degraded rangelands have focused on control of shrubs. Herbicide applications, root plowing, chaining, and burning have been widely used in desertified areas of western US rangelands. Chaining involves pulling a large chain strung between two large tractors across a landscape where trees and shrubs are uprooted by the chain. Root plowing is accomplished by a heavy steel plow that is more than 1 m in length pulled through a brush-covered area by a large tractor. Root plowing and front-end stacking used to reduce brush density had no major long-term effect on brush recovery. Grass seeding following brush removal by root plowing and stacking also had no long-term effect on brush reinfestation (Gonzalez, 1990). The rapid reinfestation of cleared areas by shrubs and the absence of any long-term effect of mechanical shrub removal have resulted in the virtual abandonment of this as a rehabilitation technique. Herbicides are generally applied aerially but some research on hand application has been conducted (Figs 10.6 and 10.7). Fire is a management tool that is limited to areas and time periods when there is sufficient vegetation cover and biomass to carry a fire. In many areas where brush cover is a problem, there are insufficient fuel loads to carry fire. Frequently the brush control measures are combined with seeding. Seed mixtures are frequently native and alien species of grass but in many instances seeding by an alien grass species has been the preferred treatment (Call and Roundy, 1991).

Rehabilitation of degraded rangelands has been attempted on extensive areas. Rehabilitation efforts are expensive and the benefits are not accrued for a sufficient period of time to recoup the expenditure. For example, between 1940 and 1981, an average of 600,000 h of brush-infested rangeland in Texas was treated annually by herbicides, fire, and mechanical means such as root-plowing, chaining, and disking (Rappole *et al.*, 1986). Despite the intensity of treatments, land managers realized only a transitory, short-term gain in productivity. Regrowth of shrubs occurred in less than a decade, and retreatment became necessary within 15 years of root plowing, 2 years after chaining and 8 years after treatment with

Figure 10.6   A mesquite (*Prosopis glandulosa*) coppice dune with an ephemeral herb, *Descurainia pinnata*, following treatment with a herbicide that killed the mesquite shrubs, Jornada Experimental Range, southern New Mexico.

Figure 10.7   Mesquite coppice dunes 10 years after treatment with a herbicide, Jornada Experimental Range, southern New Mexico.

herbicides (Rappole *et al.*, 1986). Similar results have been reported in other desert rangeland areas (Herbel *et al.*, 1983). Even when combined with exclusion of livestock grazing, root-plowing and herbicide treatment has not resulted in re-establishment of grassland (Chew, 1982; Roundy and Jordan, 1988).

Partial success in restoring some of the functions of desert grassland has been achieved by seeding nonnative grasses. These grasses have revegetated water-sheds and reduced damage from flash floods. They also provide forage for livestock at certain times of the year and on some microsites. For example, after initial screening by the US Department of Agriculture, Lehmann's lovegrass (*Eragrostis lehmanniana*) was seeded on selected sites in southeastern Arizona and southern New Mexico. Between 1940 and 1980, land managers in Arizona established Lehmann's lovegrass on 70,000 ha. Lehmann's lovegrass has since spread to an additional 130,000 ha (Cox and Ruyle, 1986; Anable *et al.*, 1992). In New Mexico where Lehmann's lovegrass was seeded along electric transmission line corridors and road verges, it has spread into some adjacent landscape units. Where Lehmann's lovegrass has spread into native black-grama (*Bouteloua erio-poda*) and drop-seed (*Sporobolus* spp.) grasslands, it appears to be displacing the native grasses. Despite the benefits of having this exotic grass provide soil cover and forage for livestock in shrub dominated ecosystems, the diversity of fauna remains severely reduced (Bock *et al.*, 1986). The spread of this exotic grass has stirred debate about the costs and benefits of restoring some of the functions of a desert grassland (Figs 10.8 and 10.9).

Figure 10.8   An example of restored desert grassland dominated by the alien Lehmann's lovegrass, *Eragrostis lehmanniana*, on the Santa Rita Experimental Range, southern Arizona.

Figure 10.9   A remnant patch of native desert grassland on the Santa Rita Experimental Range, southern Arizona.

Rehabilitation of desertified rangelands in Australia has focused on the same techniques that have been applied in North America. Blade plowing, chemical treatment and chaining were too energy-expensive to apply to large areas and were therefore not cost effective (Noble, 1997). Contour furrowing in patterned chenopod shrubland in an attempt to reduce erosion and encourage the establishment of desirable vegetation actually exacerbated the degradation (MacDonald and Melville, 1999). Because of the dispersive nature of the soils, the furrow banks were breached by the accumulation of sheet flow. The breached furrows created 'stream flow' which increased erosion. Differential leaching of cations in the furrow soils appears to be the cause of establishment of unpalatable (to livestock) plant species in the furrows. This example demonstrates the need for careful evaluation of soils, vegetation, and landscape-scale processes when designing rehabilitation projects as was emphasized by the authors. 'Care must be taken when utilizing this landscape to avoid further degradation through erosion, through changes in cation distribution, and through associated vegetation changes' (MacDonald and Melville, 1999).

Mechanical rehabilitation methods have been widely applied on degraded rangelands in South Africa. 'The degraded condition occurring on each farm or nature reserve can be regarded as a unique unit with regard to soil, climate, and degree of degradation'(Van der Merwe and Kellner, 1999). In a study of the relative success of mechanical methods (moldboard plow vs. multiple-tined rippers) the deep-furrow plow achieved the best results on 'shrink-swell' clay soils and the multiple tined ripper afforded the best results on sandy soils. The mechanical modification of the landscape changed infiltration patterns, provided seed traps, and sites for organic matter concentration. The success of rehabilitation in the South African rangelands also depended on deferment of grazing for between 1 and 3 years to allow establishment and growth of vegetation. Grazing pressure and the effects of trampling are determinants of the longevity and effectiveness of the hollows and furrows (Griffith *et al.*, 1984).

Combinations of burning and application of chemical herbicides have proven to be the most cost effective in the short-term. Strategies combining mechanical and biological treatments have proven to be effective in some instances. Goats were used to browse shrubs that were regenerating after mechanical treatment. Shrub control by goat browsing following bulldozing was significantly better than either goats or bulldozing alone (Noble, 1997).

In China a number of soil stabilization and rehabilitation techniques are used (Fullen and Mitchell, 1994). These include planting windbreaks, establishment of straw or clay checkerboards on bare dune surfaces and planting of xerophytes. In order to provide an environment where indigenous xerophytic plants can colonize and survive, it is essential to stabilize the surface material. In areas where mobile sand dunes are a problem, the Chinese have devised a strategy for establishing an artificial ecosystem on mobile dunes. The first stage in this strategy is the installation of a 'sand barrier' made from woven willow branches or bamboo to act as a windbreak. Behind the sandbarrier, straw checkerboards are established. Between 4.5 and 6 metric tons per hectare of wheat or rice straw are embedded in the sand in a grid of 1–3 m squares. The straw is inserted 15–20 cm deep so that the straw protrudes approximately 10–15 cm above the dune surface. Where available, clay is used instead of straw because of its greater durability. The checkerboards remain intact for 4–5 years. This allows time for the cultivated xerophytic plants that are placed within the squares to become established. A number of native shrubs and grasses are planted in the checkerboards. Stability of the surface is enhanced by the colonization and establishment of microbiotic crusts. These biocrusts also protect the surface from wind unless the crust is damaged by trampling or by burrowing animals.

These rehabilitation methods have changed mobile dunes to stable productive ecosystems. Twenty-five years after establishment of artificial vegetation cover, more than ten additional species have colonized the area and vertebrates have increased from two to more than 30 species (Fuller and Mitchell, 1994). Although these are impressive results, the methods are labor intensive and may be prohibitively expensive for most countries where mobile sand dunes threaten human communities and/or infrastructure.

Effective rehabilitation of desert rangelands in Tunisia was dependent upon the degree of degradation. In areas where there were sparse populations of native plant species, eliminating grazing pressure by exclosing livestock, resulted in

reestablishment of many native plant species and partial stabilization of the aeolian landscape (Aronson *et al.*, 1993). In the most degraded sites, a selection of
species that were known to be productive and resistant to livestock grazing were
replanted on a small scale. However, for replanting to be successful, wind breaks
had to be installed to stabilize the sand dunes and provide stable germination–
establishment sites. In addition, successful selection of species for rehabilitation
was based on a large body of experimental data that had been accumulated on
native species thought to be keystone species in these ecosystems (Aronson *et al.*,
1993).

In Chile, the savanna (espinal) dominated by *Acacia caven* and alien annuals
with its low soil fertility and low livestock productivity occupies the region that
was a dense mosaic of forest and shrublands prior to occupation by European
settlers (Ovalle *et al.*, 1999). In this region an aggressive intervention was
deemed necessary to improve the ecological and economic condition of the
degraded espinales. The degraded soils exhibited precipitous declines in organic
matter, water storage, duration of water storage and available nutrients. The
degraded soils require the introduction of nitrogen-fixing legumes for revegetation and amending soils with appropriate microbial inocula (Aronson *et al.*,
1993). The appropriate microbial inocula include *Rhizobium* spp. and a 'cocktail'
of vesicular–arbuscular mycorrhizae. Inoculation with both groups of symbionts
appears to be important because *Rhizobium* requires mycorrhizal association to
provide the high amounts of phosphorus needed for nodulation and nitrogen fixation (Ovalle *et al.*, 1999). In ten years of rehabilitation efforts based on planting
nitrogen-fixing legumes and improvement of soil fertility, there has been
improvement in productivity and soil fertility. Current efforts focus on re-establishment of native tree and herbaceous plants in an effort to facilitate the
spontaneous re-establishment of native plants, microbes, and animals (Ovalle *et
al.*, 1999). This case study demonstrates the need for defining achievable goals
for a rehabilitation program and the necessity for management flexibility during
the program.

In most of the arid regions of the earth, a large fraction of the land area has
been desertified or is somewhere on the trajectory towards desertification. The
vast spatial extent of the areas requiring rehabilitation (millions of hectares) and
the high costs of large-scale efforts demonstrate the need for low-cost, small spatial scale approaches to improving the condition of degraded landscapes.
Successful rehabilitation requires identification of those parts of a landscape
where low-intensity intervention will result in improved soil fertility, establishment of desirable plant species and improve habitat for native animals. In areas
where degradation has not proceeded to an undesirable end-point, modification of
land and livestock management may be all that is necessary to halt and reverse the
process. In the landscapes where removing the stress (such as excluding domestic livestock) does not result in any meaningful recovery, active intervention is
required. Rehabilitation of such areas requires a careful survey of current landscape and patch-scale processes and identification of those parts of the landscape

where inexpensive intervention will provide the most effective and long-lasting improvements in landscape function.

REFERENCES

Anable, M. E., McClaran, M. P., and Ruyle, G. B. (1992). Spread of introduced Lehmann lovegrass, *Eragrostis lehamanniana* Nees. in southern Arizona, USA. *Biol. Conserv.* **61**, 181–188.

Andrew, M. H. and Lange, R. T. (1986a). Development of a new piosphere in arid chenopod shrubland grazed by sheep. 1. Changes to the surface. *Aust. J. Ecol.* **11**, 395–409.

Andrew, M. H. and Lange, R. T. (1986b). Development of a new piosphere in arid chenopod shrubland grazed by sheep. 2. Changes to the vegetation. *Aust. J. Ecol.* **11**, 411–424

Archer, S. (1990). Development and stability of grass/woody mosaics in a subtropical savanna parkland, Texas, U.S.A. *J. Biogeogr.* **17**, 453–462.

Archer, S., Scifres, C., and Bassham, C. R. (1988). Autogenic succession in a subtropical savanna: conversion of grassland to thorn woodland. *Ecol. Monogr.* **58**, 111–127.

Arimoto, R., Duce, R. A., and Ray, B. R. (1989). Concentrations, sources and air-sea exchange of trace elements in the atmosphere over the Pacific Ocean. In J. P. Riley, R. Chester, and R. A. Duce (eds), *Chemical Oceanography*, pp. 107–149. Academic Press, San Diego.

Aronson, J., Floret, C., Le Floc'h, E., Ovalle, C., and Pontanier, R. (1993). Restoration and rehabilitation of degraded ecosystems in arid and semi-arid lands. II. Case studies in southern Tunisia, central Chile and northern Cameroon. *Restoration Ecol.* **1**, 168–187.

Arshad, M. and Rao, A. R. (1994). The ecological destruction of Cholistan Desert and its eco-regeneration. *UN Environ. Programme Desertification Control Bull.* **24**, 32–34.

Bahre, C. J. and Shelton, M. L. (1993). Historic vegetation change, mesquite increases, and climate in southeastern Arizona. *J. Biogeogr.* **20**, 489–504.

Betancourt, J. L. (1996). Long- and short-term climate influences on southwestern shrublands. In J. R. Barrow, E. D. McArthur, R. E. Sosebee, and R. J. Tausch (eds), *Proceedings: Shrubland Ecosystem Dynamics in a Changing Environment*, pp. 5–9. General Technical Report INT-GTR-338. US Department of Agriculture, Forest Service Intermountain Research Station, Ogden, UT.

Bock, C. E., Bock, J. H., Jepson, K. L., and Ortega, J. C. (1986). Ecological effects of planting African lovegrasses in Arizona. *National Geogr. Res.* **2**, 456–463.

Brady, W. W., Stromberg, M. R., Aldon, E. F., Bonham, C. D., and Henry, S. H. (1989). Response of a semidesert grassland to 16 years of rest from grazing. *J. Range Mgmt* **41**, 284–288.

Brown, J. H. and Heske, E. J. (1990). Control of a desert grassland transition by a keystone rodent guild. *Science* **250**, 1705–1707.

Buffington, L. C. and Herbel, C. H. (1965). Vegetational changes on a semidesert grassland range from 1858 to 1963. *Ecol. Monogr.* **35**, 139–164.

Call, C. A. and Roundy, B. A. (1991). Perspectives and processes in revegetation of arid and semi-arid rangelands. *J. Range Mgmt* **44**, 543–549.

Charney, J. G. (1975). Dynamics of desert and drought in the Sahel. *Q. J. R. Meteorol. Soc.* **101**, 193–202.

Charney, J. G., Stone, P. H., and Quirk, W. J. (1975). Drought in the Sahara: a biogeophysical feedback mechanism. *Science* **187**, 434–435.

Charney, J. G., Quirk, W. J., Chow, S., and Kornfield, J. (1977). A comparative study of the effects of albedo change on drought in semi-arid regions. *J. Atmos. Sci.* **34**, 1366–1385.

Chew, R. M. (1982). Changes in herbaceous and suffrutescent perennials in grazed and ungrazed desertified grassland in southeastern Arizona, 1958–1978. *Am. Midl. Nat.* **108**, 161–169.

Chew, R. M. and Whitford, W. G. (1992). A long-term positive effect of kangaroo rats (*Dipodomys spectabilis*) on creosotebushes (*Larrea tridentata*). *J. Arid Environ.* **22**, 375–386.

Cooke, R. U. and Reeves, R. W. (1976) *Arroyos and Environmental Change in the American Southwest*. Clarendon Press, Oxford.

Cox, J. R. and Ruyle, G. B. (1986). Influence of climatic and edaphic factors on the distribution of *Eragrostis lehmanniana* Nees. in Arizona USA. *J. Grassland Soc. South Afr.* **3**, 25–29.

Cox, J. R., DeAlba-Avila, A., Rice, R. W., and Cox, J. N. (1993). Biological and physical factors influencing *Acacia constricta* and *Prosopis velutina* establishment in the Sonoran Desert. *J. Range Mgmt* **46**, 43–48.

Darkoh, M. B. K. (1998). The nature, causes and consequences of desertification in the drylands of Africa. *Land Degrad. Dev.* **9**, 1–20.

Dean, W. R. J., Hoffman, M. T., Meadows, M. E., and Milton, S. J. (1995). Desertification in the semi-arid Karoo, South Africa: review and reassessment. *J. Arid Environ.* **30**, 247–264.

Derbyshire, E., Meng, X., and Kemp, R. A. (1998). Provenance, transport and characteristics of modern aeolian dust in western Gansu Province, China, and interpretation of the Quarternary loess record. *J. Arid Environ.* **39**, 497–516.

DeSoyza, A. G., Whitford, W. G., Virginia, R. A., and Reynolds, J. F. (1997a). Effects of summer drought on the water relations physiology and growth of large and small plants of *Prosopis glandulosa* and *Larrea tridentata*. In J. R. Barrow, E. D. McArthur, R. E. Sosebee, and R. J. Tausch (eds), *Proceedings: Shrubland Ecosystem Dynamics in a Changing Environment*, pp. 220–223. General Technical Report INT-GTR-338, US Department of Agriculture, Forest Service, Intermountain Research Station, Ogden, UT.

DeSoyza, A. G., Whitford, W. G., Martinez-Meza, E., and Van Zee, J. W. (1997b), Variation in creosotebush (*Larrea tridentata*) canopy morphology in relation to habitat, soil fertility and associated annual plant communities. *Am. Midl. Nat.* **137**, 13–26.

El-Baz, F. (1983). A geological perspective of the desert. In S. Wells and D. Haragan (eds), *Origin and Evolution of Deserts*, pp. 163–183. University of New Mexico Press, Albuquerque.

Fernandez, O. A. and Busso, C. A. (1999). Arid and semi-arid rangelands: two thirds of Argentina. In O. Arnalds and S. Archer (eds), *Case Studies of Rangeland Desertification: Proceeding from an International Workshop in Iceland*, pp. 41–60. Rala Report 200. Agricultural Research Institute, Reykjavik, Iceland.

Fullen, M. A. and Mitchell, D. J. (1994). Desertification and reclamation in north-central China. *Ambio* **23**, 131–135.

Gibbens, R. P. and Beck, R. F. (1988). Changes in grass basal area and forb densities over a 64-year period on grassland types of the Jornada Experimental Range. *J. Range Mgmt* **41**, 186–192.

Gibbens, R. P., Tromble, J. M., Hennessy, J. T., and Cardenas, M. (1983). Soil movement in mesquite dunelands and former grasslands of southern New Mexico from 1933 to 1980. *J. Range Mgmt* **36**, 145–148.

Glendening, G. E. (1952). Some quantitative data on the increase of mesquite and cactus on a desert rangeland in southern Arizona. *Ecology* **33**, 319–328.

Gonzalez, C. L. (1990). Brush reinfestation following mechanical manipulation. *J. Arid Environ.* **18**, 109–117.

Griffith, L. W., Schuman, G. E., Rauzi, F., and Baumgartner, R. E. (1984). Mechanical renovation of shortgrass prairie for increased herbage production. *J. Range Mgmt* **27**, 459–462.

Griffiths, D. (1901). Range improvement in Arizona. *US Department of Agriculture, Bureau of Plant Industry, Bulletin No. 4.* US Government Printing Office, Washington DC.

Grove, A. T. (1986). Desertification in southern Europe. *Clim. Change* **9**, 49–57.

Grover, H. D. and Musick, H. B. (1990). Shrubland encroachment in southern New Mexico, USA: an analysis of desertification processes in the American southwest. *Clim. Change* **17**, 305–330.

Guo, Q. (1996). Effects of bannertail kangaroo rat mounds on small-scale plant community structure. *Oecologia* **106**, 247–256.

Harper, K. T., Van Buren, R., and Kitchen, S. G. (1996). Invasion of alien annuals and ecological consequences in salt desert shrublands of western Utah. In J. R. Barrow, E. D. McArthur, R. E. Sosebee, and R. J. Tausch (eds), *Proceedings: Shrubland Ecosystem Dynamics in a Changing Environment*, pp. 58–65. General Technical Report INT-GTR 338. US Department of Agriculture, Forest Service, Intermountain Research Station, Ogden, UT.

Hastings, J. R. and Turner, R. M. (1965). *The Changing Mile*. University of Arizona Press, Tucson, AZ.

Hennessy, J. T., Gibbens, R. P., Tromble, J. M., and Cardenas, M. (1983). Vegetation changes from 1935 to 1980 in mesquite dunelands and former grasslands of southern New Mexico. *J. Range Mgmt* **36**, 370–374.

Herbel, C. H., Gould, W. L., Leifeste, W. F., and Gibbens, R. P. (1983). Herbicide treatment and vegetation response to treatment of mesquites in southern New Mexico. *J. Range Mgmt* **36**, 149–151.

Holachek, J. L. (1992). Financial aspects of cattle production in the Chihuahuan Desert. *Rangelands* **14**, 145–149.

Huenneke, L. F. and Noble, I. (1996). Ecosystem function of biodiversity in arid ecosystems. In H. A. Mooney, J.H. Cushman, E. Medina, O. E. Sala, and E.-D. Schulze (eds), *Functional Roles of Biodiversity: A Global Perspective*, pp. 99–128. John Wiley, London.

Johnson, K. H., Vogt, K. A., Clark, H. J., Schmitz, O. J., and Vogt, D. J. (1996). Biodiversity and the productivity and stability of ecosystems. *Tree* **11**, 372–377.

Karrar, G. and Stiles, D. (1984). The global status and trend of desertification. *J. Arid Environ.* **7**, 309–313

Kassas, M. (1977). Arid and semi-arid lands: problems and prospects. *Agro-Ecosystems* **3**, 185–204.

Kepner, W. G., Watts, C. J., Edmonds, C. M., Van Remortel, R. D., and Hamilton, M. E. (1999). A landscape approach for detecting and evaluating change in a semi-arid environment. *Environ. Monit. Assess.* **64**, 179–195.

Kerley, G. I. H. and Whitford, W. G. (2000). Impact of grazing and desertification in the Chihuahuan Desert: plant communities, granivores and granivory. *Am. Midl. Nat.* **144**, 78–91.

Kerley, G. I. H., Whitford, W. G., and Kay, F. R. (1997). Mechanisms for the keystone status of kangaroo rats: graminivory rather than granivory. *Oecologia* **111**, 422–428.

Kulas, C. M. (2001). The effects of Lehman lovegrass (*Eragrostis lehmanniana*) on semiarid ecosystems: biopedturbation as an ecosystem indicator. MS Thesis. New Mexico State University, Las Cruces, NM, 107 pp.

Lockwood, J. A. and Lockwood, D. R. (1993). Catastrophe theory: a unified paradigm for rangeland ecosystem dynamics. *J. Range Mgmt* 46, 282–288.

Ludwig, J. A. and Tongway, D. J. (1994). Small-scale resource heterogeneity in semi-arid landscapes. *Pacific Conserv. Biol.* **1**, 201–208.

Ludwig, J. A. and Tongway, D. J. (1995). Spatial organization of landscapes and its function in semi-arid woodlands, Australia. *Lands. Ecol.* **10**, 51–63.

MacDonald, B. C. T. and Melville, M. D. (1999). The impact of contour furrowing on chenopod patterned ground at Fowlers Gap, western New South Wales. *J. Arid Environ.* **41**, 345–357.

Miller, B., Ceballos, G., and Reading, R. (1994). The prairie dog and biotic diversity. *Conserv. Biol.* **8**, 677–681.

Milton, S. J., Dean, W. R. J., and DuPlessis, M. A. (1994). A conceptual model of rangeland degradation, the escalating cost of declining productivity. *Bioscience* **44**, 70–76

Moorhead, D. L., Fisher, F. M., and Whitford, W. G. (1988). Cover of spring annuals on nitrogen-rich kangaroo rat mounds in a Chihuahuan Desert grassland. *Am. Midl. Nat.* **120**, 443–447.

Mun, H. T. and Whitford, W. G. (1990). Factors affecting annual plant assemblages on banner-tailed kangaroo rat mounds. *J. Arid Environ.* **18**, 165–173.

Nash, M. S., Anderson, J. P., and Whitford, W. G. (1999). Spatial and temporal variability in relative abundance and foraging behavior of subterranean termites in desertified and relatively intact Chihuahuan Desert ecosystems. *Appl. Soil Ecol.* **12**, 149–157.

Noble, J. C. (1997). *The Delicate and Noxious Scrub: CSIRO Studies on Native Tree and Shrub Proliferation in the Semi-arid Woodlands of Eastern Australia*. CSIRO Division of Wildlife and Ecology, Canberra.

O'Conner, T. G. (1995). *Acacia karroo* invasions of grassland: environmental and biotic effects influencing seedling emergence and establishment. *Oecologia* **103**, 214–223.

Omar, S. A. S., and Abdal, M. (1994). Sustainable agricultural development and halting desertification in Kuwait. *UN Environ. Programme Desertification Control Bull.* **24**, 23–26.

Ovalle, C., Aronson, J., Avendano, J., Meneses, R., and Moreno, R. (1993). Rehabilitation of degraded ecosystems in central Chile and its relevance to the arid 'Norte Chico'. *Rev. Chile. Hist. Nat.* **66**, 291–303.

Ovalle, C., Aronson, J., Del Pozo, A., and Avendano, J. (1999). Restoration and rehabilitation of mixed espinales in central Chile: 10-year report and appraisal. *Arid Soil Res. Rehabil.* **13**, 369–381.

Pickup. G. (1985). The erosion cell – a geomorphic approach to landscape classification in range assessment. *Aust. Rangelands J.* **7**, 114–121.

Pickup, G. and Chewings, V. H. (1986). Mapping and forecasting soil erosion patterns from Landsat on a microcomputer-based image-processing facility. *Aust. Rangelands J.* **8**, 57–62.

Rappole, J. H., Russell, C. E., Norwine, J. R., and Fullbright, T. E. (1986). Anthropogenic pressure and impacts on marginal, neotropical, semiarid ecosystems: the case of south Texas. *Sci. Total Environ.* **55**, 91–99.

Roundy, B. A. and Jordan, G. L. (1988). Vegetation changes in relation to livestock exclusion and root plowing in southeastern Arizona. *Southwest. Nat.* **33**, 425–436.

Schlesinger, W. H., Reynolds, J. F., Cunningham, G. L., Huenneke, L. F., Jarrell, W. M., Virginia, R. A., and Whitford, W. G. (1990). Biological feedbacks in global desertification. *Science* **247**, 1043–1048.

Smith, D. A. and Schmutz, E. M. (1975). Vegetative changes on protected versus grazed desert grassland range in Arizona. *J. Range Mgmt* **28**, 433–457.

Tongway, D. J. and Ludwig, J. A. (1990). Vegetation and soil patterning in semi-arid mulga lands of eastern Australia. *Aust. J. Ecol.* **15**, 23–34.

Van Auken, O. W. (2000). Shrub invasions of North American semiarid grasslands. *Annu. Rev. Ecol. Syst.* **31**, 197–215.

Van der Merwe, J. P. A. and Kellner, K. (1999). Soil disturbance and increase in species diversity during rehabilitation of degraded arid rangelands. *J. Arid Environ.* **41**, 323–343.

Verstraete, M. M. (1986). Defining desertification: a review. *Clim. Change* **9**, 5–18.

Weltzin, J. F., Archer, S., and Heitschmidt, R. K. (1997). Small-mammal regulation of vegetation structure in a temperate savanna. *Ecology* **78**, 751–763.

Westoby, M., Walker, B., and Noy Meir, I. (1989). Range management on the basis of a model which does not seek to establish equilibrium. *J. Arid Environ.* **17**, 235–239.

Whitford, W. G. (1993). Animal feedbacks in desertification: an overview. *Rev. Chil. Hist. Nat.* **66**, 243–251.

Whitford, W. G. (1996) The importance of the biodiversity of soil biota in arid ecosystems. *Biodiversity Conserv.* **5**, 185–195.

Whitford, W. G. (1997). Desertification and animal biodiversity in the desert grasslands of North America. *J. Arid Environ.* **37**, 709–720.

Whitford, W. G., Martinez-Turanzas, G., and Martinez-Meza, E. (1995). Persistence of desertified ecosystems: explanations and implications. *Environ. Monitor. Assess.* **37**, 319–333.

Whitford, W. G., Nielson, R., and DeSoyza, A. G. (2001). Establishment and effects of establishment of creosotebush, *Larrea tridentata*, on a Chihuahuan Desert watershed. *J. Arid Environ.* **47**, 1–10.

Williams, M. A. J. and Balling, R. C. Jr. (1996). *Interactions of Desertification and Climate*. Arnold, London.

York, J. C. and Dick-Peddie, W. A. (1969). Vegetation changes in southern New Mexico during the past hundred years. In W. G. McGinnies and B. J. Goldman (eds), *Arid Lands in Perspective*, pp. 155–166. The University of Arizona Press, Tucson, AZ.

Chapter 11 | Monitoring and Assessment

The widespread continuing degradation of arid and semiarid lands has led to recognition of the need for monitoring and assessment systems to detect degradation trends and serve as early warning systems. The shift from a 'functional' ecosystem or landscape to a 'dysfunctional' ecosystem or landscape may be effectively irreversible when the system has moved past some 'threshold'. Monitoring and assessment systems need to be able to detect degradation trajectories and to provide indicators that allow identification of thresholds of irreversible change.

Despite the perception that monitoring and assessment systems are needed, there is no general agreement as to what should be monitored or how monitoring results should be reported or interpreted. What are the goals of monitoring and assessing the status of arid and semiarid lands? What are the values that society places on arid and semiarid lands? Can these values be translated into measurable goals for a monitoring and assessment program? These are questions for which there are no answers that are generally applicable to arid and semiarid lands around the world. Land managers seek monitoring and/or assessment systems that can provide information on the spectrum of goals expressed by various groups within society. The information required of monitoring and/or assessment systems is different for the various user groups in society. Each user group espouses different values for arid and semiarid lands. Monitoring and assessment systems must therefore provide sufficient data to describe the state of the system with respect to the divergent values of the user groups.

Government agencies call for assessments of the status of lands encompassed by political boundaries. Political boundaries rarely coincide with regional drainage systems or with functional landscapes. However, the status of the lands within political boundaries can only be assessed within a proper geographical context. In arid lands this is frequently a difficult problem because many of the environmental stressors affecting the sustainability of regional ecosystems originate outside the political region.

Calls for assessments are usually vague particularly with regard to the standards against which the status of the lands are to be measured. The standards against which assessments are compared are often some idealized 'pristine' or relatively 'undisturbed' concept of what the ecosystems in a region should resemble. Since all biological communities and ecosystems change both structurally and functionally over time in response to variation in climate and other abiotic variables, it is essential to factor out 'normal ecological change' from changes resulting from anthropogenic stresses imposed on ecosystems. Because of the natural temporal changes in structure and function of ecosystems, monitoring studies are essential to provide the natural background 'standard' for assessments.

11.1 DEFINITIONS

Assessment is a one-time process that relies on a statistically valid sampling within a region. An assessment system provides data on the status of the ecosystems of a region and should be used across broad geographical areas. The measurements made in an assessment system must be rapid, and sufficiently simple to be made by persons with limited experience and expertise. Assessment measurements are converted into indicators by combining a group of measurements or by calculating indicator values from the basic measurements. Indicators should be measures of ecosystem properties that are related to ecosystem processes in some documented way. A suite of such indicators provides information on how well the ecosystem(s) are performing selected functions (such as conservation of water and nutrients). If assessments are made repeatedly at the same locations, the assessment indicators can be incorporated into a monitoring system in order to ascertain the trends in ecosystems of a region. Data on landscape characteristics obtained by remote sensing have considerable potential for assessing condition.

Monitoring consists of a series of measurements designed to provide information on the trajectory of change in ecosystem structure and/or function over time. Monitoring systems are not limited by the same time constraints that are imposed on assessments. Ideally monitoring involves repeated measurements on the same location at fixed time intervals. Monitoring sites need not be selected to fit a particular statistical model. Monitoring sites should be selected on the basis of a set of criteria or questions that may differ from region to region and even among locations within a region. For example, a land manager may want to know how effective a management protocol is at restoring ecosystem function that has been recognized as degraded or lost. The manager must then decide what landscape locations will provide the information necessary to answer that question. The site locations and measurements selected should be chosen to answer one or more questions about status and trends of the ecosystems of a landscape unit.

11.2 EXISTING ASSESSMENT AND MONITORING SYSTEMS

Assessment and monitoring systems for arid rangelands are described in monitoring manuals for a variety of Australian rangelands (Tongway, 1994; Tongway and Hindley, 1995) and South African Karoo desert rangelands (Milton and Dean,

1996). Although the Australian and South African rangelands differ in a number of very important ways, they also share a number of the same attributes. The monitoring systems that have been proposed for these regions use similar indicators that are related to the same processes. In the United States, new methods to inventory, classify, and monitor semiarid and arid rangelands (lands used for livestock production) were described in the US National Research Council publication, *Rangeland Health* (National Research Council, 1994).

The Karoo Desert manual is designed primarily for the land manager (farmers) in the Karoo. The assessment system is qualitative but that allows semiquantitative evaluation of some of the most important variables. The Karoo system scores attributes of the system from 5 to 1 with 5 being highest quality and 1 lowest quality. The attributes measured or estimated include forage value, use of palatable plant species, alien weeds, the ratio of palatable:unpalatable species of seedlings of perennial plants, and soil health. The soil health variables include those that are positive (mulch, digging, shade plants) and those that are negative (capping, pedestals, rills, roots). The soil health variables are rated common to uncommon. Uncommon is good for the negative soil health variables and common is good for the positive soil health variables (Milton and Dean, 1996).

The Australian soil condition assessment manuals focus on the hypothesis that arid land ecosystems should retain water and nutrient resources. The assessment and monitoring systems are focused at the landscape scale. This scale was selected because arid and semiarid systems frequently function as 'source'–'sink' systems that are linked at the landscape scale. The indicators selected for assessing soil condition are variables that affect the retention of water and nutrients within the landscape. Vegetation is evaluated as to its obstruction value – resisting flow, reducing velocity of overland flow and causing sediment deposition. Soil surface stability and resistance to splash erosion are evaluated by a simple field measure of aggregate stability and crust characteristics (presence/absence and percentage cover of cryptogamic crusts (biological crusts composed of algae, lichens, and fungi). The soil condition assessment system can be converted into a monitoring system by establishing permanent line transects on which vegetation and soil measurements are made at a predetermined frequency (Tongway, 1994; Tongway and Hindley, 1995).

The assessment of rangeland health focused on qualitative indicators. For example, soil condition is assessed by the presence or absence of an A horizon, rills and gullies, soil pedestaling, evidence of scouring and sheet erosion, and dune formation. Biotic integrity is assessed by plant distribution, litter distribution, rooting depth, photosynthetic period, age class, plant vigor, and germination microsites (National Research Council, 1994). The rangeland health indicators are rated healthy, at risk and unhealthy by qualitative evaluations that require considerable expertise on the part of the evaluator (Table 11.1). Some of the indicators (plant vigor, rills) require completely subjective evaluation and it is nearly impossible for a group of trained observers to agree on the state of these indicators, i.e. healthy, at risk, or unhealthy. Three categories of ecosystem attributes related to

Table 11.1
The Relationship Between Ecosystem Health Attributes and the Indicators of
Health Thresholds for Arid and Semiarid Rangelands (modified from National Research Council, 1994)

Attribute	Healthy	At risk	Unhealthy
Soil stability and watershed function	No evidence of soil movement	Soil movement local Remains on site	Soil moving off site
	A horizon present and continuous	A horizon present but patchy	A horizon absent or present only in association with dominant plants
	No pedestaling of plants or rock	Some plants pedestaled but no exposed roots	Most plants and rock pedestaled, roots exposed
	Rills and gullies absent or with muted features	Small and not connected into dendritic pattern	Well-defined channels in a dendritic pattern
	No visible sheet erosion or scouring	Patches of bare soil or scours present	Bare areas coalescing, scours well developed
	No visible soil deposition	Soil accumulating around plants and obstructions	Soil accumulating in dunes or barren deposits
Distribution of nutrients and energy	Plant and litter distribution unfragmented	Plant and litter distribution developing fragmentation	Large barren areas between plant and litter patches
	Photosynthesis during entire growing season	Photosynthesis restricted during one or more seasons	Photosynthesis restricted to one season
	Roots distributed through soil profile	Roots absent from portions of the soil profile	Roots limited to one portion of soil profile
Recovery following disturbance	Diverse age class distribution	Seedlings and immature plants absent	Most plants old and dying
	Vigorous plants	Reduced plant vigor	Plant vigor poor
	Germination microsites abundant	Developing soil crust, reduction in germination microsites	Soil crusts and soil mobility inhibiting germination

health criteria were identified. These provided qualitative indicators of the thresholds for the state of the system (Table 11.1). Many of these indicators are similar to those described for assessments in the Karoo of South Africa and for arid rangelands in Australia.

11.3 DATA FOR ASSESSMENT AND MONITORING

Ecosystem processes are difficult to measure and frequently require the use of specialized instrumentation. One of the most important ecosystem processes is the cycling of nitrogen. Obtaining data on rates of nitrogen fixation, nitrogen mineralization, nitrogen immobilization, and denitrification in desert systems is complex and requires considerable investment of scientist time and access to expensive technical equipment. For example, measurement of nitrogen mineralization rates requires time for soil incubation, and expensive analytical instrumentation for the measurement of nitrate and ammonium. The analytical instrumentation requires skilled operators who have extensive knowledge of chemistry and instrumentation. Some of the most important arid ecosystem processes such as water infiltration and run-off are episodic. These processes cannot be measured inexpensively nor in a short period of time. If measurement of ecosystem processes cannot be accomplished within the time, monetary, and skill level constraints imposed by assessment and monitoring needs, what can be measured?

There are ecosystem properties that are known to be linked to ecosystem processes. For example, it is well documented that nitrogen mineralization rates vary directly with the soil organic matter content. Soil organic matter is directly related to litter accumulation, litter types and spatial distribution of root systems. Litter patches can be measured by spatially explicit techniques in order to provide information on patch characteristics and patch distribution. The cover and distribution of plant species indirectly provides information on the characteristics of root systems. The general characteristics of root systems influence the rates of nitrogen immobilization and mineralization. These readily measure properties, therefore, provide information on important nitrogen cycling processes.

There are a number of ecosystem properties that are related to infiltration and run-off: (a) aggregate stability of soil crusts; (b) proportion of soil surface area protected from direct raindrop impact by plant foliage; (c) proportion of soil surface area with cryptogamic crusts; (d) microtopography and spatial distribution of plants, litter, debris, and surface features that obstruct or slow overland flow; (e) abundance and spatial distribution of animal produced macropores. Ecosystem properties (a) through (d) are properties that are reasonably constant through time and therefore reliable indicators. The abundance and distribution of animal-produced macropores tend to be seasonal (emergence tubes of insects such as cicadas and beetles, active ant colonies, termite foraging galleries).

11.4 ECOSYSTEM FUNCTIONS

There is a long history of concern with 'integrity of community processes' (Leopold, 1948). This concern has developed into a consensus that the most important societal need is for 'ecosystem integrity ' or 'biological integrity' (Karr, 1991; West and Whitford, 1995). The elements of biological and ecosystem integrity that are most important include primary and secondary production, biogeochemical cycles (especially water), biotic and abiotic processes that affect patterns of distribution of organisms, resources, and structure in ecological systems (Shachak and Pickett, 1997). Although maintenance of ecosystem integrity may be a worthy goal, it may be difficult to design simple and inexpensive monitoring and assessment systems that provide sufficient information about ecological changes that affect ecosystem integrity to be of value.

For arid ecosystems the most essential function is the retention of water and nutrient resources. Those arid ecosystems that fail to retain water and nutrients have limited productivity. If water run-off areas are spatially extensive, the movement of water across such landscape units erodes soil and further compromises the production potential. In the extreme, arid landscapes in which the retention of water is limited are hazards to humans because they are sources of flood water during intense storms. Conservation of nutrients is frequently tied to the retention of water, since overland flow removes much of the dead plant material that is the source of mineralizable nutrients.

Productivity, as measured by plants or animals that can be harvested as a source of food for humans, is closely tied to the water–nutrient retention status of the ecosystems of a desert landscape. However, a different aspect of productivity has gained in importance in the latter half of the 20th century. As human societies shift from a subsistence life style to a commodity-driven life style, there is increased awareness of the value of open space and the biota inhabiting that space. The capacity of ecosystems to support biodiversity has become an important and contentious aspect of the function of ecosystems. Both biodiversity and productivity are properties of ecosystems that must be considered in the design of assessment and monitoring systems.

11.5 INDICATORS

Indicators are calculated from the basic set of measurements made during an assessment or monitoring session. The indicator may be an index calculated from several of the measurements or may be the actual measurement without manipulation of the numeric value (see Table 11.2). The same indicator may address more than one ecosystem function and may be related to one or more ecosystem processes. However, not all indicators are related to ecosystem processes. For both assessment and monitoring, it is necessary to have a suite of indicators that are sensitive to environmental stress, that focus on risk of degradation, and that are related to ecosystem function (Herrick *et al.*, 1995). Indicators that are meaningful for assessing and/or monitoring the health of arid and semiarid ecosystems must meet the following criteria: (1) reflect the status of a critical ecosystem process, important ecosystem property, or an economic–social value; (2) be unambiguous (i.e. the trajectory of the measure is unidirectional in response to ecosystem stressors of increasing intensity); (3) be applicable to the range of ecosystems encountered in the arid and semiarid landscapes; and (4) be readily and inexpensively measured.

Table 11.2

The Condition or Health of Arid and Semiarid Ecosystems Includes the Capacity of the System(s) to Perform Selected Functions and Maintain Those Functions Following a Stress-Disturbance. Each of the indicators listed can be used to test for both criteria depending on when the system is examined relative to the most recent stress/disturbance event(s) (data from Herrick *et al.*, 1995)

Ecosystem function	Indicator
Soil stability and watershed function (*indicators related to wind and water erosion potential*)	1. Total vegetation cover and average height of vegetation 2. Size of unvegetated patches (mean, median, skewness of distribution, weighted skewness (skewness × mean)) 3. Spatial distribution and orientation of unvegetated patches 4. Surface stability (aggregate measure – slake test) 5. Cryptogamic crust cover 6. Litter and rock cover 7. Infiltration capacity (single-ring, unconfined) 8. Size and spatial distribution of litter patches 9. Penetration resistance (compaction) (cone penetrometer) 10. Root density and depth based on species composition and cover of the vegetation. 11. Soil disturbance by animals (percent surface area disturbed, volume moved/unit area, soil horizon origin of ejected soil 12. Predictability of annual plants (rainfall seasonality) 13. Ratio of long-lived to short-lived grasses 14. Ratio of seed-reproducing grasses/vegetative-reproducing grasses
Productivity (energy flow)	Indicator 1 15. Rainfall use efficiency (based on biomass production estimates by harvest or dimension analysis) 16. C_3/C_4 plant cover ratio vs. rainfall seasonality
Animal production (including wildlife)	17. Palatability index for each animal species (Sum index over all species of animals (index = proportion of cover × months useable/12 × palatability rating)) 18. Forage value index (Sum over all species of plants) Index = (palatability index × nutritive value × biomass production rating))
Nutrient cycling	Indicators 1, 2, 3, 8, 12, 14

There are ecosystem properties that provide information on the status of most ecosystem processes (Table 11.2). Ecosystem properties such as structural characteristics of the vegetation, spatial distribution patterns of plant species, morphological characteristics of plants, and physical and biological character-istics of the soil surface do not change rapidly over time (in unstressed ecosystems) and are frequently directly related to one or more ecosystem processes (Table 11.2). Potential indicators need to be evaluated for their sen-sitivity to disturbance and stress. Disturbance is defined as a variable to which the ecosystem has been exposed over evolutionary time and from which recov-ery to the predisturbance state occurs in ecological time (years to a few decades). In arid and semiarid ecosystems, drought is a disturbance (rainless time periods are characteristic of arid lands), grazing and browsing of vegeta-tion are also disturbances when large herbivores are free-ranging. Overgrazing by livestock that are confined in paddocks or limited to small areas can be an ecosystem stress.

A sensitive indicator is one that yields very different quantitative values when measured at locations that are known to be degraded and at locations that are known to be in 'healthy' or good condition (DeSoyza et al., 1997). In addition, a sensitive indicator of a healthy ecosystem is one that rebounds quickly after a dis-turbance. In a stressed ecosystem (high impact by cattle), most short life-span grasses died during a drought (disturbance) but in a low-stress system (light graz-ing by cattle) most of the short-life-span grass clumps had some surviving tillers at the end of the drought. Grass clumps with surviving tillers quickly recovered following rainfall (Whitford et al., 1999b). In this example, life-span character-istics of grasses, is an indicator of resistance and resilience to disturbance and how this indicator responds to stress.

Sensitive indicators yield very different quantitative values when measured at locations known to have been exposed to different levels of environmental stress. Indicators that are sensitive to stress and resistant/resilient to disturbance are also applicable to the evaluation of the success of restoration efforts. Several of the rangeland health indicators were found to be related to the relative success of restoration efforts (Whitford et al., 1998).

Tests of the sensitivity of indicators for arid and semiarid lands used for live-stock production in the southwestern US showed that not all of the potential indicators were suitable for monitoring or assessment (Whitford et al., 1998; DeSoyza et al., 2000a, b). The sensitive indicators that applied to most desert areas tested were those related to wind and water erosion (average size of bare patches (fetch lengths) and an index of soil stability (related to soil aggregate sta-bility). In the Chihuahuan Desert rangelands the following indicators were judged sufficiently robust to be included in assessment and monitoring programs: aver-age size of bare patches, cover of long-lived grasses, a palatability index (season and duration of palatability to livestock), and a soil surface stability index (meas-ure of aggregate stability). Other indicators, where values above a threshold are indicators of rangeland health, include: cover of alien species, cover of species toxic to livestock, and cover of increaser species (native plant species that rapidly spread into stressed environments).

Biodiversity is thought to affect many ecosystem processes, including but not limited to, productivity and nutrient cycling (Tilman, 1999). Tilman (1999) argues that diversity must be added to composition, disturbance, nutrient dynamics and climate as determinants of ecosystem structure and function. However, for many groups of organisms in many biological communities there appears to be a few 'functional groups' composed of a number of ecologically equivalent species. In such communities it is argued that some species may be lost with little or no effect on ecosystem processes (the redundancy hypothesis, Walker, 1992). At present the interactions between biodiversity and ecosystem services are poorly understood. Many ecosystem services and processes may be unaffected by small losses in biodiversity. However, ecosystem services may deteriorate rapidly when most of the species in a functional group are gone (Balvanera et al., 2001). For arid ecosystems there are limited data on the relationship between sustainability and biodiversity and the effects of degradation/desertification on biodiversity.

One indicator that appears to be useful in identifying vegetation and soil indicator values at or near the threshold for irreversible degradation is the presence/absence of a keystone species (Krogh et al., 2001). Studies at 117 sites in the northern Chihuahuan Desert found that the banner-tailed kangaroo rat (*Dipodomys spectabilis*) did not occur at sites with shrub cover greater than 20%. The banner-tailed kangaroo rat is considered a keystone species in the desert grasslands because of its burrow mounds that are habitat for several relatively rare plant species. This species also affects desert grassland dynamics by harvesting immature seed head tillers, excavating cache pits that form 'safe' sites for seeds, and creating small nutrient-rich patches. The loss of this species may be one of the main factors affecting the irreversibility of degradation. By documenting the vegetation and soil indicators associated with sites where banner-tailed kangaroo rats cannot maintain viable populations, those indicator values can be used as threshold values (Krogh et al., 2001).

Indicators based on abundance and species richness of plants and animals were not sensitive to environmental stress. These included species richness, abundance and functional groups of ants (Formicidae), species richness and abundance of breeding birds, and perennial plant species richness (Whitford et al., 1998). The abundance, species composition, and species richness of ants were sensitive to dominance of degraded lands by an alien grass (*Eragrostis lehmanniana*) but not to dominance by shrubs (Whitford et al., 1999a). The value of ants as early warning bioindicators of rangeland degradation was also tested in Australian rangelands. It was concluded that 'neither overall ant abundance and richness, nor the abundances of the most common species and functional groups responded significantly to grazing in this local-scale experiment, which rendered ants of limited use as early warning indicators of unsustainable management' (Read and Andersen, 2000). The general rangeland health indicators were also useful in evaluating the relative success of various rehabilitation efforts (herbicide treatment, root-plowing, and leveling of coppice dunes).

Tests of suite of indicators used in the Chihuahuan Desert provided very different results when tested in the Great Basin Desert in Utah, eastern Oregon, and Idaho. The Great Basin Desert is a strongly seasonal precipitation desert (winter precipitation as snow) and is a cold desert. The most robust indicators of condition in Great Basin ecosystems were percent vegetation cover (including cryptogams), percent cover of life-forms, percent cover of sagebrush (*Artemisia* spp.) and percent cover of resilient species. There were sufficient differences in plant community composition and range of bare patch sizes in different parts of the Great Basin, that the sensitive indicators were found to be location specific. It was concluded that assessments and monitoring should be based on a suite of indicators. Another difference between the Chihuahuan Desert and Great Basin is the reliability of annual plant communities. In the Great Basin, spring annuals occur nearly every year. An alien annual species, cheat grass (*Bromus tectorum*) fills the intershrub spaces in the Great Basin in many years. Cheat grass provides sufficient fuel to carry devastating fires. Great Basin shrublands are not fire-adapted and burned areas are subject to severe erosion and slow recovery of perennial vegetation cover (DeSoyza *et al.*, 2000a).

Studies of potential animal indicators of the condition of arid rangeland have yielded mixed results. Soil microarthropod community structure appears to be a good indicator of soil quality. The abundance of soil mites increased as the intensity of disturbance by grazing livestock decreased. Soil microarthropods responded to a complex of factors including soil compaction, depth to an impervious soil layer, root biomass variation with depth, and residual effects of herbicide. All plots, except those treated with herbicide, were dominated by mites of the family Nanorchestidae. The numerical responses of nanorchestid mites provided a sensitive indicator of ecosystem health in the Chihuahuan Desert (Kay *et al.*, 1999). Soil-nesting ants were not sensitive to ecosystem degradation except in areas dominated by an alien grass, *E. lemanniana* (Whitford *et al.*, 1999a). Studies on rodents concluded that population variation and dispersal confound the interpretation of environmental stress on abundance and species richness (Douglass, 1989).

There has been considerable investment in the potential of remote sensing as a method for conducting assessment and for monitoring arid and semiarid rangelands. The most cost-effective imagery that is widely available for monitoring or assessment is from the US National Oceanic and Atmospheric Administration's polar orbiting satellites' Advanced Very High Resolution Radiometer (AVHRR). The AVHRR system has coarse spatial resolution (1100 m at nadir), high temporal resolution (twice daily coverage), high radiometric resolution (1024 gray levels) and a synoptic view (scanning ± 55° from satellite orbital track. The most useful information on vegetation cover and productivity is derived from the visible red (0.58–0.68 μm), near-infrared (0.725–1.10 μm), and thermal-infrared

(10.5–11.5 μm). Because growing vegetation strongly reflects near-infrared (NIR) and absorbs red (RED) energy, these data are used to compute a normalized difference vegetation index (NDVI) for a series of dates. The NDVI signatures for a season can be analyzed to provide geographically explicit changes in productivity or cover (Minor et al., 1999). Applying NDVI to estimate green vegetation cover in Namibia provided a poor relationship between NDVI and actual green vegetation cover because NDVI did not detect some of the variability in green vegetation cover (DuPlessis, 1999). A study of semiarid and arid rangelands of the state of New Mexico utilizing AVHRR imagery for five consecutive years, detected 14 geographic areas with significant NDVI change. Among the causes for these changes were, adoption of different range management strategies, cross-country military training activities, and increases in area of irrigated cropland (Minor et al., 1999). A more detailed analysis of vegetation and soils using AVHRR was conducted to identify areas where rangelands were irreversibly degraded (creostebush, (*Larrea tridentata*, on eroded soils, and mesquite, *Prosopis glandulosa*, coppice dunes and open shrubland). Ground-truth quantitative analysis of vegetation cover and bare soil revealed that the most accurate assessment of pixels was from a four-year block of images during which precipitation patterns did not produce high ground cover of winter–spring annuals. The reflectance from winter–spring annuals and summer annuals contributed considerable variation in the NDVI signatures of pixels from areas with little or no grass cover (Eve et al., 1999). The regression between percent bare soil (from NDVI) and the average size of unvegetated patches (the most robust on-ground indicator of degradation) accounted for less than 10% of the variance in the relationship (DeSoyza et al., 2000b). If the average size of bare patches cannot be discerned from remote sensing imagery, identifying areas at risk for further degradation cannot be accomplished.

Finer spatial-scale evaluation of the condition of arid and semiarid lands can be made using the Landsat Thematic Mapper (Landsat TM) with pixel size of 30 meters and the use of a moving standard deviation index (MSDI) derived by passing a moving standard deviation filter across the Landsat TM red band. Degraded and unstable landscapes exhibited higher MSDI values than the undisturbed, stable reference areas. There was a significant correlation between NDVI and MSDI (Tanser and Palmer, 1999). Degradation could not be detected in areas smaller than 8100 m^2 because of the size of the filter. The studies of remote sensing imagery for regional assessments, although showing some promise, fail to provide reliable, interpretable indicators of condition. Further research on remote sensing image analysis may develop tools for deriving the required indicator values from the imagery. Because use of remote sensing imagery and geographic information systems would provide a cost-effective means of obtaining monitoring and assessment data for large areas, it is important that this research effort be continued.

REFERENCES

Balvanera, P., Daily, G. C., Ehrlich, P. R., Ricketts, T. H., Bailey, S-A., Kark, S., Kremen, C., and Pereira, H. (2001). Conserving biodiversity and ecosystem services. *Science* **291**, 2047.

DeSoyza, A. G., Whitford, W. G., and Herrick, J. E. (1997). Sensitivity testing of indicators of ecosystem health. *Ecosyst. Health* **3**, 44–53.

DeSoyza, A. G., Van Zee, J. W., Whitford, W. G., Neale, A., Tallent-Hallsel, N., Herrick, J. E., and Havstad, K. M. (2000a). Indicators of Great Basin rangeland health. *J. Arid Environ.* **45**, 289–304.

DeSoyza, A. G., Whitford, W. G., Turner, S. J., Van Zee, J. W., and Johnson, A. R. (2000b). Assessing and monitoring the health of western rangeland watershed. *Environ. Monit. Assess.* **64**, 153–166.

Douglass, R. J. (1989). Assessment of the use of selected rodents in ecological monitoring. *Environ. Mgmt* **13**, 355–363.

DuPlessis, W. P. (1999). Linear regression relationships between NDVI, vegetation and rainfall in Etosha National Park, Namibia. *J. Arid Environ.* **42**, 235–260.

Eve, M. D., Whitford, W. G., and Havstadt, K. M. (1999). Applying satellite imagery to triage assessment of ecosystem health. *Environ. Monit. Assess.* **54**, 205–227.

Herrick, J. E., Whitford, W. G., DeSoyza, A. G., and Van Zee, J. W. (1995). Soil and vegetation indicators for assessment of rangeland ecological condition. In A. G. Cededonio (ed.), *North American Workshop on Monitoring for Ecological Assessment of Terrestrial and Aquatic Ecosystems*, pp. 157–166. USDA Forest Service, Rocky Mountain Forest and Range Experiment Station, Ft. Collins, Colorado.

Karr, J. R. (1991). Biological integrity: a long neglected aspect of water resource management. *Ecol. Appl.* **1**, 66–84.

Kay, F. R., Sobhy, H. M., and Whitford, W. G. (1999). Soil microarthropods as indicators of exposure to environmental stress in Chihuahuan Desert rangelands. *Biol. Fertil. Soils* **28**, 121–128.

Krogh, S. N., Zeisset, M. S., Jackson, E., and Whitford, W. G. (2001). Presence/absence of a keystone species as an indicator of rangeland health. *J. Arid Environ.* In Press.

Leopold, A. (1948). *A Sand Country Almanac.* Oxford University Press, Oxford, UK.

Milton, S. J. and Dean, W. R. J. (1996). *Karoo Veld: Ecology and Management.* ARC Range and Forage Institute, Lynn East, South Africa.

Minor, T. B., Lancaster, J., Wade, T. G., Wickham, J. D., Whitford, W. G., and Jones, K. B. (1999). Evaluating change in rangeland condition using multitemporal AVHRR data and geographic information system analysis. *Environ. Monit. Assess.* **59**, 211–223.

National Research Council (1994). *Rangeland Health.* National Academy Press, Washington, DC.

Read, J. L. and Andersen, A. N. (2000). The value of ants as early warning bioindicators: responses to pulsed cattle grazing at an Australian arid zone locality. *J. Arid Environ.* **45**, 231–251.

Shachak, M. and Pickett, S. T. A. (1997). Linking ecological understanding and application: patchiness in a dryland system. In S. T. A. Pickett, R. S. Ostfeld, M. Shachak, and G. E. Likens (eds), *The Ecological Basis of Conservation: Heterogeneity, Ecosystems, and Biodiversity*, pp. 108–119. Chapman and Hall, New York.

Tanser, F. C. and Palmer, A. R. (1999). The application of remotely-sensed diversity index to monitor degradation patterns in a semi-arid, heterogeneous, South African landscape. *J. Arid Environ.* **43**, 477–484.

Tilman, D. (1999). The ecological consequences of changes in biodiversity: a search for general principles. *Ecology* **80**, 1455–1474.

Tongway, D. J. (1994). *Rangeland Soil Condition Assessment Manual.* CSIRO Publishing, Melbourne.

Tongway, D. J. and Hindley, N. (1995). *Manual for Soil Condition Assessment of Tropical Grasslands.* CSIRO Ecology and Wildlife, Canberra.

Walker, B. (1992). Biological and ecological redundancy. *Conserv. Biol.* **6**, 18–23.

West, N. E. and Whitford, W. G. (1995). The intersection of ecosystem and biodiversity concerns in the management of rangelands, pp. 72–79. In *Biodiversity on Rangelands – Natural Resources and Environmental Issues*, Volume IV, Utah State University, Logan, Utah.

Whitford, W. G., DeSoyza, A. G., Van Zee, J. W., Hierrick, J. E., and Havstad, K. M. (1998). Vegetation, soil, and animal indicators of rangeland health. *Environ. Monit. Assess.* **51**, 179–200.

Whitford, W. G., Van Zee, J., Nash, M. S., Smith, W. E., and Herrick, J. E. (1999a). Ants as indicators of exposure to environmental stressors in North American desert grasslands. *Environ. Monit. Assess.* **54**, 143–171.

Whitford, W. G., Rapport, D. J., and DeSoyza, A. G. (1999b). Using resistance and resilience measurements for 'fitness' tests in ecosystem health. *J. Environ. Mgmt* **57**, 21–29.

Chapter 12 | Desert Ecosystems in the Future

The value and future of natural ecosystems is a growing concern (Costanza *et al.*, 1997). The goods and services provided by ecological systems and the natural capital that produces them are critical to the functioning of the earth's life-support system. In their analysis of global ecosystem services and natural capital, the contributions of arid and semiarid ecological systems was not estimated because of the lack of sufficient data (Costanza *et al.*, 1997). Despite the inability of Costanza *et al.* (1997) to place a monetary value on the services and capital of arid and semiarid ecological systems, many of the ecosystem goods and services listed in their article are of great importance to the human occupants and visitors to arid regions (Table 12.1). If desert ecosystems are to provide these goods and services, the occupants and managers of the lands in arid regions will have to adopt management strategies that conserve the desired functional and structural attributes of the ecosystems and landscapes within a region.

The sustainable production of food and raw materials from arid regions will require a shift from traditional agriculture (irrigated crops and husbandry of three or four species of domesticated animals). At the end of the 20th century we cultivate only 100 species on a large scale and 90% of the food needs of the human population come from 12 crops (primarily grains such as maize, wheat, and rice). None of the significant food crops are xerophytes and there are no significant fiber crops that are xerophytic (Hinman, 1984). The potential for food and fiber production in semiarid and arid lands has been recognized for some time. Light and temperature regimes in subtropical regions provide growing conditions for most of the year. Seasonal growing constraints are generally due to lack of soil moisture rather than temperature and light constraints. However, the search for xerophytes that can provide the basis for commercial agriculture has attracted little research support and very little support for commercial development of those species that have been identified as potential commercial crops (Hinman, 1984).

There is a real need to develop nontraditional crops and to domesticate or manage arid-adapted animals for meat production. In every arid region of the world, there are many plants and animals that have been used for food and fiber by hunter–gatherer societies in the past. The large and rapidly growing human populations in many of the arid (geopolitical) regions of the world are currently sustained by imported food and fiber. As agricultural surpluses become less available due to increased costs of supply, and increased local demand, food and raw

Table 12.1

Ecosystem Goods and Services Provided by Semiarid and Arid Ecological Systems with Examples of How These Goods and Services Impact Human Populations That Reside Outside Arid Regions as Well as the Impacts on Humans Living Within Arid Regions (Modified from Costanza et al., 1997)

Goods/service	Impact	Ecological process
Atmospheric composition	Photosynthesis, respiration, N cycling, volatilization of compounds on leaves	CO_2/O_2 balance, UVB absorption, potential allergens
Climate regulation	Changing albedo, atmospheric dust content, water vapor inputs, and CO_2 inputs	Changing energy balance and water-holding capacity of air, changing rainfall patterns, possible sink for anthropogenic CO_2 from other regions
Disturbance regulation	Run-off from watersheds, survival during droughts, ephemeral stream capacity	Flooding, dust storms
Water supply	Storage and retention of water	Sediment concentration in water into reservoirs (reservoir life expectancy), reducing hydrostatic head for perennial streams
Erosion control and sediment retention	Retaining soil within an ecosystem or landscape unit	Reservoir sedimentation, reduced capacity for plant production
Soil formation	Soil turnover, biological and abiotic weathering of rock	Long-term effect on productivity of arid lands
Nutrient cycling	Internal cycling of nutrients, processing input nutrients, storage of nutrients	Potential sink for some atmospheric pollutants
Refugia	Habitat for resident and transient species	Reducing or degrading winter habitat for migratory species, loss of critical habitat for marginal species
Waste treatment	Nutrient cycling, water retention	Arid regions used for disposal of sewage sludge and other wastes from municipalities, changes in suitability for commodity production (e.g. domestic stock)
Food production	Production of harvestable plants, energy and nutrient supply for domesticated herbivores (energy flow)	Loss of potential food and fiber plants, reduction in livestock production
Raw materials	Maintenance of biodiversity	Loss of potential medicinals from plants and animals, loss of plant species with potential as food or fiber
Recreation	Maintenance of biodiversity	Loss of aesthetically pleasing open space, loss of recreational opportunities such as bird watching
Genetic resources	Biodiversity	Potential loss of genetic materials that could improve resistance to pathogens or insect pests, loss of potential medicinals

materials must be harvested from local regions and processed by the human populations within the region. This is a critical issue for countries where most of the land area is arid or semiarid. One of the more successful programs for development of new crops for arid zones has been at the Ben-Gurion University of the Negev. That program was designed to attract settlement of the desert by providing means of production that offered sufficient income to support a reasonable standard of living within the constraints of limited water supply (Pasternak *et al.*, 1986). Some desert plants are the basis for cottage industries in a number of countries. Among the most economically important species are the Cactaceae. One species (*Opuntia ficus-indica*) was an important commercial plant in the Mediterranean region of Europe and North Africa during the 18th century where it was cultivated to grow cochineal insects for a source of dye. The genus *Opuntia* is widely used as fodder, forage, fruit, and as a green vegetable especially in Mexico but also in many other parts of the world (Russell and Felker, 1987).

Cactaceae use crassulacean acid metabolism which can be up to five times more efficient in converting water into dry matter than the most efficient grasses. Some of the domesticated strains of *Opuntia* have fresh fruit yields of 8000–12,000 kg ha^{-1} yr^{-1} and dry matter production of 20,000–50,000 kg ha^{-1} yr^{-1} (Russell and Felker, 1987). The fruits of *Opuntia* cultivars are sold in markets in the US, Chile, Mexico, Brazil, North Africa, and Mediterranean countries. The young cladodes (pads) are used extensively as a fresh green vegetable in Mexico and the border states of the US. Research on methods of processing cactus fruits and cladodes has documented a number of food-processing technologies that can increase the shelf life of cactus products. The fruits can be processed into juices, marmalades, gels, and liquid sweetener. The cladodes can be pickled and processed into food additives (Saenz, 2000). Other uses of cactus fruits and cladodes, such as food coloring and medical uses, are currently being explored.

In Mexico and Argentina, *Opuntia* spp. are planted in plantations, often on marginal lands, where they are harvested for human food and for animal fodder. Cactus production was found to be feasible in a direct browsing system on a 100 ha plantation in a 300-mm rainfall region and on a 50 ha plantation in a 400-mm rainfall region. It was estimated that the establishment costs of such plantations would increase the investment costs of the stockman by 7.4–10% (Guevara *et al.*, 1999). Local dairy operators consider cactus pad fodder important for lactation of their cows and for the quality of the milk they produce (Russell and Felker, 1987). These authors suggest that prickly pear (*Opuntia* spp.) can be grown as a fodder crop on lands that are marginal for grain production. *Opuntia* can be successfully intercropped with plants, such as mesquite (*Prosopis* spp.) that enrich the soil and increase the productivity of the prickly pear. Because prickly pear are important forage for game animals, such as deer (*Odocoileus virginianus*), javelina (*Pecari tajacu*) and several species of game birds in arid and semiarid regions, they provide the basis for recreational hunting. In Texas, prickly pear–mesquite rangeland produces two to three times more income per hectare from recreational hunting permits than from cattle grazing (Russell and Felker, 1987).

There are many other desert plants that could be cultivated to produce food and fiber in arid environments. Becker (1986) identified several plants native to the Sahel that are important as 'wild' food for the humans in the region. These plants provide important sources of vitamins and minerals. Many species of trees and shrubs produce fruits and seeds that are edible. These species are more reliable as sources of food than are annual plants. Several of the species listed by Becker (1986) appear to be good candidates for plantation or orchard crop plants.

The large volume of brackish or saline water that is available in some arid regions is a resource that can potentially be utilized for production of food crops (O'Leary, 1987). The most promising 'new' crops that could be produced using saline water are halophytic seed crops. The foliage of halophytic species such as *Atriplex* spp. contain salt as a high percentage of the dry mass. Livestock production utilizing the foliage of halophytic species requires fodder produced by fresh water in order to balance the use of the halophytic foliage. However, the salt content of halophyte seeds is no higher than that of seeds of nonhalophytic plants.

There are many species of arid zone plants that produce fibers with desirable characteristics. For example, hard fibers are tissues obtained by decorticating leaves of a variety of monocotyledonous plants. These fibers have traditionally been used in the manufacture of rope and burlap but are also pulped and used in the manufacture of specialty papers including currency, tea bags, and other products that require high tensile strength. The fiber characteristics of several genera of *Agavaceae* that occur in North American hot deserts were examined to determine their paper-making qualities (McLaughin and Schuck, 1991). Several species had fibers that were as long as or longer and smaller in diameter than the fibers of sisal (primary fiber for rope manufacture). Species in two genera, *Hesperaloe* and *Yucca* appeared to be most suitable for manufacturing paper. Another example of an arid land plant that provides fiber for a native plant industry is the palmilla (*Nolina* spp.). The leaves are split, trimmed, dried, and bundled for sale to broom-manufacturing companies. Leaves may be harvested from the same population of plants every 18–24 months which provides a sustainable harvest of an arid land plant (Nabhan and Burns, 1985).

The potential for meat production in arid and semiarid lands has not been realized because of the focus on husbandry of cattle and sheep. These domesticated species have few if any physiological adaptations to arid environments. The one exception is the camel (*Camelus* spp.) which has been used extensively as a source of milk products, fiber for clothing, meat, and for transporting materials and humans. The desertification problem is attributable in large part to the replacement of native herbivores with domestic species. This problem was eloquently addressed by Fraser Darling (1960) who said in reference to African deserts and savannas, 'to exchange the wide spectrum of 20–30 hoofed animals, living in delicate adjustment to their habitat, for the narrowed spectrum of three ungulates exotic to Africa – cattle, sheep and goats – is to throw away a bountiful resource and marvelous ordering of nature.' Fraser Darling's statement applies (with modification of number of native herbivores) to the arid lands of the America's, Australia, and Asia. There have been no organized efforts to domesticate the several species of African antelopes that are physiologically adapted to arid environments. There is growing evidence that much greater meat and fiber production can be obtained from arid-adapted animals with much less stress on the arid ecosystems than from present-day agricultural enterprises. One example is from an introduction of a desert-adapted species (the South African gemsbok, *Oryx gazella*) as a source of recreational hunting in the Chihuahuan Desert of New Mexico. From a few animals introduced into the Tulorosa Basin in the 1970s, these animals now number in the thousands and have spread into large areas off the reserve where they were originally introduced. Because *Oryx* do not use large amounts of water for thermoregulation, they are able to forage over large areas where surface water is not available. There is increasing evidence that *Oryx* in New Mexico have higher rates of reproduction and higher survivorship of calves than the source populations in Southern Africa. Arid-adapted animals like *Oryx* are not constrained by distance from surface water and therefore are able to distribute their grazing–browsing activity more evenly across the landscape than domestic cattle or sheep. Despite the success of this species in Chihuahuan Desert shrublands, there are increasing calls for elimination of the species from grazing lands because they are 'exotic'. The rationale is that *Oryx* compete with domestic livestock (which are really exotic animals in a desert environment). Because of cultural bias, it is difficult to convince the public to consider cattle and sheep as exotics in desert environments.

In Africa, cattle account for approximately two-thirds of the biomass of domestic livestock. Cattle because of their Eurasian origin have neither the physiology nor grazing behavior adapted to the arid and semiarid pastoral lands of Africa. This is especially apparent during drought when cattle are malnourished because they overgraze areas that are within their range from water (Kay, 1997). Because cattle cannot range far from water and are dependent on drought-susceptible grasses rather than on the more drought-resistant trees and shrubs, overgrazing contributes to desertification. Many of the native African herbivores are capable of moving long distances from water and are facultative browsers when grass biomass is low.

Despite the documentation of the adaptations of red and gray kangaroos to life in arid environments, most of the arid and semiarid rangelands of Australia are still devoted to cattle and sheep pastoralism. Several scientists have made the case for replacing sheep and cattle pastoralism with management and harvest of native kangaroos and other native marsupials. These animals, even in high numbers, have a much lower impact on the arid and semiarid rangelands than sheep or cattle.

Because arid-adapted animals are not tied to water sources, there is reduced impact of these herbivores in the area around water points. A reduction in activity that produces piospheres around water points is an important consideration in reducing or stopping desertification. Unless the meat production industry in the arid regions of the world dramatically changes their management systems, degradation of rangelands will continue unabated. Exchanging cattle and sheep that are cool, mesic climate-adapted species, for native or nonnative herbivores that are desert adapted could be an important step in halting and even reversing the degradation trends in the semiarid regions of the world.

In most deserts, termites appear to process a large fraction of the total litter input. Because termites are social insects, their numbers and biomass far exceeds that of other detritivorous macroarthropods in most deserts. High biomass of subterranean and epigeic termites have been reported for semiarid savanna in southern Africa (Ferrar, 1982) and large populations have been reported from Saudi Arabia (Badawi *et al.*, 1984). Johnson and Wood (1980) report on a number of species of termites that occur in the arid regions of Africa and the Arabian Peninsula. Their summary of the feeding habits of the species shows that in most of these regions, dead roots and above-ground herbaceous and woody vegetation are utilized by one or more of the species. I have observed large numbers of harvester termites (*Hodotermes mossambicus*) cutting stems of dead grasses or collecting stems and leaves of grasses in the Kalahari Desert and in parts of the Karoo Desert in South Africa. The abundance of soil dumps and foraging entrances in these areas suggest that termites probably process a large fraction of the standing dead grass and detritus in these deserts.

Because such a large fraction of the primary production in most deserts is converted into termite biomass, it would seem logical to develop ways in which termites can be converted into food for humans. The high nutrient content of termites and high rates of population growth by some species led to a proposal that termites be harvested to provide high protein supplement for chickens or other domestic stock (French, 1982). French (1982) proposed that termites could be attracted into baits that could be removed from epigeic mounds or collected from the soil surface. The termites could be emptied from the bait containers into the chicken feed trays. French's proposal could become an important component of food production in arid regions where only a small fraction of the net primary production can be converted into consumable biomass by domesticated animal species.

In South Africa where domesticated ostrich (*Struthio camelus*) are an important part of the agricultural industry, termites have been shown to be a suitable protein supplement to the vegetable matter diet typically used for raising ostrich chicks (Milton *et al.*, 1993). Utilizing termites to feed chickens or other domesticated animals can provide a mechanism for harvesting a larger fraction of the net primary production and for converting unusable cellulose into food for human consumption.

Sustainable use of arid lands by humans requires thinking 'outside the box' and a willingness to explore nontraditional uses of arid lands. Sustainable use requires that the ecosystem services provided by arid lands are maintained and not sacrificed to short-term goods production. Management of arid lands for sustainability requires an understanding of ecosystem properties and processes and how these interact with the unique conditions of arid environments and arid landscapes.

REFERENCES

Badawi, A., Faragalla, A. A., and Dabbour, A. (1984). Population studies of some species of termites in Al-Kharji oasis, central region of Saudi Arabia. *Z. Angewandte Entomol.* **97**, 253–261.

Becker, B. (1986). Wild plants for human nutrition in the Sahelian Zone. *J. Arid Environ.* **11**, 61–64.

Costanza, R., d'Arge, R., de Groots, R., Farber, S., Grasso, M., Hannon, B., Limburg, K., Naeem, S., O'Neil, R. V., Paruelo, J., Raskin, R. G., Sutton, P., and ven den Belt, M. (1997). The value of the world's ecosystem services and natural capital. *Nature* **387**, 253–260.

Ferrar, P. (1982). Termites of a South African savanna. II. Densities and populations of smaller mounds, and seasonality of breeding. *Oecologia* **52**, 113–138.

Fraser Darling, F. (1960). *Wildlife in an African Territory*. Oxford University Press, Oxford, UK.

French, J. (1982). Termites can be chicken feed. *Grass Roots* **32**, 50–51.

Guevara, J. C., Estevez, O. R., and Stasi, C. R. (1999). Economic feasibility of cactus plantations for forage and fodder production in the Mendoza plains (Argentina). *J. Arid Environ.* **43**, 241–249.

Hinman, C. W. (1984). New crops for arid lands. *Science* **225**, 1445–1448.

Johnson, R. A. and Wood, T. G. (1980). Termites of the arid zones of Africa and the Arabian Peninsula. *Sociobiology* **5**, 279–293.

Kay, R. N. B. (1997). Responses of African livestock and wild herbivores to drought. *J. Arid Environ.* **37**, 683–694.

McLaughlin, S. P. and Schuck, S. M. (1991). Fiber properties of several species of Agavaceae from the southwestern United States and northern Mexico. *Econ. Bot.* **45**, 480–486.

Milton, S. J., Dean, W. R. J., and Linton, A. (1993). Consumption of termites by captive ostrich chicks. *South Afr. J. Wildl. Res.* **23**, 58–60.

Nabhan, G. B. and Burns, B. T. (1985). Palmilla (*Nolina*) fiber: a native plant industry in arid and semi-arid U.S./Mexico borderlands. *J. Arid Environ.* **9**, 97–103.

O'Leary, J. W. (1987). Halophytic food crops for arid lands. In E. F. Aldon, C. E. Gonzales-Vincente, and W. H. Moir (eds), *Strategies for Classification and Management of Native Vegetation for Food Production in Arid Zones*, pp. 1–5. General Technical Report RM 150, USDA Forest Service, Rocky Mountain Forest and Range Experiment Station, Ft. Collins, Co.

Pasternak, D., Aronson, J. A., Ben-Dov, J., Forti, M., Mendlinger, S., Nerd, A., and Sitton, D. (1986). Development of new arid zone crops for the Negev Desert of Israel. *J. Arid Environ.* **11**, 37–59.

Russell, C. E. and Felker, P. (1987). The prickly-pears (*Opuntia* spp., Cactaceae): a source of human and animal food in semiarid regions. *Econ. Bot.* **41**, 433–445.

Saenz, C. (2000). Processing technologies: an alternative for cactus pear (*Opuntia* spp.) fruits and cladodes. *J. Arid Environments* **46**, 209–225.

Index

A

Aardvark (*Orycteropus afer*), 100
Acacia, 22, 38, 103, 189, 190, 206, 254, 260
  see also Mulga woodland
Acacia aneura, 34, 200
  see also Mulga woodland
Acacia caven, 254, 277, 300
Acacia constricta, 113
Acacia erioloba, 101
Acacia karoo, 287
Acacia schaffneri, 104
Acari, soil meso/microfauna, 221
  anhydrobiosis/cryptobiosis, 247
  decomposition activities, 248
  soil quality indicator, 314
Actinomycete nitrogen fixation, 260
Adaptations, 123–151
  amphibians, 126–129
  annual plants, 124–125
  arthropods, 131–133, 142–143
  avoidance of extremes, 123–133
  birds, 131, 150–151
  carnivores, 150
  mammals, 129–131, 143–150
  perennial plants, 125–126, 133–151
  physiological/morphological, 133–151
  reptiles, 129, 141–142
  soil mesofauna, 143
Advanced Very High Resolution Radiometer
  (AVHRR) remote sensing system,
  314–315
Aedes, 223
Aeolian features, 37–38
  classification, 38
Aepyceros melampus (impala), 189
Agavaceae, 322
Agavae, 105
Agave deserti, 138
Algal crusts
  nitrogen fixation, 259
  resistance to wind erosion, 68

  soil nutrient enrichment, 259–260
  see also Crytogamic crusts
Alkaloids, 185, 186, 192, 195
Alluvial fans, 25–30
Ambrosia, 140, 174
Ambrosia deltoidea, 107
Ambrosia dumosa, 105, 107
Ammonia volatilization, 262–263
Amphibians, 223, 224
  adaptations, 126–129
Anabasis, 22, 138
Anhydrobiosis, 143, 247
Animal activities
  disturbance patches, 110–117, 184, 214
    nutrient heterogeneity, 253
    temporal changes, 118
  pits as seed traps/germination sites, 98–99,
    100–101, 184
  rainfall infiltration effects, 82–84, 184
  seasonality of reproduction, 48
  seed dispersal, 101–102
  soil disturbance, 110
    macropore formation, 82–84, 110
    wind erosion susceptibility, 68–69, 70
Animal adaptations *see* Adaptations
Animal indicators for monitoring/assessment,
  314
Annual above ground net primary production,
  157
  comparisons with mesic ecosystems,
    160–161
  landscape relationships, 166–170
  measurement, 158–159
  models, 163–166
  rainfall relationship, 163–164, 165, 166,
    167, 168, 170, 171–172
Annual plants
  adaptations, 124–125
  rainfall seasonality relationships, 50
  rooting systems, 254
  seed dormancy polymorphism, 124–125

Antelope, 323
Anthyllis henoniana, 140
Antioxidant feeding deterents, 186
Ants, 198, 199, 201–202, 217, 268
  behavioural adaptations, 133
  granivory, 207, 208, 211, 212, 213
  impact on ecosystem structure/processes,
    213–215
  indicators for monitoring/assessment, 313,
    314
  lizard predation, 218
  omnivory, 185
  plant productivity impact, 113
  soil effects, 113, 114–117
    macropore formation, 84, 110
    nutrients, 110, 113
Aplomodo falcon (*Falco femoralis*), 295
Apparent competition, 185
Arabian desert, 21–22
  aridity index (I_b), 46
  insect communities, 198, 199
  seasonality, 49, 51
Arachnids, 133
Arenivaga, 131
Arid regions, 1
  definition, 12
  water balance parameters, 46–48
Aridisols, 15
Aridity
  classification, 12
  definition, 11–12
Aridity index (I_b), 46–47
Aristida, 141
Aristida pungens, 166
Arroyos see Ephemeral streams
Artemisia, 22, 314
Artemisia campestris, 140
Artemisia herba-alba, 140, 168, 169
Artemisia monosperma, 37
Artemisia tridentata (sagebrush), 107, 188
Arthropods
  adaptations, 131–133
  canopy insect communities, 192–198
  herbivores
    leaf chewing, 190
    sucking, 193, 194
  mycorrizal spore dispersal, 268
  plant productivity impact, 113
  secondary production, 182
  soil effects, 113, 114–117
    macropore formation, 87, 110
  soil fauna, 220–221
  soil microfauna

decomposition activity, 247–248, 251
  nitrogen mineralization, 264
  soil quality indicator, 314
  surface active, 198–202
Asphodelus ramosus, 196
Assemblages, 10
Assessment systems *see*
    Monitoring/assessment systems
Astragalus, 22, 186, 260
Atalaya hemiglauca, 35
Atmospheric layer, hydrological landscape, 73
Atriplex, 29, 76, 322
Atriplex canescens, 266, 269
Atriplex confertifolia, 53, 149, 162
Atriplex polycarpa, 29
Atriplex vesicaria, 281
Australian desert
  aridity index (I_b), 48
  desertification, 276, 279, 287
  desertified landscape rehabilitation, 298
  drought, 44–45
  dune field vegetation, 39
  lizard communities, 215
  monitoring/assessment systems, 306, 307,
    309
  nutrient availability, 236
  rodent granivory, 209, 210, 211
  seasonality, 49
  surface active arthropods, 199–200
Australian dotterel, 151
Australian netted dragon (*Cnemidophorus
    nuchalis*), 217
Autoecological hypothesis, 12–13, 14
Azobacter, 259
Azospirillum, 259, 261
Azotobacter, 261

B
Baccharis, 195
Bacterial decomposition, 249
Badger (*Taxidea taxus*), 206
Bajadas, 25–30
Banded vegetation patterns, 33, 118–119
Banner-tail kangaroo rat (*Dipodomys
    spectabilis*), 129, 288, 313
Barkhan (crescent-shaped) dunes, 38
Basins, 21, 22, 33–36
Bat-eared fox (*Otocyon megalotis*), 100
Bedouin goat, 147
Beetles, 133, 198, 199, 217
Behavioural adaptations, 123
  arthropods, 131–133
  birds, 131

mammals, 129–131
Below-ground food webs, 220–222
Below-ground productivity, 162
Bettongia lesuer, 111
Bioclimatic aridity indices, 11
Biodiversity, 310, 313
    adaptations to regional climate, 45
    desertification impact, 293–295
Biomass estimates, 157–159
    root biomass, 162
Biome, 2
    adaptations to seasonality, 45
Biosphere, 2
Biotic interactions, 6–7, 17
Biotic–abiotic interactions, 9–10, 17
Birds, 313
    adaptations
        behavioural, 131
        physiological, 150–151
    granivory, 207
    nomadism, 131
    omnivory, 185
    secondary production, 181
    seed/fruit dispersal, 99, 101, 206
Black grama (*Bouteloua eriopoda*), 69, 117, 141, 235–236, 284, 297
Black-tail jackrabbit (*Lepus californicus*), 186, 188
Body temperature lability, large mammals, 144–145
Bootettix punctatus, 181
Boulder-strewn surfaces, 22, 36
Bouteloua eriopoda (black grama), 69, 117, 141, 235–236, 284, 297
Brachystomella arida, 200
Bradyhizobium, 260
Brain countercurrent heat-exchange system, 145
Brickellia lanciniata, 32, 257
Brittlebush (*Encelia farinosa*), 126
Bromus rubens, 288
Bromus tectorum (cheat grass), 287, 314
'Brousse tigree' vegetation, 33
Browsing, 186, 188–190, 191
    plant responses, 204
Bruchid beetles, 207
Bufo cognatus, 224
Bugs, 217
Bunch grasses, 5, 6
    shield/platform desert, 22
    weather interface, 5
Burr sage (*Ambrosia*), 105, 107
Burr sage (*Franseria*), 29, 76

Burrowing activities, 110, 184
    amphibians, 126
    arthropods, 131
        spiders/scorpions, 200
    mammals, 129–130
    nutrient transport, 110, 253
Buttes, 24

C
C_3 photosynthetic pathway, 134–135, 138, 141
C_4 photosynthetic pathway, 134–135, 138, 141
Cache pits, 100, 111, 214
Cactaceae, 137, 138
    commercial exploitation, 321
Cactus wren, 101, 102
Calcium, 117
Calcrete, 36
CAM (crassulacean acid metabolism)
        photosynthetic pathway, 134, 137, 138, 321
Camel (*Camelus*), 143, 144, 146–147
Camel cricket, 198
Canis latrans (coyote), 150, 185, 206, 288
Canis lupus (lobo wolf), 288
Canopy
    insect communities, 192–198
    morphology, 5
Cape fox (*Vulpes chama*), 100
Cape ground squirrel (*Xerus inauris*), 130
Cape hare, 186
Capsodes infuscatus, 196
Carbohydrate exudation, 5
Carbon–nitrogen ratios, 238
Carnegia, 138
Carnegia gigantea (saguaro), 27, 29, 105, 107, 109, 288
Carnivores
    adaptations, 150
    omnivory, 185
    seed dispersal, 206
Cassida, 195
Casuarina cunninghamiana, 260
Catenas
    fluvial transport, 84–85
    see also Watersheds
Cell, 2
Celtis pallida, 99
Cenchrus ciliare, 288
Centipedes, 132, 217
Ceratoides lanata (winterfat), 53, 162, 188
Cercidium (palo verde), 105, 173
Cercidium microphyllum, 29, 80, 107, 109
'Chaco' woodland, 166

Cheat grass (*Bromus tectorum*), 287, 314
Chenopod shrubland, banded vegetation
  patterns, 33
Chihuahuan Desert
  bird-mediated seed dispersal, 101–102
  decomposition processes, 244, 245
  desertification changes, 284, 286
  ephemeral plant seed production, 124
  ephemeral pond/lake fauna, 223, 224, 225,
    226
  ephemeral streams, 32, 92
  granivory, 207–208
  herbivory, 204
  indicators for monitoring/assessment,
    312–313
  multi-species patches, 104–105
  nitrogen availability, 235
  productivity, 161, 163, 166, 177
  rainfall, 56, 57
    spatial variability, 52
  seasonality, 46, 50, 51
    animal reproduction, 48, 50
  secondary production/herbivory, 182, 186,
    196
  seed densities, 98
  surface active arthropods, 198, 199, 200,
    201
  termite activities, 83, 117, 249, 250
  vegetation, 32, 74, 79–80
  wind erosion, 69, 71
Chilopsis linearis (desert willow), 32, 257
China
  desertified landscape rehabilitation, 299
  drought, 44
Chiromantis, 127
Chottes *see* Ephemeral ponds/lakes
Chrysothamnus viscidiflorus (green
  rabbitbush), 188
Chuckwalla (*Sauromalus obesus*), 141, 142
Cladodes, 137
Clam shrimp (*Eulimnadia texana*), 223
Climate, 43–61
  seasonalality, 45–46, 48–52
  spatial effects, 52–53
  water balance parameters, 46–48
  see also Rainfall
Climate change, desertification processes,
  290–291
Clostridium, 261
Cnemidophorus, 217
Cnemidophorus tigris, 216
Cockroaches, 198, 199, 217
Cocoon formation, 127

Collared peccary (*Pecari tajuca*), 206, 321
Collembola, 143, 200, 221
  anhydrobiosis, 247
Colorado River, 37
Commensalism, 107, 109
  nurse plant relationships, 107, 109
Community, 2–3, 7
Compensatory growth, 188, 191
Competition, 10, 17, 106–110
  apparent, 185
  patch structure effects, 97
Consumers, 181–185
Convectional heat flux, 144
Convectional rainfall, 52
Coppice dunes, 276, 286
Corvus cryptoleucus (wite-necked raven), 150
Cottontails, 186
Coyote (*Canis latrans*), 150, 185, 206, 288
Creosotebush (*Larrea tridentata*), 3, 29, 51,
  57, 70, 71, 74, 75, 76, 80, 103, 105, 106,
  107, 109, 117, 198, 199, 201, 207, 217,
  235, 238, 257, 284, 290, 315
  adaptations, 135–137
  feeding deterents, 186
  herbivory, 186, 188, 193, 196, 197
  primary productivity, 173–174, 175
Crepuscular activity cycle, 130
Crescent-shaped (barkhan) dunes, 38
Crickets, 198, 199, 217, 268
Cryptobiosis, 143
Cryptogamic crusts
  disruption by animal activities, 82
  nitrogen fixation, 259
  rainfall infiltration effects, 81
  soil nutrient enrichment, 259–260
  see also Surface crusts
Cryptopygus ambus, 200
Ctenophorus isolepis, 217
Ctenophorus nuchalis (Australian netted
  dragon), 217
Ctenotus, 216, 217
Cyanobacterial nitrogen fixation, 258, 259
Cynomys ludovicianus (prairie dog), 288, 289

D
Dalea, 260
Dark septate fungal–root associations,
  268–269
Dasyochloa (Erioneuron) pulchella, 141
Decomposition, 237–252
  conceptual models, 251–252
  fragmentation processes, 237, 238, 243, 251
  litter, 238–239

rates/mechanisms of mass loss, 239–244
soil processes, 245–247, 251
mineralization processes, 237–238, 243
nitrogen mineralization, 264–265
photodegradation processes, 240, 251
rainfall relationship, 240, 243, 244, 245, 247
root materials, 245–247, 251, 264
soil microfauna activities, 247–249, 251
spatial variation, 244
termite activities, 249–251
Definition of deserts, 11–12
Defoliation, 196
compensatory growth response, 202–203, 204
Degu (*Octodon degus*), 214
Denitrification, 262, 263–266
Desert cottontail (*Sylvilagus auduboni*), 188
Desert iguana (*Dipsosaurus dorsalis*), 141, 215
Desert tortoise (*Gopherus agassizii*), 142
Desert willow (*Chilopsis linearis*), 32, 257
Desertification, 9, 275–301, 322, 324
biodiversity impact, 293–295
climate change, 290–291
definition, 275
dynamic conceptual models, 282–283
ecosystem/landscape consequences, 282–289
environmental stressors, 275, 283
grazing-related changes, 278, 279, 280–282, 283–284
history, 277–282
landscape rehabilitation, 295–301
landscape resistance/resilience, 289–290
landscape structure changes, 280
loss/replacement of forage species, 276–277, 284, 286, 287
monitoring/assessment systems, 305
social/economic consequences, 291–293
spread of nonnative species, 287–288
tree/shrub harvesting, 282
Detritus-based food webs
sand dune arthropods, 198
soil, 221, 222
Dinothrombium, 133
Dipodomys (kangaroo rat), 45, 145, 149, 213, 217–218
Dipodomys merriami, 210, 213
Dipodomys microps, 149
Dipodomys ordii, 149, 213
Dipodomys spectabilis (banner-tail kangaroo rat), 129, 288, 313

Dipsosaurus dorsalis (desert iguana), 141, 215
Dolichotis patagonum, 206
Domestic livestock, 322, 323
seed dispersal into grassland, 206
Dominant species, 3, 5, 9
longevity, 105
physiological threshold limits, 55
Dorcas gazelle (*Gazella dorcas*), 190
Dromarius novaehollandiae (emu), 206
Drought, 43, 312
change in habitat use, 45
decomposition processes, 244
El Niño–Southern Oscillation, 44
impact on humans, 276
recovery period, 5–6
simulation experiments, 175–176
tropical sea surface temperature relationship, 43, 44
Drought-deciduous perennials, 125–126
Dryfall, 261–262
Dust storms, 65

E
Earthworms, 220
soil macropore formation, 82
Echinocereus, 109
Economic impact of desertification, 276, 277, 291–293
Ecosystem, 2, 3
definition, 6–9
desertification consequences, 282–289
function, 7–8
health indicators, 307, 308–309
integrity, 310
structure, 2, 3–4, 7
Ecosystem engineering, 184
Ecosystem processes, 6, 16–17
consumer effects, 184–185
human management issues, 8
monitoring/assessment, 309, 310
scaling problems, 18
El Niño–Southern Oscillation
long-term records, 45
rainfall effects, 43–45
Embryonic diapause, 147
Emu (*Dromarius novaehollandiae*), 206
Encelia farinosa (brittlebush), 126
Energy flow, 6, 7, 16, 181
consumer effects, 184, 185
Entomobrya, 200
Ephedra torreyana (Torrey ephedra), 186
Ephedra trifurca (mormon tea), 186

Ephemeral plants
  nitrogen-fixing legumes, 255
  patchy distribution, 99
  productivity, 167, 169
  rainfall seasonality correlations, 50
  seed germination delay/inhibition, 124–125
  see also Annual plants
Ephemeral ponds/lakes, 90–92
  aquatic fauna, 222–226
  flooding cycle, 92
  food webs, 92
Ephemeral streams, 30–32, 37, 166
  stream bed vegetation, 31–32
  transmission losses, 31, 32
Episodic events, 88–90
Eragrostis lehmanniana (Lehmann's love
    grass), 141, 213, 214, 281, 294, 297, 314
Eremias lineoocellata, 215, 216
Eremias lugubris, 215, 216
Erioneuron pulchellum (fluff-grass), 235
Eucalyptus (mallee), 73, 75, 76, 267
Eulimnadia texana (clam shrimp), 223
Euphorbiaceae, 137
Evaporative heat loss, 144
Evaporative water loss, granivorous rodents,
    149

F
Facilitation, 109–110
Fagonia, 22
Falco femoralis (Aplomodo falcon), 295
Fallugia paradoxa, 257
Fat stores, 142, 144
Feedbacks, 8
Feeding deterents, 185–186, 189, 192, 193, 195
Felis concolor (mountain lion), 288
Ferocactus acanthodes, 107, 138
Festuca, 259
Festuca pallescens, 172
Fiber production, 322
Field capacity, 15
Fire, 288
Fire ants (*Solenopsis xyloni*), 208
Flatlands, 33–36
Flood, 88–89
Floodplains, 34, 37
Flourensia cernua (tarbush), 33, 73, 74, 76,
    186, 190, 192, 193, 196, 257
Fluff-grass (*Erioneuron pulchellum*), 235
Fluvial transport, 84–85
Food crops production, 319, 321–322
Food webs, 11, 185
  below-ground, 220–222

ephemeral ponds/lakes, 92
Foquieria, 105
Foquieria splendens (ocotillo), 125, 257
Foraging pits, 100, 111, 253
Formica perpilosa, 113
Fox, 130
  omnivory, 185
Fragmentation processes, 237, 238, 243
  photodegradation, 240
Frankia, 260
Franseria (burr sage), 29, 76, 190
Franseria dumosa, 126
Frontal rainfall, 52
Frugivory, 206
Functional groups, 313
  primary production, 172–177
Fungal decomposition, 249

G
Gazella dorcas (Dorcas gazelle), 190
General circulation models, 44
Geoffroea decorticans, 277
Geolycosa rafaelana, 200
Geomorphology, 17, 21
  productivity effects, 168
  see also Landform
Germination delay/inhibition, 124
Gilgai relief, 36
Gnathamitermes tubiformans, 133, 249
Goannas (varanid lizards), 101
Gobi Desert, orographic effects, 53
Goods and services provision, 8, 319, 320
Gopherus agassizii (desert tortoise), 142
Granivory, 206–212
  ant–rodent interactions, 213, 214–215
  impact on ecosystem structure/processes,
    213–215
Grasses
  adaptations, 141
  dune vegetation, 38
  herbivory, 190
    compensatory growth response, 202–203,
      204
    impact of kangaroo rat seed consumption,
      213–214
  photosynthetic pathways, 141
  rainfall infiltration/run-off, 79, 80
  rooting systems, 254
Grasshopper, 190–191, 193, 217, 268
Grasshopper mouse (*Onychomys leucogaster*),
    209
Grassland
  banded vegetation patterns, 33

conversion to coppice dunes, 276
granivory, 208
invasion by shrubs (desertification changes),
276–277, 278, 279, 280–281, 282, 284,
286, 287
rehabilitation efforts, 295, 297
seed germination–establishment sites, 99
surface active arthropods, 199
Gravel plains, 36
Grayia, 140
Grazing, 202–206
associated desertification changes, 278, 279,
280–282, 283–284
optimization conceptual model, 202–203
soil effects, 202, 204–205
species preferences, 203, 204
vegetation effects, 202–204
Great Basin Desert
desertification changes, 287–288
herbivory, 188
indicators for monitoring/assessment, 314
orographic effects, 53
productivity, 163
rodent-related seed traps/germination sites,
99
seasonal climate, 46
vegetation, 53, 162
Green rabbitbush (*Chrysothamnus
viscidiflorus*), 188
Grey fox (*Pseudalopex griseus*), 206
Grey fox (*Urocyon cinereoargenatus*), 288
Gross rainfall, 72
Gular flutter, 151
Gutierrezia (snakeweed), 76, 186
Gypcrete, 36

H
Hairy-nosed wombat (*Lasiorhinus latifrons*),
111
Halogeton glomeratus, 287
Haloxylon, 138
Hamadas, 36
Hammada, 22, 138, 167, 168
Hare, 130
Herbivory, 17, 185
canopy insect communities, 192–198
foliage quality determinants of abundance,
197
compensatory plant growth, 188, 191
foliage chewers/browsers, 185–191
large grazers, 202–206
patch temporal changes, 118
physiological adaptations

mammals, 143, 144–145
reptiles, 141, 142
plant deterrent spinesence, 189
plant tissue water content relationship, 188,
189, 190
plant toxins/feeding deterents, 185–186,
189, 192, 193, 195
protective nurse plant associations, 109
rainfall infiltration/run-off effects, 82
seed dispersal, 206
sucking arthropods, 193, 197
Hesperaloe, 322
Heuweltjies, 111
Hilaria mutica (tabosa grass), 33, 69
Hilaria rigida, 107, 109
Hillslopes, 22–24
Hodotermes mossambicus, 324
Horned lizard (*Phrynosoma*), 129, 218, 220
Hydraulic lift, 138–139
Hydric stress, soil surface/subsurface, 58, 61
Hydrological landscape, 73
Hyper-arid regions, 1
ant communities, 201–202
decomposition of buried materials, 245
definition, 12
ephemeral stream vegetation, 32
productivity, 170–171
Hyperolius, 127
Hystrix austro-africanae (porcupine), 100

I
Ibex, 100
Ifloga spicata, 99
Impala (*Aepyceros melampus*), 189
Indian desert
drought, 44
seasonality, 49
Indian rice-grass (*Oryzopsis hymenoides*), 99
Indicators, 306, 310–315
abundance/species richness, 313–314
animals, 314
criteria, 311
qualitative, 307, 308–309
sensitivity to disturbance/stress, 312
Indus, 37
Infiltration of water, 72, 78–84
animal activities/macropore formation,
82–84, 110
arthropod activity effects, 113
termites, 251
catenas/watersheds, 85
ephemeral stream channels, 92
measurement, 309

Inputs, 6, 7
Interception rainfall, 72, 73–76
Interception zones, 118–119
Intermediate disturbance hypothesis, 185
Intraguild predation, 185
Invertebrate adaptations, 142–143
Iranian Desert, aridity index (I_b), 46
Ironwood (*Olyneya tesota*), 29, 109
Irrigation experiments, 175–176
Isomeris aborea, 126

J
Jackrabbit (*Lepus*), 100, 130, 138, 186, 188, 206
Jatropha dioica, 105
Javelina (*Pecari tajacu*), 206, 321
Jet stream flow, 44

K
Kalahari Desert
  bird-mediated fruit dispersal, 101
  dune vegetation, 38
  lizard communities, 215
  seasonality, 49, 51
  wind erosion, 67–68
Kangaroo, 100, 147, 323
Kangaroo rat (*Dipodomys*), 45, 145, 213, 217–218
reproduction, 149
Karoo
  animal-produced pits, 100
  ant granivory, 211
  monitoring/assessment systems, 306, 307, 309
  seasonality, 49, 51
  shrub herbivory, 189
  shrub invasion, 287
Kelso sand dunes, 38
Keystone predation, 185
Keystone species, 294, 313
Kit fox, 150
Krameria, 140
Krameria parvifolia, 175
Kudu (*Tragelaphus strepsiceros*), 189

L
Lactation, 147, 149
Lagostomus maimus, 206
Lake Eyre, 37
Landform, 17, 21–40
Landform stability, 40
Landscape, 2, 3, 4, 5, 9–10

desertification consequences, 282–289
  hydrological, 73
  nutrient ditribution patterns, 252–257
  water balance, 72–73
Landstat Thematic Mapper, 315
Larrea, 140, 190, 193
Larrea cuneifolia, 186, 193
Larrea tridentata see Creosotebush
Lasiorhinus latifrons (hairy-nosed wombat), 111
Leaf adaptations, 123, 136
Leaf growth, 140
Leaf nutrient resorption, 257–258
Leaf size, 123
Leaf succulents, 138
Legume nitrogen fixation, 260
Lehmann's love grass (*Eragrostis lehmanniana*), 141, 213, 214, 281, 294, 297, 314
Lepidium lasiocarpum, 124
Lepus (jackrabbit), 100, 130, 138, 186, 188, 206
Lepus californicus (black-tail jackrabbit), 186, 188
Lichen crusts
  disruption by animal activities, 82
  nitrogen fixation, 259
  rainfall infiltration effects, 81
  soil nutrient enrichment, 259–260
  see also Crytogamic crusts
Life form, 5
Life history, 5
  adaptation to seasonality, 45
Lignin, 238, 239, 240, 245, 249
Linear (longitudinal) dunes, 38, 39
Litter
  accumulation around shrubs, 70–71, 103
  dams/microterraces caused by run-off, 86–87
  decomposition, 238–239
    rates/mechanisms, 239–244
    soil processes, 245–247
  measurement for monitoring/assessment, 309
  termite removal, 249
  water content, 58
Lizards, 129
  desert communities, 215
  herbivores, 215
  patch refugia, 54
  physiological adaptations, 141
  predators, 215–220, 252
    prey, 216–217

soil enrichment, 252
  rainfall relationships, 217–218
  secondary production, 181, 183
Lobo wolf (*Canis lupus*), 288
Local extinction, 4
Loess accumulation, 39
Long-term measurements, 90
Longitudinal (linear) dunes, 38, 39
Lophocereus schottii (senita), 27
Lotus, 260
Lupinus, 260
Lycium, 22, 140, 174
Lygus, 193

M
Macropores, 82–84, 87, 110, 251
Macropus robustus, 147
Macropus rufus (red kangaroo), 147
Magnesium, 117
Mallee (*Eucalyptus*), 73, 75, 76, 267
Mammalaria, 109
Mammals
  adaptations
    behavioural, 129–131
    physiological, 143–150
  burrows, 129–130
  crepuscular activity cycle, 130
  granivory, 207, 208–210
  omnivory, 185
  patch refugia, 54
  secondary production, 181, 182, 183
  soil disturbance patches (pedturbation), 110,
    111–112
Management strategies, 319
Markovian Dynamics, 105–106
Marsubium vulgare, 281
Marsupial reproduction, 147
Meat production, 319, 322, 323, 324
Mephitis (skunk), 288
Meriones crassus, 212
Mesas, 24
Mesquite *see Prosopis*; *Prosopis glandulosa*;
    Prosopis juliflora
Metabolic water production, 145
  granivorous rodents, 149
Microclimate, 58–61
  soil, 247
Migratory locust (*Schistocerca gragaria*),
    196
Millipedes (myriopods), 131
Mineralization processes, 237–238, 243
  soil microfauna activities, 247–249
Mites *see* Acari

Moina wierzejski, 223, 225
Moisture index (*I*), 46
Mojave Desert
  alien species expansion, 288
  herbivory, 186
  productivity, 164, 165, 174, 175
  seasonal climate, 46
  seasonality of animal reproduction, 48,
    50
  seed densities, 98
  vegetation, 107, 109, 140, 162
Mole rats, 111
Mollisols, 15
Moloch horridus, 220
Monitoring/assessment systems,
    305–315
  data, 309
  definitions, 306
  existing systems, 306–309
  indicators, 306, 307, 308–309,
    310–315
Mormon tea (*Ephedra trifurca*), 186
Morphological variation, single-species
    patches, 103–104
Mosaics, 3, 5, 9, 97, 104–106
  grazing effects, 204
Mosquitoes (*Aedes*), 223
Mountain and basin deserts, 21, 22
Mountain lion (*Felis concolor*), 288
Mountains, 22–24
  orographic effects/rainfall spatial variability,
    53
Muhlenbergia porteri, 3
Mulga woodland, 73, 75, 101, 200
  arthropod activities, 110
  banded vegetation patterns, 33, 34
  ephemeral lakes, 92
  productivity, 160–161
Mulhenbergia porterii, 109
Mycorrhizae, 266–269
Myriopods (millipedes), 131
Myrmecocystus, 218

N
Namib Desert, 40
  ant granivory, 211
  aridity index (*I*$_b$), 47
  decomposition of buried materials, 245
  herbivory, 186
  productivity, 171
  rainfall, 56
  seasonality, 49
  surface active arthropods, 201

Negev Desert
decomposition processes, 243–244
episodic events frequency, 88
herbivory, 190, 196
loess accumulation, 39
productivity, 167, 168
rainfall, 52, 56
surface active arthropods, 200–201
vegetation, 51
Nematodes, 221, 264
adaptations, 143
anhydrobiosis, 247
decomposition activity, 248, 251
Neotoma (woodrat), 186, 189
New Mexico
desertification changes, 278
orographic effects/rainfall spatial variability,
53
rainfall predictability, 54
wind dispersal of seeds, 98
Nile, 37
Nitrogen availability, 14, 16–17, 235–236, 264
arthropod activity effects, 113, 117
mycorrhizal associations, 268
nitrogen-fixing legumes, 254–255
primary productivity relationship, 175, 176
soil organic matter relationship, 255, 257
Nitrogen cycle, 258–266
ammonia volatilization losses, 262–263
denitrification, 262, 263
measurement of processes, 309
nitrogen mineralization/immobilization
rates, 264–266
Nitrogen fixation, 258–262
atmospheric inputs, 261–262
free-living bacteria, 258, 261
root nodule symbionts, 254, 258, 259, 260
Nitrogen mineralization, 264–266, 309
Nitrogen resorption, 257, 258
efficiency, 136
Nitrogen-fixing legumes, 254–255, 260
Nolina (palmilla), 322
Nomadism in birds, 131
North Africa, aridity index (I_b), 46, 47
North American desert
desertification, 278–279
desertified landscape rehabilitation,
295–297
ephemeral lakes, 90, 92
lizard communities, 215
rodent granivory, 210, 211
Notiosorex crawfordii, 209
Nurse plants, 107, 109–110

environmental modification, 105
Nutrient levels
animal disturbance of soil, 110
arthropod activity effects, 113, 117
distribution patterns, 252–257
Nutrient limitation, 235–236
productivity modelling, 164
rain use efficiency, 162
Nutrient resorption, 257–258
drought deciduous perennials, 125–126
Nutrient retention, 310
Nutrient/materials cycling, 6, 8, 16

O
Oberlander index (aridity index; I_b), 46–47
Ocotillo (*Foquieria splendens*), 125, 257
Octodon degus (degu), 214
Odocoileus virginianus (white-tailed deer),
206, 321
Olyneya tesota (ironwood), 29, 109
Omnivory, 185
Onychomys leucogaster (grasshopper mouse),
209
Onychomys torridus, 209
Opuntia, 101–102, 105, 138, 206
commercial exploitation, 321
Opuntia ficus-indica, 321
Opuntia fulgida, 109
Opuntia leptocaulis (pencil cholla), 109
Opuntia polyacantha, 56
Opuntia rastera, 105
Opuntia streptacantha, 104, 105
Opuntia violacea, 188
Organ-pipe cactus (*Stenocereus thruberi*), 27
Organic material redistribution
accumulation in animal-produced pits,
100
ant activities, 113
ephemeral lakes, 92
run-off, 86–87
termite activities, 117
wind erosion, 70
Organism, 2
Organism–environment interactions, 8
Organization levels, 2
Orographic effects, 53
Orycteropus afer (aardvark), 100
Oryx (*Oryx gazella*), 143, 144, 186, 323
Oryzopsis hymenoides (Indian rice-grass), 99
Osteospermum sinuatum, 189, 195
Ostrich (*Struthio camelus*), 151, 324
Otocyon megalotis (bat-eared fox), 100
Outputs, 6, 7

P

P (precipitation)/ET (mean annual
evapotranspiration) aridity index, 11–12
Palmilla (*Nolina*), 322
Palo verde (*Cercidium*), 29, 80, 105, 107, 109,
173
Panicum, 22
Panicum obtusum, 222
Panicum turgidum, 37
Pans *see* Ephemeral ponds/lakes
Panting in birds, 150
Paper making, 322
Patches, 3, 4, 5, 6, 7, 9, 17, 97–119
animal-produced, 110–117, 184
competition, 106–107
grazing effects, 204
limiting resources location, 21
Markovian Dynamics model, 105–106
multi-species, 104–106
nutrient heterogeneity, 253
refugia, 54
scaling-up problems, 18
seed germination–establishment sites,
97–102
single species, 102–103
morphological variation, 103–104
temporal dynamics/feedbacks, 117–119
Pavements, 36–37
Pecari tajuca (collared peccary; javelina), 206,
321
Pencil cholla (*Opuntia leptocaulis*), 109
Percolation, 72
Perennial plants
adaptations, 133–151
drought-deciduous, 125–126
functional groups, 137
root systems, 254
Pericopsis, 260
Pheidole, 208
Pheidole rugulosa, 213
Phenolic compounds, 192, 238
Phosphorus availability, 236, 266
arthropod activity effects, 113, 117
mycorrhizal associations, 268
resorption efficiency, 257
Photodegradation processes, 240
Photosynthesis, 139
pathways, 123, 134
Phreatic layer, 73
Phrynosoma (horned lizard), 129, 218, 220
Phrynosoma cornutum, 129, 218, 220
Phrynosoma modestum, 218
Phyllomedusa, 128

Physical conditions, 7
Physiological threshold limits, 54–55
Piedmonts, 24–30
Piospheres, 204–205, 323
Plant adaptations *see* Adaptations
Plant toxins/feeding deterents, 185–186, 189,
192, 193
Plant–weather interactions, 5
Plaruroctonus mesaenis, 201
Platform desert vegetation, 22
Playas *see* Ephemeral ponds/lakes
Plectrachne schinzii, 39
Pogonomyrmex, 129, 218
Pogonomyrmex barbatus, 113
Pogonomyrmex desertorum, 208
Pogonomyrmex occidentalis, 268
Pogonomyrmex rugosa, 113
Polypedilum vanderplanki, 143
Polyphenols, 189
Population, 2, 3
Porcupine (*Hystrix austro-africanae*), 100
Porcupine grass (spinifex), 39
Potassium
reptile physiological adaptations, 141, 142
soil levels, 113
Potential evapotranspiration (PET):mean
rainfall ratio, 51–52
Prairie dog (*Cynomys ludovicianus*), 288, 289
Predators, 17, 185
lizards, 215–220
poisoning campaigns for livestock
introduction, 288
rodent prey, 212
surface active arthropods, 199–200, 201
Primary production, 157–177, 310
below-ground, 162
comparisons with mesic ecosystems,
160–161
desertification effects, 276
extreme deserts, 170–171
grazing response, 202–203
landscape relationships, 166–170
net, 157
measurement, 157–159
nutrient limitation, 162, 235–236, 237
rainfall relationship, 51–52, 163–164, 165,
166, 167, 168, 170, 171–172
experimental approaches, 175–176
species/functional groups, 172–177
see also Annual above ground net primary
production
Prong-horned antelope, 186
Prosopis, 103, 105, 173, 254, 260

Prosopis flexuosa, 166, 206
Prosopis glandulosa (mesquite), 32, 33, 39, 51, 73, 76, 99, 107, 113, 118, 140, 177, 189, 191, 196, 198, 206, 207, 255, 257, 258, 276, 280, 284, 290, 294, 315
Prosopis juliflora, 29
Prostigmatid mites, 221
    decomposition activity, 248
Protozoa, soil microfauna, 221, 264
    decomposition processes, 247, 251
Pruning activities, 186, 188
Psammomys obesus, 149
Pseudalopex griseus (gray fox), 206
Psocoptera, 221
Psorothemnus spinosus, 260
Pterocarpus, 260
Pteronia pallens, 195
Pulse-reserve paradigm, 12–14, 16, 17, 157, 170, 223

Q
Qualitative indicators, 307, 308–309

R
Rabbit, 130, 186, 188
Radiative energy flux, 144
Rain use efficiency, 161–162
Rain-splash, 77–78, 103
Rainfall, 14, 17
    aridity index (I_b), 46–47
    autoecological hypothesis, 12–13, 14
    characteristics, 17
    convectional, 52
    decomposition relationship, 240, 243, 244, 245, 247
        soil microfauna activities, 247
    duration, 55–58
    episodic storms, 88, 89
    frequency/return time, 17, 55–58
    frontal, 52
    gross, 72
    infiltration, 72, 78–84
    intensity, 17, 55–58
    interception, 72, 73–76
    nitrogen availability relationship, 14, 264
    nutrient deposition, 262
    partitioning, 72
    percolation, 72
    predictability (variability), 53–55
    productivity relationship, 163–164, 165, 166, 167, 168, 171–172
        experimental approaches, 175–176
        extreme deserts, 170–171

pulse-reserve model of ecosystem function, 13, 14, 16
    quantity, 17
    redistribution, 4, 71–77
        exchanges among landscape units, 84–86
        exchanges within landscape units, 86–88
        following loss of grass cover, 280, 284, 286
    run-off, 72, 78–84
    seasonality, 17, 45, 48–50, 51
    spatial variability, 52–53
    stemflow, 72, 73–76
    throughfall, 72, 76–77
    see also Wet season
Ranching lifestyle, 292–293
Rangelands, 1, 22, 202
    ecosystem dynamics, 282–283
    monitoring/assessment systems, 306–307
    ranching lifestyle, 292–293
Rantherium, 22
Rantherium sanveolens, 140
Red kangaroo (*Macropus rufus*), 147
Regs, 36
Remote sensing, 306, 314–315
Reptiles
    adaptations, 129
        physiological, 141–142
    fat stores, 142
    salt glands, 141
    thermoregulatory behaviour, 129
Resilience, 9, 10–11
    desertified landscape, 289–290
Resistance, 9
    desertified landscape, 289–290
Retama retam, 166
Rhizobium, 254, 258, 260, 300
Rhizosheaths, 260–261
Rhus microphylla, 257
Rills, 35
Rio Grande, 37
Rivers, 37
Roadrunner, 151
Rock-strewn surfaces, 36
    rainfall infiltration/run-off, 84
Rodents
    abundance patterns, 210, 212
        following livestock introduction, 288–289
    desertification vegetation changes, 281
    granivory, 207, 208–210, 211–212, 213
    impact on ecosystem structure/processes, 213–215
    metabolic rates, 149–150

physiological adaptations, 148–150
seasonality of reproduction, 48, 50
soil disturbance, 111
seed germination sites, 98–99
Rooting patterns/root systems, 123, 137, 138, 140
measurement for monitoring/assessment, 309
nutrient status effects, 253–254
Roots
biomass estimates, 162
decomposition, 239, 251, 264
soil processes, 245–247
depth, 138, 140, 177
Rope making, 322
Rotifers, 224
Run-off, 72, 78–84, 161, 310
catenas/watersheds, 85
ephemeral lake filling, 90, 92
exchanges among landscape units, 84–86
exchanges within landscape units, 86–88
measurement, 309
microtopographic effects, 85, 86
organic matter redistribution, 86–87
productivity relationship, 164, 166, 167
Runnels, 35, 37

S
Sagebrush (*Artemisia tridentata*), 107, 188
Saguaro (*Carnegia gigantea*), 27, 29, 105, 107, 109, 288
Sahara
ephemeral stream transmission losses, 32
lizard communities, 215
nutrient availability, 236
seasonality, 49, 51
Sahel
drought, 44
nutrient availability, 235, 236
seasonality, 49, 51
social/economic impact of desertification, 276, 277, 291–292
termite activity, 82–83
wild food plants, 322
Saline environments, 138
Salsola, 287
Salsola vermiculata, 140
Salt glands, 141, 151
Sand dunes, 37–40
detritivores, 198
microphytic crusts, 81
productivity, 177
rainfall infiltration, 78, 81

stabilization with artificial vegetation, 299
surface active arthropods, 198
wind-related effects, 67, 68
Sand features, 37–40
Sand partridge, 151
Sand plains, 22, 34–35
Sand seas, 22, 37
Sarcobatus, 29
Sauromalus obesus (chuckwalla), 141, 142
Scaling, 18
Scaphiopus (spadefoot toad), 126–127, 128, 129, 226
Scaphiopus bombifrons, 224, 225
Scaphiopus couchi, 223, 224, 225
Scaphiopus (hammondi) multiplicatus, 127, 128, 223, 224, 225
Scarabeus christatus, 198
Schistocerca gragaria (migratory locust), 196
Scleropogon brevifolia, 3, 69
Scolopendra, 132
Scorpio marus palmatus, 201
Scorpions, 198, 200–201
Scouring events, 89, 90
Sea surface temperature, 43, 44
Seasonality, 45–46, 48–52
animal reproduction, 48, 50
predictability, 53
productivity modelling, 163
vegetation effects, 50–51
Sebkas *see* Ephemeral ponds/lakes
Secondary production, 181–185, 310
Seed consumption *see* Granivory
Seed dispersal
animals, 101–102
birds, 99, 206
carnivores, 206
herbivores, 206
desertification vegetation changes, 281, 284
wind, 97–98
Seed dormancy polymorphism, 124–125
Seeds
emphemeral plants, 124
germination delay/inhibition, 124
germination–establishment sites, 97–102, 105, 214
nurse plant protective effects, 109
Seidlitzia, 138
Seira bipunctata, 200
Semiarid regions, 1
definition, 11, 12
Senita (*Lophocereus schottii*), 27
Sheep, 143, 144

Shield deserts, 21
  vegetation, 22–23
Shrubs
  alluvial fans/bajadas, 29
  chewing-browser herbivory, 186, 188–190,
    191
  deep-rooted, 138–139, 140, 177
  drought deciduous, 125–126
  ephemeral stream channel, 32
  flatlands, 34, 35
  grassland invasion (desertification changes),
    276–277, 278, 279, 280–281, 282, 284,
    286, 287
    rehabilitation efforts, 295, 297
    resistance/resilience, 289–290
  hydraulic lift, 138–139
  leaf growth phenology, 140
  litter accumulation, 70–71, 103
  morphological impact on stemflow, 103
  patch development, 99
  physiological adaptations, 138
  productivity, 172, 173
  rainfall infiltration/run-off, 80–81, 86
  root systems, 254
  seed dispersal by herbivores, 206
  shield/platform desert, 22
  soil nutrient accumulation ('fertile islands'),
    102, 252–253
    desertification processes, 284, 286
  spatial distribution, 107
Silcrete, 36
Silver fish, 217
Simmondsia chinensis, 191
Skunk (*Mephitis*), 288
Smithurides pumilis, 200
Snakes, 129
Snakeweed (*Gutierrezia*), 76, 186
Social impact of desertification, 276, 277,
  291–293
Soil, 4, 14–16
  alluvial fans/bajada vegetation, 27–30
  animal disturbance patches, 110, 214, 253
  arthropod activity-related enrichment, 113,
    114–117
  bacteria
    denitrification, 262–263
    nitrogen fixation, 259, 261
  desert types, 15–16
  erosion, 5
    cohesive resistance to wind, 68–69
    raindrop splash, 77–78
  food webs, 220–222
  gradients along catenas/watersheds, 85

  grazing effects, 202, 204–205
  macrofauna, 220
  mesofauna, 221
    anhydrobiosis, 143
  microclimate, 247
  microfauna, 221
    decomposition/mineralization activities,
      247–249, 264
    soil quality indicator, 314
  mites *see* Acari
  moisture, 5, 15, 16
    water balance parameters, 46
    wind erosion susceptibility relationship,
      66
  nutrient accumulation beneath shrubs
    ('fertile islands'), 102, 252, 253
  organic matter, 16, 309
    nutrient availability relationship, 255, 257
    termite abundance relationship, 250–251
  rainfall partitioning, 72
    infiltration, 78–84
    macropores, 82–83
    run-off, 86
    stemflow water, 74–76
  surface/subsurface temperature, 58, 61
  texture, 16
  vadose water, 73
  water potential, 15, 16, 73
Solar radiant energy, 58
Solenopsis xyloni (fire ants), 208
Solfugids, 200
Sonoran Desert
  alien species expansion, 288
  herbivory, 190, 191
  nitrogen fixing legumes, 260
  plant community dynamics, 105–106
  productivity, 163, 177
  seasonality, 46, 50, 51
  seed densities, 98
  vegetation, 27, 29, 107, 109, 140
South African desert
  aridity index (I_b), 47, 48
  desertified landscape rehabilitation,
    298–299
South American desert
  aridity index (I_b), 47, 48
  desertification processes, 287
  desertified landscape rehabilitation, 300
  lizard communities, 215
  rodent granivory, 210
  seasonality, 50, 51
South Asian desert, aridity index (I_b), 47
Southern Oscillation, 44

Spadefoot toad (*Scaphiopus*), 126–127, 128, 129, 226
Spatial scale, 2, 4
Spatial variation, 11
  climate, 52–53
  decomposition processes, 244
  desertification processes, 286–287
  plant distribution, 106–107
Spiders, 199, 200, 217
Spinesence, 189
Spinifex grassland, 39
Splash erosion-kinetic energy of raindrops, 77–78
Sporobolus, 141, 297
Sporobolus flexuosus, 6
Stability, 9, 10, 11
Stem adaptations, 136
Stem photosynthesis, 123
Stem succulents, 138
Stemflow, 5, 72, 73–76
Stenocereus thruberi (organ-pipe cactus), 27
Stipagrostis amabilis, 38
Stipagrostis obtusa, 38
Stomatal adaptations, 123
Stony surfaces, 36–37
Storms, 88, 89, 310
  debris flows, 89
  fluvial flows, 89
  primary production modelling, 163, 164, 166
  watershed scouring events, 89, 90
Streams, 37
  see also Ephemeral streams
Streptocephalus texanus, 223
Struthio camelus (ostrich), 151, 324
Suaeda, 29, 138
Subhumid zone, 12
Succulents, 123, 137–138
  Australian semi-arid regions, 44–45
  leaf, 138
  saline environments, 138
  stem, 138
Sucking herbivorous arthropods, 193, 194
Surface active arthropods, 198–202
Surface crusts
  measurement for monitoring/assessment, 309
  rainfall infiltration effects, 81–82
  resistance to wind erosion, 68
  see also Algal crusts; Cryptogamic crusts; Lichen crusts
Surface layer, hydrological landscape, 73
Sylvilagus auduboni (desert cottontail), 188

Symbiosis, 10
Symbiotic nitrogen fixation, 254, 258, 260

T
Tabosa grass (*Hilaria mutica*), 33, 69
Tadpole shrimp (*Triops longicaudatus*), 223
Tail shading behaviour, 130
Tannins, 186, 189, 192, 195
Tarbush (*Flourensia cernua*), 33, 73, 74, 76, 186, 190, 192, 193, 196, 257
Taxidea taxus (badger), 206
Tegeticula yuccasella (yucca moth), 204
Temperature
  organic materials decomposition, 240
  seasonal timing, 45
  soil surface/subsurface, 58, 61
Temporal heterogeneity, 11
  desertification processes, 286–287
Tenebrionid beetles, 142–143, 198, 199
Termites, 84, 87, 220–221, 294
  behavioural adaptations, 133
  decomposition activities, 249–251
  exploitation for food production, 324
  gut symbionts, 249
  lizard predation, 216–217, 252
  plant productivity impact, 113
  soil effects, 113, 114–117
    burrowing disturbance patches, 110, 111
    macropore production, 82–83, 110, 251
    nutrient transport, 110
  soil organic matter relationship, 250–251
  vegetation effects, 117
Themeda triandra, 117
Themnocephalus platyurus, 223
Thermal energy balance, 123, 143–144
  birds, 150
Thermoregulatory behaviour, reptiles, 129
Throughfall, 72, 76–77
Time scales, 2
Tolerance thresholds, 3, 4
Topography
  productivity effects, 168, 170
  rainfall spatial variability, 53
Torrey ephedra (*Ephedra torreyana*), 186
Traganum, 138
Tragelaphus strepsiceros (kudu), 189
Transmission losses
  ephemeral lakes, 92
  ephemeral streams, 31, 32
Transverse dunes, 38

Trees
alluvial fans/bajadas, 29
Australian semi-arid regions, 45
banded patterns, 33
hydraulic lift, 138–139
physiological adaptations, 138
productivity, 173
rainfall infiltration/run-off, 80–81, 86
root systems, 254
seed dispersal by herbivores, 206
shield/platform desert, 22
soil nutrient accumulation ('fertile islands'),
252
Trimerotropis pallidipennis, 190, 191
Trinervitermes trinervoides, 117
Triodia basedowii, 39, 217
Triodia pungens, 39
Triops longicaudatus (tadpole shrimp),
223
Trophic relationships, 11, 16–17, 181
Tunisian desert
desertified landscape rehabilitation,
299–300
productivity, 169
rainfall, 52, 53
vegetation, 140

U
Urocyon cinereoargenatus (grey fox), 288
Uromastix, 142, 215
Uromastix acanthinurus, 141
Uta stansburiana, 181, 216

V
Vadose layer/vadose water, 73
Varanus gouldii, 200
Variable stem death, 140
Vegetation
alluvial fans/bajadas, 27–29
arcs, 33
banded patterns, 33
catena/watershed sediments, 85
ephemeral streams, 31–32
grazing effects, 202–204
platform deserts, 22
rainfall distribution effects, 73
infiltration/run-off, 78, 79–81, 85, 86
kinetic energy reduction, 78
rainfall seasonality effects, 50–51
rainfall spatial variability effects, 53
rodent/ant seed consumption impact,
213–215
sand dunes, 38, 39

seasonality adaptations, 45
shield deserts, 21, 22–23
termite mounds, 117
wind erosion reduction, 67–68, 70
Ventilago viminalis, 34
Vertebrate predators, 215–220
Vesicular–arbuscular endomycorrhiza, 266,
267–268, 300
Vulpes chama (Cape fox), 100

W
Wadis, 22
see also Ephemeral streams
Water balance
birds, 151
carnivorous mammals, 150
landscape, 72–73
parameters, 46–48
large herbivorous mammals, 145–147
rodents, 148–149
Water infiltrating landscapes, 73
Water loss reduction, 123
amphibians, 126–127
Water retention, 310
Water spreading landscapes, 73
Water use efficiency, 136, 160
see also Rain use efficiency
Watersheds, 4, 9, 17
fluvial transport, 84–85
scouring events, 89, 90
Wet season, 48
potential evapotranspiration (PET):mean
rainfall ratio, 51–52
productivity, 51–52
White-necked raven (*Corvus cryptoleucus*),
150
White-tailed deer (*Odocoileus virginianus*),
206, 321
Wilting point, 15
Wind dispersal of seeds, 97–98
Wind erosion, 65–71
animal activities impact, 68–69, 70
erosion cells model, 70
fetch length/threshold velocity relationship,
69–70
organic material redistribution, 70
soil cohesive resistance, 68–69
threshold wind speed, 66
vegetation-mediated reduction, 67–68,
70
Wind storms, 88
Winterfat (*Ceratoides lanata*), 53, 162, 188
Woodrat (*Neotoma*), 186, 189

X
Xerophytic food crops, 319
Xerus inauris (Cape ground squirrel), 130

Y
Yucca, 87, 105, 133, 195, 322
Yucca elata, 102, 204

Yucca moth (*Tegeticula yuccasella*), 204

Z
Zinnia acerosa, 238
Zygogramma tortusa, 190, 196
Zygophyllum dumosum, 167, 168, 169